Emerging Technologies in Solar Energy

Emerging Technologies in Solar Energy

Editor: Catherine Waltz

⟨M⟩ MURPHY & MOORE

www.murphy-moorepublishing.com

Murphy & Moore Publishing,
1 Rockefeller Plaza,
New York City, NY 10020, USA

Visit us on the World Wide Web at:
www.murphy-moorepublishing.com

ISBN: 978-1-63987-188-9 (Hardback)

Cataloging-in-Publication Data

Emerging technologies in solar energy / edited by Catherine Waltz.
    p. cm.
Includes bibliographical references and index.
ISBN 978-1-63987-188-9
1. Solar energy. 2. Solar energy--Technological innovations.
3. Renewable energy sources. I. Waltz, Catherine.
TJ810 .E44 2022
621.47--dc23

Table of Contents

Preface

Solar energy is a renewable form of energy which is produced by conversion of solar irradiance. The heat and light energy of the sun can be converted into electrical and thermal energy. Solar technologies are categorized into passive and active systems. Passive systems don't use any external devices, but use heat and light directly from the sun. An active system employs an external device to convert solar energy into electricity. Some of the common techniques used to harness solar energy are photovoltaics, solar heating and cooling, and concentrating solar power. Some of the major applications of solar power are cooking, heating, lighting and transportation. This book includes some of the vital pieces of work being conducted across the world, on various topics related to solar energy. Also included herein is a detailed explanation of the various concepts and applications of solar energy. As this field is emerging at a rapid pace, the contents of this book will help the readers understand the emerging technologies in this field.

This book is a comprehensive compilation of works of different researchers from varied parts of the world. It includes valuable experiences of the researchers with the sole objective of providing the readers (learners) with a proper knowledge of the concerned field. This book will be beneficial in evoking inspiration and enhancing the knowledge of the interested readers.

In the end, I would like to extend my heartiest thanks to the authors who worked with great determination on their chapters. I also appreciate the publisher's support in the course of the book. I would also like to deeply acknowledge my family who stood by me as a source of inspiration during the project.

Editor

.

1

Silicon Solar Cells: Recombination and Electrical Parameters

Saïdou Madougou[1], Mohamadou Kaka[1] and Gregoire Sissoko[2]
[1]University Abdou Moumouni of Niamey, BP 10 963 - Niamey
[2]Université Cheikh Anta Diop de Dakar,
[1]Niger
[2]Senegal

1. Introduction

Nowadays, the world's energy needs are growing steadily. However, the conventional sources of energy are limited.

Solar energy such as photovoltaic energy (PV) is the most available energy source which is capable to provide this world's energy needs. The conversion of sunlight into electricity using solar cells system is worthwhile way of producing this alternative energy. The history of photovoltaic energy started in 1839 when Alexandre-Edmond Becquerel discovered the photovoltaic effect (S.M. SZE 1981, W. Shockley 1949). Photovoltaic system uses various materials and technologies such as crystalline Silicon (c-Si), Cadmium telluride (CdTe), Gallium arsenide (GaAs), chalcopyrite films of Copper-Indium-Selenide (CuInSe$_2$), etc (W. Shockley 1949, W. Shockley et al. 1952). Now, silicon solar cells represent 40 % of the world solar cells production and yield efficiencies well higher than 25 % (A. Wang et al 1990). In solar technology, the main challenge of researchers is to improve solar cells efficiency. Due to this challenge, several investigations have been developed to characterize the solar cells by the determining their parameters. Indeed, it is important to know these parameters for estimating the degree of perfection and quality of silicon solar cells.

This chapter first describes the device physics of silicon solar cells using basic equations of minority carriers transport with its boundary conditions, the illumination mode and the recombination mechanisms. Then, a silicon solar cells recombination and electrical parameters are presented (S. Madougou et al 2005a, 2007b). Finally, some methods of determination of these parameters are described.

2. Overview of silicon material

In most cases, solar cells are manufactured on a silicon material. Its proportion represents 40% of world-wide semiconductor solar cells production. Pure silicon material is founded directly in solid silica by electrolysis. The production of silicon by processing silica (SiO$_2$) needs very high energy and more efficient methods of synthesis. Also, the most prevalent silicon solar cell material is crystalline silicon (c-Si) or amorphous silicon (a-Si).

Crystalline silicon can be separated into multiple categories according to its crystallinity and its crystal size. These include: monocrystalline silicon, poly or multicrystalline silicon, Ribbon silicon and new structures.

Compared to amorphous silicon, crystalline silicon absorbs the visible part of the solar spectrum more than infrared portion of spectrum. Crystalline silicon has a smaller band gap (Eg = 1.1 eV) than amorphous silicon (Eg = 1.75 eV) in live with Shockley-Hall-Read's recombination process experiment (W. Shockley 1949).

Crystalline silicon solar cells generate approximately 35 mA/cm² of current, and voltage 550 mV. Its efficiency is above 25 %. Amorphous silicon solar cells generate 15 mA/cm2 density of current and the voltage without connected load is above 800 mV. The efficiency is between 6 and 8% (S. W. Glunz et al. 2006).

But, all solar cells require a light absorbing material contained within the cell structure to absorb photons and generate electrons (G. Sissoko et al. 1996).

3. Device physics of silicon solar cells

3.1 Silicon solar cells

Commonly, most silicon solar cells are configured in N-P junctions or vice versa (S.M. SZE 1981) in one side and N+-N-P+ structure (or vice versa) for double sides named bifacial silicon solar cell (S. Madougou et al. 2004, 2005a, 2005b, 2007a et 2007b). Silicon solar cells have all contacts on the back of the cell. Figure 1 shows an example of silicon solar cell with its contacts.

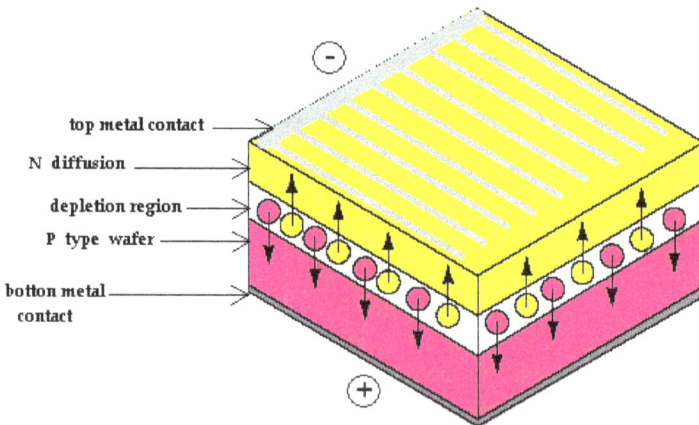

Fig. 1. Silicon Solar cell with its contacts

In this section, we will study the structure and the operation of N-P junction (monofacial and bifacial silicon solar cells).

3.1.1 Monofacial silicon solar cell

N-P junction or a P-N junction is a one side solar cell (W. Shockley 1949). When a P-type is placed in intimate contact with an N-type, a diffusion of electrons occurs from the region of high electron concentration (N-type) into the region of low electron concentration (P-type). Figure 2 shows the N-P junction and its forward biased with its corresponding diode schematic symbol and its I-V characteristic curve.

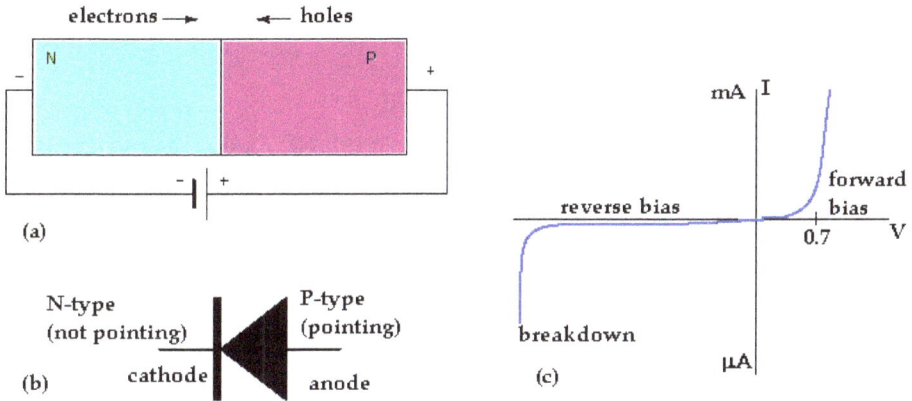

Fig. 2. N-P junction: (a) Forward biased N-P junction, (b) Corresponding diode schematic symbol (c) silicon diode I versus V characteristic curve.

3.1.2 Bifacial silicon solar cell

Bifacial silicon solar cell is a double sided silicon solar cell with N^+-N-P^+ or N^+-P-P^+ structure or vice versa (S. Madougou et al. 2004).

A bifacial silicon solar cell with N^+-P-P^+ structure has an N^+-P front side (surface) and P^+-P back side (surface). This back surface (P^+-P) is an important seat of a Back Surface Field (J. D. Alamo et al. 1981). In some solar cells, the front surface doping density ranges from 10^{17} to 10^{19} cm^{-3}. In the base, the doping ranges 10^{15} to 10^{17} cm^{-3}. The bifacial silicon solar cell can be illuminated from the front side, the back side or simultaneously from both sides as shown in figure 3.

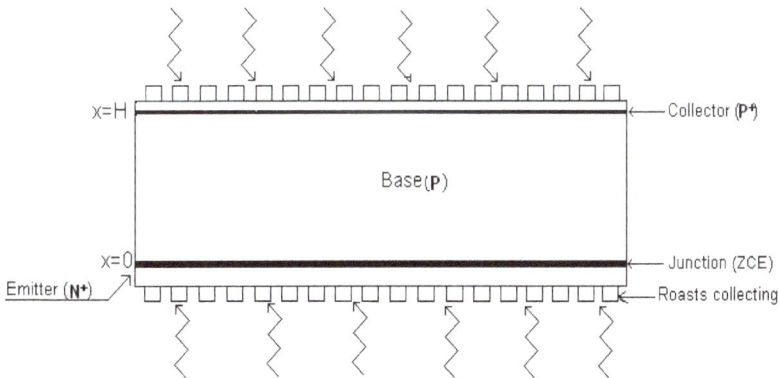

Fig. 3. Bifacial silicon solar cell with n+-p-p+ structure.

3.2 Recombination mechanisms

In a conventional silicon solar cell, recombination can occur in five regions (W. Shockley 1949):

- at the front surface;
- at the emitter region (N^+);
- at the junction (the depletion region of the junction);
- at the base region (P);

- at the back surface.

They are mainly two types of recombination: the recombination mechanisms in bulk (volume) and the surface recombination.

3.2.1 Recombination mechanisms in bulk of the silicon solar cell

In a bulk of the silicon solar cell, three fundamental recombination mechanisms are produced.

- *Auger recombination.* We have Auger recombination when the energy of the electron which falls in the valence band is transferred as kinetic energy to:
  - another free electron which will be transferred to a higher level in the conduction band;
  - a hole on a deep level of the valence band.
- *Shockley-Read-Hall recombination (through defect).* They are twofold:
  - those due to defects in the crystal lattice or chemical impurities;
  - those that occur when a deep level captures an electron.
- *Radiative or band-to-band recombination.* When a junction is forward biased, the recombination can be radiative. It is the opposite of the absorption phenomenon. The free carriers go directly from the conduction band to the valence band by emission of photon. This recombination is related by the lifetime τ of the excess minority carriers. They also intervene in the diffusion of charge carriers through the term $U(x)$ given by the expression:

$$U(x) = \frac{\delta n(x)}{\tau} \tag{1}$$

Where, $\delta n(x)$ is the excess minority carriers density, and τ is its lifetime.

3.2.2 Surface recombination

The many faults that characterize the surface of a semiconductor disrupt its crystalline structure. The Surface recombination corresponds to a phenomenon where excited electrons in the conduction band recombine with holes in the valence band via defect levels at the surface, called surface states. These surface states are the result of the abrupt discontinuity of a crystalline phase at the surface, which forms unsatisfied dangling silicon bonds.

3.3 Illumination mode

The solar cells can be under monochromatic light (single wavelength), constant multispectral light or variable and intense light concentration (more than 50 suns) or under other mode. The optical generate rate $G(x)$ of monochromatic light is given by (S. Madougou 2007b):

$$G(x) = \alpha(1 - R)\Phi_0 \exp(-\alpha x) \tag{2}$$

Where, R is the reflectivity coefficient, α is the optical absorption coefficient and Φ_0 is the incident flux of monochromatic light.

When the solar cell is illuminated by the multispectral light the generated rate $G(x)$ is given by (S. Madougou 2007a):

$$G(x) = \sum_{i=1}^{3} a_i \left[\exp(-b_i x) \right] \tag{3}$$

Where, a_i and b_i are the coefficients deduced from the modelling of the generation rate considered for over all solar radiation spectrums.

Figure 4 shows an experimental set-up of silicon solar cell illuminated with a direct light from a lamp.

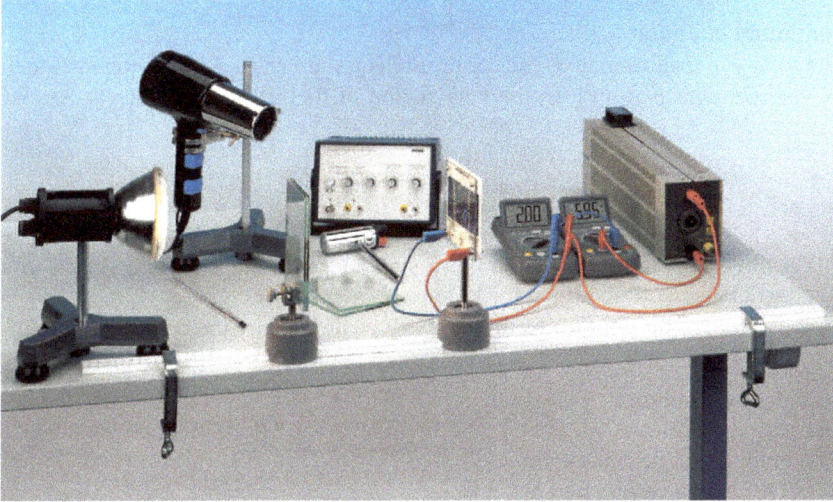

Fig. 4. Experimental set-up of silicon solar cell illuminated with a direct light from a lamp.

3.4 Basic equations of minority carriers transport

The basic equations describe the behaviour of the excess minority carriers in the base of the solar cell under the influence of an electric field and/or under illumination; both cause deviations from thermal equilibrium conditions. These equations can be expressed on one or two dimensions. In the following, we will work in one dimension.

3.4.1 Poisson equation

The Poisson equation relates the gradient of the electric field E to the space charge density ρ. According to W. Shockley(1949), it is given as:

$$- \frac{d^2\phi(x)}{dx^2} = \frac{dE(x)}{dx} = \frac{\rho}{\varepsilon_0 \varepsilon} \qquad (4)$$

Where, ϕ is the electrostatic potential, ε_0 is the permittivity of free space and ε is the static relative permittivity of medium.

In the same conditions, the electrons' current density I_n and the holes' current density I_p are obtained as follows:

$$I_n = +q\,\mu_n\,n(x)E(x) + qD_n\frac{dn(x)}{dx} \qquad (5)$$

$$I_p = -q\,\mu_p\,p(x)E(x) + qD_p\frac{dp(x)}{dx} \qquad (6)$$

Where, q is the elementary charge, μ_n and μ_p are the mobility's of electrons and holes, D_n and D_p are the diffusion constants related through the Einstein relationships: $D_n = \dfrac{kT}{q\,\mu_n}$; $D_p = \dfrac{kT}{q\,\mu_p}$.

k is the Boltzmann constant.

3.4.2 Continuity equation
When the solar cell is illuminated, the continuity equation related to photogenerated excess minority carriers density $\delta n(x)$ in the base region of the cell is given by (G. Sissoko et al 1996):

$$\frac{\partial^2 \delta n(x)}{\partial x^2} - \frac{\delta n(x)}{L^2} + \frac{G(x)}{D} = 0 \tag{7}$$

Where, D is the excess minority carriers diffusion constant and L is their diffusion length. $G(x)$ is the carriers generation rate in the base.
The solution $\delta n(x)$ of the continuity equation is well defined by the boundary conditions.

3.4.3 Boundary conditions
According to G. Sissoko et al. (1996), the boundary conditions defined by the minority carriers recombination velocities are:
-     The emitter-base junction at x = 0:

$$\left.\frac{\partial \delta n(x)}{\partial x}\right|_{x=0} = \frac{Sf}{D}\delta n(0) \tag{8}$$

-     The back-surface of the base at x = H:

$$\left.\frac{\partial \delta n(x)}{\partial x}\right|_{x=H} = -\frac{Sb}{D}\delta n(H) \tag{9}$$

Where, Sb is the minority carriers recombination velocity at the back-surface and Sf is the minority carriers recombination velocity at the junction.

3.5 Equivalent circuit of the solar cell
To understand the electronic behaviour of a solar cell, it is useful to create its model which is electrically equivalent at the solar cell. Because no solar cell is ideal, a shunt resistance and a series resistance component are therefore added to the model to have the equivalent circuit. This equivalent circuit of the solar cell is based on discrete electrical components. Figure 5 shows an example of an equivalent circuit of a solar cell with one diode.
For the practical analysis of the solar cell performance the dark current-voltage (I-V) characteristics curve is shifted down by a light generated current I_L resulting in the illuminated I-V characteristics.
The $I - V$ characteristic of a single-junction P-N under illumination can be written as follows:

$$I = I_0\left[\exp\left(\frac{qV}{kT}\right) - 1\right] - I_L \tag{10}$$

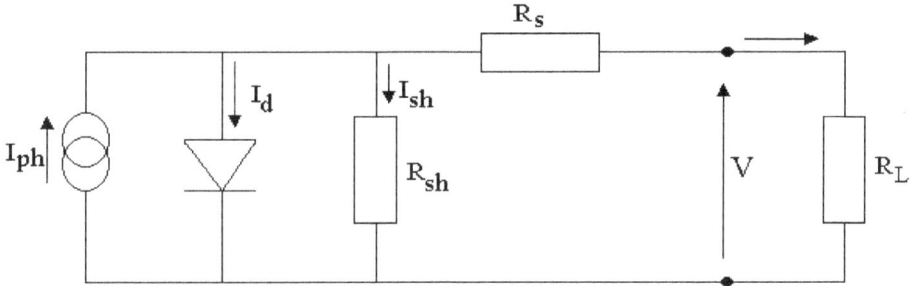

Fig. 5. Equivalent circuit of the solar cell with one diode.

And the dark current density of the P-N junction by:

$$I_{dark} = I_0 \left[\exp\left(\frac{qV}{kT}\right) - 1 \right] \tag{11}$$

Where, I_0 is the reverse saturation current density, V the voltage and T is the absolute temperature.

Figure 6 below shows the dark and illuminated current-voltage (I-V) characteristics.

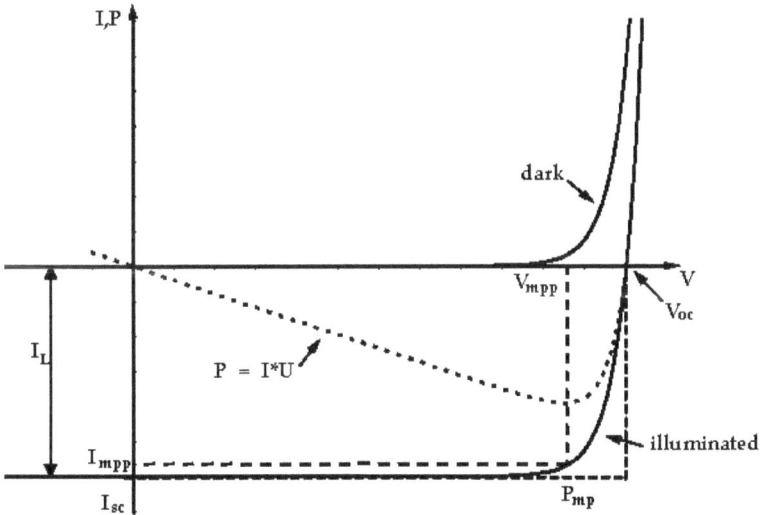

Fig. 6. Dark and illuminated current-voltage I-V curves of silicon solar cell.

Solar cells can be also usually connected in series in modules, creating an additive voltage. Connecting cells in parallel will yield higher amperage. Modules are then interconnected, in series or parallel, or both, to create the desired peak DC voltage and current.

4. Silicon solar cells electrical and recombination parameters

The solar cells have two categories of parameters: electrical parameters and recombination parameters.

4.1 Electrical parameters
4.1.1 Photocurrent
The photocurrent density I at the junction of solar cell is obtained from the excess minority carriers density of each illumination mode as follows (S. Madougou et al. 2004 et 2007a):

$$I = qD\frac{\partial \delta n(x)}{\partial x}\bigg|_{x=0} \tag{12}$$

Where q and D are constants defined above.

4.1.2 Voltage
By means of Boltzmann's relation, the voltage V can be expressed as (J.F. Phyllips 1997):

$$V = V_T Log\left(N_B\frac{\delta n(0)}{n_i^2} + 1\right) \tag{13}$$

Where, $V_T = \dfrac{kT}{q}$ is the thermal voltage, n_i the intrinsic carriers density and N_B the base doping density.

4.1.3 Power
The Power generated for the cell is given by:

$$P = VI \tag{14}$$

Where, I and V are the photocurrent and the voltage defined above.

4.1.4 Fill factor
The fill factor is given by (W. Shockley 1949):

$$FF = \frac{V_{max}I_{max}}{V_{co}I_{sc}} \tag{15}$$

Where, V_{max} and I_{max} are voltage and current at maximum power point respectively. V_{CO} is the open-circuit voltage and I_{SC} is the short-circuit current.

4.1.5 Series and shunt resistances
The series resistance is given by (M. Wolf et al. 1963):

$$R_s = \frac{V_{co} - V}{I} \tag{16}$$

The shunt resistance is given by (M. Wolf et al. 1963):

$$R_{Sh} = \frac{V}{I_{SC} - I} \tag{17}$$

I, V, V_{CO} and I_{SC} are defined in sections 4.1.3 and 4.1.4 above.

4.1.6 Capacitance
The capacitance is given by (Edoardo Barbisio 2000; S. Madougou et al 2004):

$$C = \frac{qn_0}{V_T}\exp\left(\frac{V}{V_T}\right) \qquad (18)$$

Where, $n_0 = \frac{n_i^2}{N_B}$, V and V_T are the voltage and the thermal voltage defined above.

4.1.7 Internal quantum efficiency
The internal quantum efficiency η_{IQE} is expressed by (S. Madougou et al. 2007b):

$$\left(\eta_{IQE}\right) = \frac{I_{SC}}{q[1-R]\Phi_0} \qquad (19)$$

I_{SC}, R and Φ_0 are defined above.

4.2 Recombination parameters
The different recombination parameters include: the diffusion length L, the minority carriers lifetime τ and the recombination velocities (Sf, Sb).

4.2.1 Diffusion length
The excess minority carriers diffusion length is given by:

$$L = \sqrt{\tau D} \qquad (20)$$

4.2.2 Lifetime
The excess minority carriers lifetime is given by:

$$\tau = \frac{L^2}{D} \qquad (21)$$

4.2.3 Recombination velocities (Sf, Sb)
When Sb is higher, the photocurrent tends towards the open-circuit current. Thus, we have the relationship:

$$\left(Sf\right) = \left[\frac{\partial J}{\partial Sf}\right]_{Sb \to +\infty} \qquad (22)$$

When, Sf is higher, the photocurrent density tends towards its maximum value (short-circuit current). Thus, we have the following relationship:

$$\left(Sb\right) = \left[\frac{\partial J}{\partial Sf}\right]_{Sf \to +\infty} \qquad (23)$$

5. Methods for determining the solar cells parameters

Research indicates many techniques for determining electrical and recombination parameters of solar cells (S. Madougou et al. 2007b, 2005a; G. Sissoko et al 1996; S. K. Sharma et al. 1985).
In this part, we will present some methods of electrical and recombination parameters determination.

5.1 Methods of electrical parameters determination
For determining solar cells electrical parameters, several methods exist.

5.1.1 The method based on the current-voltage characteristics
In these categories of methods, the authors use an algorithm for extracting solar cell parameters from I-V-curve using a single or double exponential model (S. Dib et al. 1999; C.L. Garrido Alzar 1997). They also use a non linear least squares optimization algorithm based on the Newton model using the measured current-voltage data and the subsequently calculated conductance of the device (M. Chegaar et al. 2001). To extract the solar cells parameters, the authors also utilize an analysing method of the current-voltage (I-V) characteristics of silicon solar cells under constant multispectral illumination and under magnetic field (S. Madougou et al 2007a).

5.1.2 The analytical method
In this method, researchers use the data of some solar cells parameters (measured data of current, short circuit current, current at maximum power point, voltage, open circuit voltage, voltage at maximum power point) to determine others parameters.

5.1.3 Others methods
In those methods, we have:
- Methods using numerical techniques approaches;
- The vertical optimisation method;
- Etc...

All of these techniques can be extended or modified adequately to cover many cases of solar cells.

5.2 Methods of recombination parameters determination
Here, we present two methods of recombination parameters determination. The first method is a linear fit of Internal Quantum efficiency (IQE) reverse curves versus light penetration depth. The second one is a programming method. For each method, after determining the minority carriers diffusion length, we calculate the excess minority carriers lifetime and recombination velocities.
In this approach, new analytical expressions of recombination parameters were established.

5.2.1 Fitting method
It is a linear fit of the IQE reverse curves versus light penetration depth. From this linear fit, we extract the minority carriers diffusion length (S. Madougou et al. 2007b). Figure 7 presents an example of a linear fit of measured IQE reverse curves versus light penetration depth.

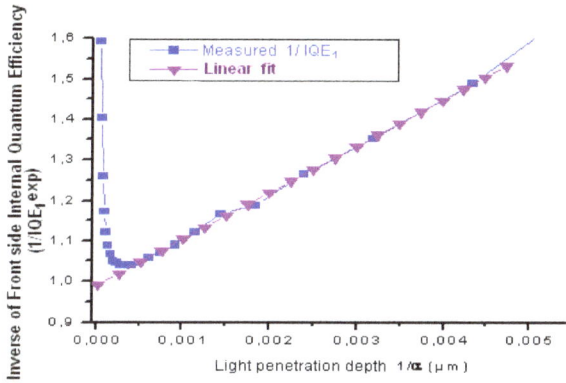

Fig. 7. Linear fit of the reverse of IQE with light penetration depth.

5.2.2 Programming method

Some researchers develop an algorithm to calculate the excess minority carriers diffusion length. This algorithm searches the weakness square type value between theoretical and measured internal quantum efficiency data. At this weakness square type value, we determine the corresponding diffusion length value of the solar cell (S. Madougou et al. 2007b).

6. Conclusion

This study shows that most of silicon solar cell is configured in N-P junctions. In these solar cells, recombination mechanisms occur in five regions and are mainly two types: recombination mechanisms in bulk and surface recombination. In practice, the solar cells can be illuminated using several illumination modes: under monochromatic light, under multispectral light, under intense light concentration or under other modes. Basic equations describe the behaviour of the excess minority carriers generated in the base of the solar cells. To understand the electronic behaviour in the study, the solar cell is modelled in an equivalent circuit containing a shunt resistance and a series resistance.

Silicon solar cells have two categories of parameters (electrical parameters and recombination parameters) which, the knowledge is very important to ameliorate the efficiency of the solar cells. Nowadays, many determination techniques of electrical and recombination parameters of solar cells exist. It is a great challenge for researchers to find a way to improve the solar cells efficiency. If this challenge is won, solar energy through photovoltaic energy can reveal itself to be a unique opportunity to solve energy and environmental problems simultaneously.

7. References

Alamo J. D., J. Eguren, and A. Luque (1981): "Operating limits of Al-alloyed high-low junctions for BSF solar cells". *Solid-State Electronics*. Vol. 24, pp. 415-420. ISSN: *0038-1101*.

Chegaar M., Z. Ouennoughi, and A. Hoffmann (2001): A new method for evaluating illuminated solar cell parameters, *Solid-state electronics*, Vol. 45, pp. 293. ISSN: 0038-1101.

Dib S., M. de la Bardonne, A. Khoury (1999), "A new method for extraction of diode parameters using a single exponential model". *Active and Passive Electronic Components*. Vol. 22, PP. 157. ISSN *0882-7516*.

Edoardo Barbisio (2000): "Diffusion capacitance identification of PV cells". *Proc. 16th European photovoltaic solar energy conference.* ISBN *1 902916 18 2.* Date 1 – 5 May 2000, Glasgow, UK, pp 919 - 922.

Garrido-Alzar C. L. (1997): "Algorithm for extraction of solar cell parameters from I-V-curve using double exponential model". *Renewable Energy.* Vol. 10, pp. 125. ISSN: 0960-1481.

Glunz S. W., S. Janz, M. Hofmann, T. Roth, and G. Willeke (2006): "Surface passivation of silicon solar cells using amorphous silicon carbide layers". *Proc. 4th WCPEC,* pp. 1016-1019. *ISBN*: 1-424-40017-1.

Madougou S., Nzonzolo, S. Mbodji, I. F. Barro, G. Sissoko (2004): Bifacial silicon solar cell space charge region width determination by a study in modelling: Effect of the magnetic field. *Journal des Sciences.* Vol.4, N°3, pp. 116-123. ISSN 0851 – 4631.

Madougou S., B. Dieng, A. Diao, I. F. Barro, G. Sissoko (2005a): Electrical parameters for bifacial silicon solar cell studied in modelling: space charge region width determination. *Journal des Sciences pour l'Ingénieur.* Volume 5, pp. 34 – 39. (Senegal). ISSN 0851 4453.

Madougou S., I.F. Barro, G. Sissoko (2005b): Effect of magnetic field on bifacial silicon solar cell studied in modeling: space charge region width determination. *Proc. 2005 Solar World Congress.* PaperN°1979. ISBN 089553 1771. Date 6-12 August 2005, Orlando (Florida, USA).

Madougou S. F. Made, M. S. Boukary, G. Sissoko (2007a): I –V characteristics for bifacial silicon solar cell studied under a magnetic field. *Advanced Materials Research: Trans Tech Publications Inc.* Volume 18 until 19 pp. 303 - 312; (Zurich – Switzerland). ISSN 1022 6680.

Madougou S., F. Made, M. S. Boukary, G. Sissoko (2007b): Recombination Parameters Determination by Using Internal Quantum Efficiency (IQE) Data Of Bifacial Silicon Solar Cells. *Advanced Materials Research. Trans Tech Publications Inc.* Volume 18 until 19 pp. 313 - 324; (Zurich – Switzerland). ISSN 1022 6680.

Phyllips J.E., T. Titus and D. Hofmann (1997): Determining the voltage dependence of the light generated current in CuInSe2-Based solar cells using I-V measurements made at different light intensities. *Proc 26th IEEE PVSC conference.* isbn 0080438652. Sept. 30 - Oct. 3, Anaheim, CA, USA.

Sharma S. K., S.N. Singh, B.C. Chakravarty, and B.K. Das (1986): "Determination of minority carrier diffusion length in a p-silicon wafer by photocurrent generation method" *Journal of Appl. Phys.* Vol 60, N°10. pp 3550 – 3552. ISSN: 0021-8979.

Shockley W., J. Bardeen and W. Brattain, "The Theory of P-N Junctions in Semiconductors and P-N Junction Transistors," *Bell System Technical Journal,* Vol 28: 1949, pp. 435 Semiconductive Materials Transistor. ISSN : 0005-8580.

Shockley W. and W. T. Read (1952): "Statistics of the recombination of holes and electrons," *Physical Review.* Vol. 87, pp. 835-842. ISSN: 0031-9007.

Sissoko G., C. Museruka, A. Correa, I. Gaye, A. L. Ndiaye (1996): "Light spectral effect on recombination parameters of silicon solar cell". *Proc. World Renewable Energy Congress.* Part III, pp 1487-1490. ISBN 0-7918-3763-7. Date 15 – 21 June. Denver – USA (1996).

Sze S.M. (1981): "*Physics of semiconductors devices*"; 2nd Edition, Wiley Interscience, Editor: John Wiley and Sons (WIE). ISBN-10 0471 0566 18. New York 11 p. 802.

Wang A., J. Zhao, and M. A. Green (1990): "24% efficient silicon solar cells," Applied Physics Letters, vol. 57, pp. 602-604. *ISSN* 0003-6951.

WOLF M. and Hans RAUSCHENBACH (1963): Series resistances effects on solar cell measurement. *Advanced energy conversion.* ISSN *0196-8904.* Volume 3, pp 445 - 479. Pergamon Press 1963. ISSN: 0196-8904.

2

Aerostat for Solar Power Generation

G. S. Aglietti, S. Redi, A. R. Tatnall, T. Markvart and S.J.I. Walker
University of Southampton
United Kingdom

1. Introduction

One of the major issues in the use of ground based photovoltaic (PV) panels for the large scale collection of solar energy is the relatively low energy density. As a result a large area is required on the ground to achieve a significant production. This issue is compounded by the fact that the power output of the devices is strongly dependent on the latitude and weather conditions. At high latitudes the sun is relatively low on the horizon and a large part of the solar energy is absorbed by the atmosphere. Countries situated at high latitudes, with climates such as the UK, are therefore challenged in their exploitation of solar energy as the average number of Peak Solar Hours (PSH - numerically equal to the daily solar irradiation in kWh/m²) is relatively low. In Europe, typical annual average PSH values for horizontal surfaces range from about 2.5 h in northern England to 4.85 h in southern Spain (Markvart & Castañer, 2003). As, roughly speaking, the cost of the energy produced is inversely proportional to the average PSH, northern European countries are at a considerable economical disadvantage in the exploitation of solar energy with respect to other regions. On the other hand, areas with high ground solar irradiations (e.g. African deserts, see Kurokawa, 2001) are remote from most users and the losses over thousands of miles of cables and the political issues entailed in such a large project, severely reduce the economic advantages.

A different approach to address most of the shortcomings of ground based solar energy production was proposed by Glaser et al., 1974 and his idea has captured the imagination of scientists up to this day. The basic concept was to collect solar energy using a large satellite orbiting the Earth. This satellite would be capable of capturing the full strength of the solar radiation continuously and transmit it to the ground using microwave radiation. The receiving station would then convert the microwave radiation into electric energy for widespread use.

The original concept was revisited in the late 90's (Mankins, 1997) in view of the considerable technological advances made since the 70's and research work on this concept is still ongoing. However a mixture of technical issues (such as the losses in the energy conversions and transmission), safety concerns (regarding the microwave beam linking the satellite with the ground station) and cost have denied the practical implementation of this concept. The latter is a substantial hurdle as the development of Satellite Solar Power (SSP) cannot be carried out incrementally, in order to recover part of the initial cost during the development and use it to fund the following steps, but it requires substantial funding upfront (tens of billions of dollars according to Mankins, 1997) before there is any economical return.

As a compromise between Glaser's (SSP) and ground based PV devices it is possible to collect the solar energy using a high altitude aerostatic platform (Aglietti et al., 2008a, b). This approach allows most of the weather related issues, except for very extreme weather conditions, to be overcome as the platform will be above the cloud layer. As the platform is also above the densest part of the troposphere, the direct beam component from the sun will travel through considerably less air mass than if it was on the ground (in particular for early morning and evening) and this will further improve the energy output. Therefore this method enables considerably more solar power to be collected when compared to an equivalent ground based system. In addition, the mooring line of the platform can be used to transmit the electric energy to the ground in relative safety and with low electrical losses. Although this approach would capture between 1/3 and 1/2 of the energy that could be harvested using a SSP, the cost of the infrastructure is orders of magnitude lower, and this approach allows an incremental development with a cost to first power that is a few orders of magnitudes smaller than that necessary for SSP.

Most researchers up till now have proposed harvesting energy at high altitude by exploiting the strong winds existing in the high atmosphere such as the jet streams (Roberts et al., 2007). This would be achieved using Flying Electrical Generators, that are essentially wind turbines collecting wind power at altitudes from few hundred meters (www.magenn.com) to over 10 km.

The extraction of this energy using the type of machines proposed by Roberts et al. 2007, although feasible and most probably economically viable, is relatively complex in mechanical terms. One of the issues is that in low wind the machine (that is heavier than air) needs to reverse its energy flow and take energy from the ground to produce enough lift to support itself and the tether. Alternative designs like the MAGENN (www.magenn.com) overcome this problem using a lighter-than-air approach so that the buoyancy keeps it in flight all the time. However the mechanical complications are still considerable.

The exploitation of solar energy at high altitude may therefore be simpler in engineering/mechanical terms, and provide a very predictable/reliable source. One of the crucial steps to demonstrate the viability of the concept is a reliable calculation of the solar energy available as a function of the altitude. After a brief introduction on aerostatic platforms, the energy available at different altitudes is investigated. The concept of the Aerostat for Solar Power Generation (ASPG) is then described together with the equations that link its main engineering parameters/variables, and a preliminary sizing of an ASPG, based on realistic values of the input engineering parameters is presented.

2. Aerostatic platforms

Lighter-than-air craft (aerostats) have been progressively neglected by the main stream research in Aerospace Engineering during the second half of the past century after having made remarkable technological progress that culminated in the 1930's with the construction of over 200m long airships (Dick & Robinson, 1992, Robinson, 1973). There have been some developments of historical interest (Kirschner, 1986) but little of significance.

However, in the last few years, aerostats have attracted a renewed interest. Their typical market niches (scientific ballooning, surveillance/reconnaissance (Colozza & Dolce, 2005)) are expanding and more researchers have proposed several different applications, ranging from high altitude aerostats as astronomical platforms (Bely & Ashford, 1995) to infrastructures for communication systems (Badesha, 2002).

Amongst the most recent achievements in scientific ballooning are the Ultra-High Altitude Balloon (UHAB) developed for NASA (launched in 2002 with a volume of nearly 1.7 million cubic metres and reached an altitude of 49 km) and the ultra-thin film high altitude balloon constructed by the Institute of Space and Astronautical Science (ISAS) of Japan, which successfully carried a 10 kg payload to a world-record altitude of 53 km.

Tethered aerostats are limited to lower altitudes due to the weight of the tether, which increase linearly with height. Commercial aerostats fly up to 8km but various studies have been conducted to prove that considerably greater altitudes can be reached. For example the Johns Hopkins University Applied Physics Laboratory (JHU/APL) has conducted a successful feasibility study (although not experimentally demonstrated) on a high altitude (20 km) tethered balloon-based space-to-ground optical communication system (Badesha, 2002). The US Airforce has made extensive use of aerostats as a surveillance system, and there are aerostats available on the market like the Puma Tethered Aerostat (www.rosaerosystems.pbo.ru) or the TCOM's 71M (www.tcomlp.com) that can fly up to approximately 5 km tethered with payloads of 2250 kg and 1600 kg respectively. These aerostats have a mooring cable (i.e. their tether) that supplies the aerostat onboard systems and the payload with electric power, and they are designed to be able to withstand lightning strikes and strong winds.

Concerning the size, today's airships are considerably smaller than those constructed in the 1930's and that is mainly due to the economics of their typical functions. However, from a technical point of view, the state of the art in the relevant technologies would allow the construction of aerostats much larger than those currently in operation.

The possibility of using solar power as source of energy for the airship propulsion and/or to supply energy to on board systems has been investigated by Khoury and Gillett, 2004. The "sunship" that he proposed was a very simple and conventional envelop design, filled with helium, with thin film solar arrays covering appropriate areas of the external surface. The electrical power produced by the cells was then used for the propulsion and on board electrical system, with part of the energy stored in suitable units with high storage energy to weigh ratio. Notwithstanding the quality of the case made by the author, the "sunship" was never built.

However changes in the economy, driven by politics and/or technical factors (limitation of resources or scientific advances) transform the markets and the viability of certain technologies may change as a result. A typical case is that of wind turbines, whose technology has been available for decades, but only in the last few years have become a viable method to produce large quantities of electric energy.

3. High altitude solar radiation

The first step in the development of the ASPG concept is to evaluate how much solar energy is available as a function of altitude. This allows a direct assessment of the potential of the ASPG when compared to an equivalent ground based system.

This section presents a set of calculations to enable this comparison and considers the possible influence of cloud layers at different heights above the ground (up to 12 km). The results obtained are based on existing models for ideal clear sky conditions and are integrated with experimental data acquired by scientific instrumentation at a specific site in the south of the United Kingdom. Although this makes the analysis very location specific, the general conclusions about the potential of high altitude solar collectors can be extended to other countries at similar latitudes and with similar climatic conditions. Moreover it must be noted that the assumptions made in the calculation process are quite conservative in order to avoid a possible overestimate of the potential of the ASPG concept.

The attenuation that a solar beam experiences as it travels through the atmosphere is called extinction and it is mainly due to two different kinds of processes, absorption and scattering, the former being the conversion of photon energy into thermal energy and the latter involving the deflection of the photons after the interaction with atmospheric molecules or larger particles suspended in the air. These two processes cause the radiation falling on a surface to be divided into two components: direct (or beam) and diffuse. The contribution of the diffuse radiation becomes proportionally more important when the collector is located at lower altitudes, in particular under cloudy sky conditions.

Since several models that describe the characteristics of the clear atmosphere at various altitudes have been developed in the past, existing publications can be considered when dealing with these conditions. Some of these models have also been embedded in specific software (such as SMARTS (Donovan & Van Lammeren, 2001) hosted by the National Renewable Energy Laboratory), which are widely used now as design tools in the PV industry. For the present purposes the sky is considered clear above an altitude of 12 km, since it is rare to find clouds at higher layers of the atmosphere and the impact on the final results is assumed to be minimal. Therefore, the clear sky radiation falling on a sun tracking surface at 12 km can be directly determined with the use of SMARTS (Gueymard, 1995) using the site location and time of year as inputs. For the initial calculations only the beam contribution of the solar radiation is included in the analysis.

The influence of clouds is incorporated into the calculation process using the extinction parameter, determined from experimental data. The data used was acquired by radar/lidar systems (www.cloud-net.org) at a station located at Chilbolton Observatory (51.1445 N, 1.4370 W) in the South of the UK. These experimental measurements have been elaborated (see Redi 2009) to provide the extinction parameter profile in actual sky conditions, which relates the attenuation of the solar radiation to its path through the atmosphere, considering the possible presence of cloud layers. The observations were performed almost everyday of the year (from April 2003 to September 2004), 24 hours a day, in the height range between 0 and 12 km. Averaging the data for the different layers of the atmosphere and in different months of the year, it is possible to obtain the extinction parameter profile at various altitudes above the ground. As an example the values obtained for the month of March is presented in Fig. 1.

Having determined the extinction parameter the Lambert-Beers' attenuation law (Liou, 2002) can be used to calculate the loss of intensity of a solar beam as it travels through the atmosphere. By dividing the atmospheric path along the vertical in segment of length Δh_i and defining the extinction parameter for each segment α_i (m^{-1}), the variation of intensity I (the irradiance in W/m^2) can be expressed as:

$$I(\vartheta_Z) = I_{12km} \exp\left(-AM_{REL}(\vartheta_Z) \cdot \sum(\alpha_i \cdot \Delta h_i)\right) \qquad (1)$$

where I_{12km} is defined as the solar irradiance at 12 km estimated with SMARTS and $AM_{REL}(\vartheta_Z)$ is the relative air mass, which describes the path length relative to that at the Zenith and is therefore a function of the solar Zenith angle ϑ_Z.

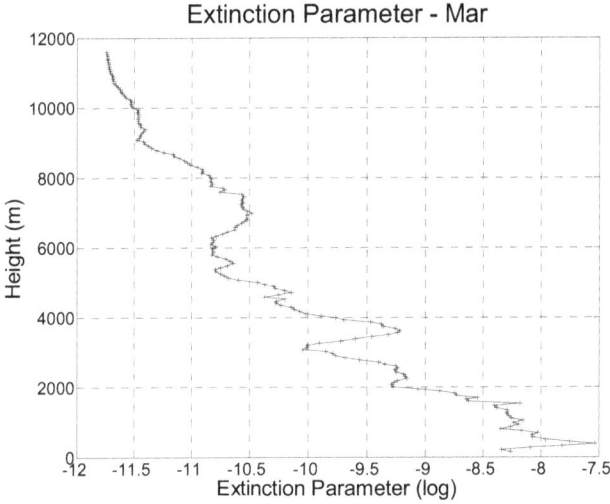

Fig. 1. Daily Mean Extinction (log) for the month of March - Chilbolton Observatory (51.1445 N, 1.4370 W)

Models like MODTRAN (Berk et al., 1989) or LibRadtran (Mayer & Kylling, 2005) that are able to integrate information about the cloud structure with the clear sky data, could be considered as a more accurate alternative to the one proposed here. However these models are quite sophisticated and they are generally more oriented towards atmospheric physics studies rather than engineering ones.

Starting from the irradiance values obtained with SMARTS at 12 km, Eq. 1 is applied to the extinction parameter values in actual sky for different months, in order to get an estimate of the irradiance below an altitude of 12 km in atmospheric conditions including possible clouds. The results obtained for the month of March at 12 km, 6 km and on the ground are presented in Fig. 2.

As a final step, it is necessary to integrate the irradiance values during the day (from sunrise to sunset) to calculate the total beam energy (beam irradiation E_B) falling on the high altitude solar collector:

$$E_B(h) = \int_{SR}^{SS} I_B(h)dt \qquad (2)$$

where SS and SR are the time of Sunset and Sunrise.

The total beam energy can now be used to evaluate the potential of the ASPG system when compared to an equivalent ground based PV array.

Considering an altitude of 6 km, and integrating graphs like the ones shown in Fig. 3 gives value of about 3600 kWh/m^2 for the energy reaching the sun tracking platform in one year. It must be noticed that due to various conservative assumptions in the calculations this shoud be a rather conservative estimate. A general overview of the beam energy that can be collected by a platform located at an altitude up to 12 km is provided in Table 1.

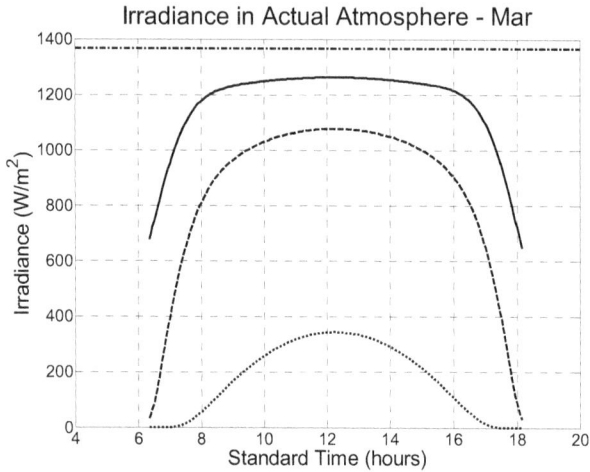

Fig. 2. Irradiance at 12 km (solid line),and at 6 km (dashed line) and on the ground (dotted line) including possible clouds, solar constant (dash-dotted line) - Chilbolton Observatory (51.1445 N, 1.4370 W)

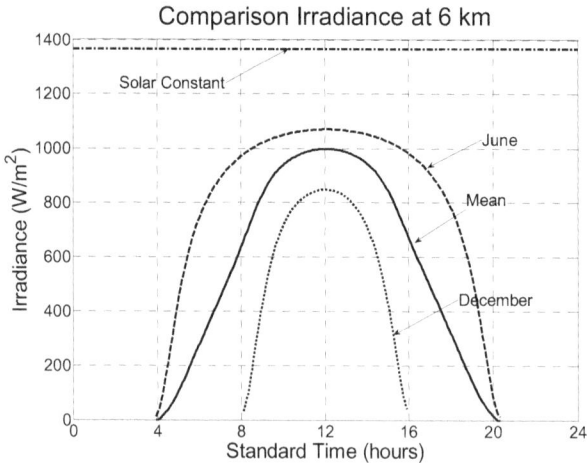

Fig. 3. Comparison between irradiance at 6km for different monthly means - Chilbolton Observatory (51.1445 N, 1.4370 W)

Altitude [km]	Total Year Beam Irradiation (including possible clouds/) [kWh/m^2]	Total Year Beam Irradiation (clear sky) [kWh/m^2]
6	3600	4530
9	4710	4800
12	5310	5310

Table 1. Year Beam Irradiation at different altitudes - Chilbolton Observatory (51.1445 N, 1.4370 W)

The study has been focused so far on the evaluation of the beam (direct) component of the solar radiation that can reach the surface. However, the contribution of the diffuse part of the radiation can be not negligible. Having considered this, the contribution of the diffuse radiation is estimated at different altitudes, with the use of SMARTS in ideal clear sky conditions. This assumption is expected to be extremely conservative since clouds can increase significantly the amount of this component which can become therefore higher particularly if actual sky conditions are considered.

The ratio between diffuse and global irradiation is estimated to range from about 5% (12 km) to 6% (6 km). This contribution is then summed to the values presented in Table 1, leading to the determination of the total radiation reaching a sun pointing surface located at an altitude up to 12 km which is presented in Table 2. Here, the comparison between the results obtained and the typical irradiation value expected for a PV array based on the ground (facing south and tilted at a fixed angle close to the latitude of its location) is given. The value of the energy collected by the solar power satellite (Glaser et al., 1974) is also included as a limit solution.

In addition to the contribution of the diffuse component of the solar radiation, some considerations about the albedo flux should be made. Considering that the albedo radiation factor can reach a value of 10 % for a satellite in Low Earth Orbit (Jackson, 1996), this component can significantly contribute to the total radiation estimate. In our specific case it is difficult during this preliminary analysis to give an estimate of this component with a sufficient degree of confidence. It must be noticed though, as a caveat, that the values presented in Table 2 are conservative and they are expected to increase when the albedo is included, especially in presence of cloud layers below the platform when the sun is at high zenith angles.

Altitude [km]	Year Global Irradiation including clouds (conservative estimate) [kWh/m²]	Year Global Irradiation (clear sky) [kWh/m²]
Ground Based (Chilbolton)	1150	
6	3830	4819
9	4985	5080
12	5590	5590
Solar Power Satellite	12000	

Table 2. Comparison between the year global irradiation (including diffuse) values at different altitudes for Chilbolton Observatory (51.1445 N, 1.4370 W)

The results obtained can give a preliminary idea of the gain that a high altitude solar generator could bring in terms of energy collected, if compared to the same generator located on the ground.

Due to the very conservative assumptions the values in the first column can be considered as "minimum" values, and the one in the last column as maximum, although the inclusion of the albedo should give results even higher.

Moreover a collector placed at an altitude of 12km could collect around 45% of what could be collected by the same PV system in a geostationary position (i.e. Solar Power Satellite).

The study presented is preliminary and it involves several assumptions that have been made to simplify the analysis and provide useful results to support the following phases of the project. In particular the extinction parameter data used are acquired in a defined and limited time period and they are relative to a precise location in the south of the UK, which is relatively well placed to collect solar energy on the ground. For this reason and the other discussed previously, the figures presented are conservative and the advantage with respect to ground installations is expected to increase as installations at northern latitude are considered and the values concerning the estimate of the diffuse and albedo contributions are revised.

4. High altitude winds

The knowledge of the mean wind speed at a certain altitude (and its statistical properties) is essential to calculate the aerodynamic forces acting on the aerostat and in particular to determine the forces along the mooring cable.

The wind speed data described in this section were provided by the Natural Environment Research Council (NERC), from the Mesosphere-Stratosphere-Troposphere (MST) Radar station located at Capel Dewi (52.42°N, 4.01°W), near Aberystwyth in west Wales (UK). This facility can provide vertical and horizontal wind speed data, covering an altitude range from 2 to 20 km, with 300 m resolution. However for this study only the data up to 10km were processed. The particular set of data described here covers the period January-December 2007 and measurements were acquired every day continuously. The radar is located near the coast, where the wind speed is expected to be slightly higher than inland and therefore the estimate should be conservative.

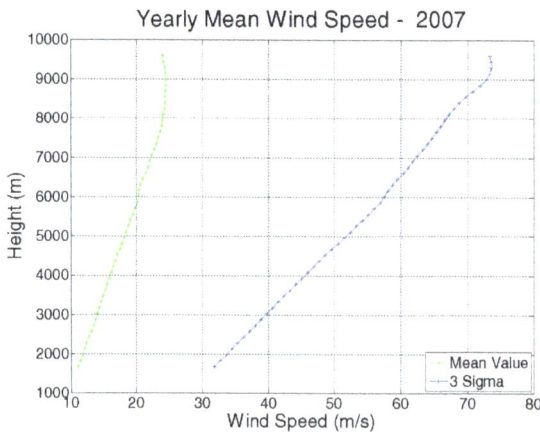

Fig. 4. Wind speed variation with altitude, year mean and 3 sigma values shown

Fig. 4 shows the mean wind speed and the 3 sigma value, and it is possible to notice that at 6km altitude these values are 20m/s and 55m/s respectively.

Another factor to be considered is that, in the time domain the wind speed is quite variable with relatively rapid transients (i.e. gusts). As discussed in Aglietti, 2009 the tethered aerostat is a non-linear system, with considerable damping, and therefore rapid transients of the input in reality produce a response that is significantly lower than what is calculated simply using the maximum wind speed in a static analysis.

5. Concept design

5.1 Tethered aerostat

For an aerostat to support PV devices, this has to be able to produce enough lift via its buoyancy to overcome its weight, the weight of the solar cells plus any control system and that of the tether, still leaving enough margin to produce an appropriate tension in the tether to avoid excessive sag.

Neglecting any aerodynamic lift which could be generated by the shape of the Aerostat, the lifting force due to the aerostat buoyancy is:

$$B = (\rho_{air} - \rho_{gas})gVol \tag{3}$$

where *Vol* is the volume of the aerostat and ρ_{air} and ρ_{gas} the densities of air and gas (helium or hydrogen can be used) filling the aerostat envelop at the specific conditions of operations (e.g. pressure, altitude), and g is the gravity acceleration (9.81m/s^2). Here it is assumed that there is a negligible pressure differential between inside and outside the aerostat envelope, and for simplicity is also taken that the whole volume of the envelope is occupied by the gas (i.e. ballonets for altitude control completely empty).

Fig. 5. Schematic configuration of an Aerostat for Electrical Power Generation, as a gimballed tethered balloon - the grey area represents PV cells cladding.

Typically aerostats have streamlined bodies to reduce the aerodynamic drag, however when such shaped aerostats are moored they then tend to rotate to the oncoming flow direction like a weathervane. Here the aerostat is required to maintain its orientation towards the Sun, therefore a spherical shape pointed through a system of gimbals seems more appropriate (see Fig. 5). A spherical aerostat generates more aerodynamic drag and clearly would require a more substantial structure and tether, but these issues can be tackled by its structural design (Miller & Nahon, 2007). A tethered sphere also suffers substantial vortex induced vibrations (Williamson & Govardhan, 1997). However a previous study (Aglietti, 2009) has shown that due to the non-linearity of the structural problem (mainly the sag of the tether) and the very slow frequency response characterized by a high value of damping, the force oscillations in the tether line (produced by relatively rapid force transients, e.g. gusts) are relatively small. The resultant rotations of the aerostat are only a few degrees which in turn produces a drop in the energy production of less than 1%.

Given its spherical shape, from the volume it is possible to calculate the surface area, and from this, taking an appropriate material area density, it is possible to estimate the weight of the envelope. The area density of the material for the skin can then be increased by 33% as suggested in (Khoury & Gillett, 2004), to account for the weight of various reinforcements, support for the payload etc.

With a similar approach the weight of the PV cells can be estimated by the surface covered, and assuming that a fraction γ of the whole aerostat envelope is covered by the cells, knowing the area density of the cells (also here including wiring etc), it is possible to estimate the weight of the cells. Therefore the weight of aerostat and PV devices can be written as:

$$W_{Aero} = (1.33\delta_{aero} + \delta_{cells}\gamma)g\,4\pi R^2 \qquad (4)$$

where δ_{aero} and δ_{cells} are the area density of the envelope material and PV cells respectively, γ is the fraction of the envelope surface that is covered by the PV cells, g is gravity acceleration and R is the radius of the balloon.

To assess the weight of the tether it is necessary to estimate the weight of the electrical conductors (taken as aluminum for this high conductivity over mass ratio) plus that of the strengthening fibers (e.g. some type of Kevlar). The size of the required conductor can be estimated from the electrical current (that is the ratio between the power generated by the PV devices on the aerostat and the transmission voltage) and setting the electrical losses permitted in the cable to a specific value. Therefore the cross section of the conductor will be

$$A_{cond} = r_{Al}\frac{S}{\eta_{trans}}\frac{P_{gen}}{V^2} \qquad (5)$$

where r_{Al} is the resistivity of the aluminum, S is the overall length of the conductor, η_{trans} is the ratio between the power lost in the cable and that generated by the PV devices (that is P_{gen}), and V is the voltage.

The power generated by the PV system can be estimated from the area covered by the cells (that is a fraction γ of the whole aerostat surface), their efficiency (η_{cells}), an efficiency parameter (η_{area}) that considers that the cells are on a curved surface and therefore the angle of incidence of the sun beam varies according to the position of the cells and finally the solar flux Φ at the aerostat operational altitude that is the irradiance discussed in the previous sections:

$$P_{gen} = 4\pi R^2 \gamma \eta_{cells} \eta_{area} \Phi \tag{6}$$

Finally the weight of the conductor will be its cross sectional area multiplied by length and by its specific weight (density times g), so substituting equation (6) in (5) the weight of the conductor can be written as:

$$W_{cond} = \delta_{Al} g r_{Al} \frac{S^2}{\eta_{trans}} \frac{4\pi R^2 \gamma \eta_{cells} \eta_{area} \Phi}{V^2} \tag{7}$$

The weight of the reinforcing fibres can be calculated from the strength necessary to keep the aerostat safely moored.
The maximum tensile force on the tether can be calculated as:

$$T = \sqrt{(B - W_{Aero})^2 + D^2} \tag{8}$$

where D is the aerostat drag force, equal to:

$$D = \frac{1}{2}\rho_{air} v^2 C_d \pi R^2 \tag{9}$$

In the above expression v is the maximum wind velocity and C_d is the drag coefficient. From the maximum expected tension in the tether, knowing the fibres strength (σ_u) it is possible to calculate the required cross section and from that the weight of the reinforcing fibres.

$$W_{fib} = \delta_{fib} S_T \frac{T}{\sigma_u} \tag{10}$$

where δ_{fib} is the density of the fibers and S_T the length of the tether.
So that the overall weight of the tether will be:

$$W_{Tether} = \delta_{cond} g r_{cond} S^2 \frac{\eta_{cells} \eta_{area} A_{cells} \Phi}{\eta_{trans} V_{gen}^2} + \delta_{fib} S_T \frac{T}{\sigma_u} \tag{11}$$

Aerodynamic forces will also act on the tether line, and they will produce further sagging (see Aglietti, 2009). However this effect does not modify significantly the maximum tension in the tether that will still be at the attachment between the balloon and the tether (equation 8).

5.2 Engineering parameters
In the previous sections, the equations that govern the preliminary sizing of the aerostatic platform have been derived and these equations can be combined, for example, to design a facility with a specified power output. Overall, as the lift and the weight are proportional to the aerostat volume and surface respectively, it will always be possible to design an aerostat large enough to "fly". Here the volume as been set at 179,000 m³ (that is a 35 m radius sphere) which gives a suitable ratio between lift and drag (using helium as a gas filler would give a buoyancy of 1 MN).

In order to reduce the interference with the aviation industry and international air traffic the maximum altitude will be set to 6 km.

The values of the specific engineering parameters which appear in the equations (like for example the area density of the skin) have a crucial role in defining the size of the aerostat. In this section realistic and sometimes conservative values for these parameters will be discussed and utilized in the equations to size a viable platform.

Starting with the solar cells, there are various types available on the market. These range from light weight amorphous silicon triple junction cells (with an efficiency of up to 7%) that could be directly integrated on the skin (see for example Amrani et al., 2007), with a mass penalty that could be as low as 25 g/m^2, to heavier but more efficient cells (e.g. Triple-Junction with Monolithic Diode High Efficiency Cells (www.emcore.com) efficiency 28%), which require some rigid backing and could be used with a mass penalty that can be in the region of 850 g/m^2. These types of cells could be mounted on light weight carbon fibre reinforced plastic tiles that would be used to clad part of the aerostat envelope. Although amorphous silicon cells seem more appropriate, judging by the efficiency over area density ratio, there are issues concerning the ease of installations, repairs, amount of surface available and finally costs that have to be considered. In this study, an efficiency of 15% and an overall area density of the PV cells δ_{cells} (including connectors etc.) of 1 kg/m^2 will be considered.

Taking a maximum peak solar irradiation of 1.2 kW/m^2, from the equations in the previous section it is possible to calculate W_{Aero} (18.9×10^3 kg) and the peak power generated (~0.5 MW). However to size the conductor in the tether it is necessary to set a transmission voltage V, and this should be high enough in order to reduce the losses in the cable. One option is to connect the solar arrays to obtain a voltage in the region of 500V DC and use a converter to bring it up to a few kV. However the converter will introduce some electrical losses and its weight might be an issue as it has to be supported by the aerostat (although the weight of the converter might be compensated by a lighter cable). The other option is to "simply" connect identical groups of solar arrays in series, to maintain the same current and obtain a DC voltage in the region of 1.5-3kV. The solar panels would be provided with bypass and blocking diodes and other circuitry that might be necessary to protect the elements of the system. Setting the transmission voltage at 3kV and allowing for 5% electrical losses in the cable (i.e. η_{trans} = 0.05) enables the cross section of the aluminium conductor and its weight to be calculated as 388 mm^2 and 13.0×10^3 kg respectively.

Using the results in the previous section and taking a maximum wind speed of 55 m/s (3 sigma value) and using equation 9 it is possible to calculate the weight of the fibres as 2.7×10^3 kg, so that the overall weight of the tether will be 15.7×10^3 kg. It should be stressed that the 55 m/s value is quite conservative, in fact this corresponds to the peak wind speed during a gust and due to the highly non linear behaviour of the tether system (see Aglietti, 2009) the force in the tether will be considerably smaller. On the other hand this level of conservatism is more than justified by the catastrophic effect that the tether rupture would have.

6. Conclusion

This chapter has investigated the possibility of using a high altitude aerostatic platform to support PV modules to increase substantially their output by virtue of the significantly enhanced solar radiation at the operating altitude of the aerostat.

Although the figures presented for the analysis of the radiation have been obtained for a specific set of data relative to a well defined location in the UK (and the calculations presented involve some approximations, justified by the preliminary character of the analysis). The results obtained illustrate the advantages, in terms of irradiation, of collecting solar energy between 6km and 12 km altitude, rather than on the ground. The general conclusions can be extended, with a certain degree of approximation, to other countries at the same latitude and with similar climates.

Based on realistic values for the relevant engineering parameters that describe the technical properties of the materials and subsystems, a static analysis of the aerostat in its deployed configuration has been carried out. The results of the computations, although of a preliminary nature, demonstrate that the concept is technically feasible.

As the AEPG requires minimum ground support and could be relatively easily deployed, there are several applications where these facilities could be advantageous respect to other renewables.

It is acknowledged that the concept mathematical model and its concept design are of a preliminary nature. However they do indicate that there is the potential for a new facility to enter the renewable energy market, and further work should be carried out to investigate this possibility more in depth.

7. References

Aglietti, G.S. (2009). Dynamic Response of a High Altitude Tethered Balloon System. AIAA Journal Of Aircraft, Vol. 46, No. 6,   page 2032-2040, doi: 10.2514/1.43332, November–December 2009.

Aglietti, G.S.; Markvart, T.; Tatnall, A.R. & Walker S.J.I. (2008). Aerostat for Electrical Power Generation Concept feasibility. *Proceedings of the IMechE Part G; Journal of Aerospace Engineering*, Vol 222, pp 29-39, ISBN0954-4100; DOI: 10.1243/09544100JAERO258 - Feb 2008.

Aglietti, G.S.; Markvart, T.; Tatnall, A.R. & Walker S.J.I. (2008) Solar Power Generation Using High Altitude Platforms Feasibility and Viability. *Progress in Photovoltaics: Research and Applications*, Vol. 16 pp 349-359, 2008.

Badesha, S.S. (2002). SPARCL: A high-altitude tethered balloon-based optical space-to-ground communication system, *Proceedings of the SPIE - The International Society for Optical Engineering*, Volume 4821, 2002, ISSN: 0277-786X.

Bely, P., & Ashford, R. L. (1995). High-altitude aerostats as astronomical platforms, *Proceedings of SPIE - The International Society for Optical Engineering*, Volume 2478, 1995, Pages 101-116.

Berk, A.; Bernstein L.S. & Robertson, D.C. (1989). MODTRAN: A moderate resolution model for LOWTRAN 7, Report GL-TR-89-0122, Geophysics Laboratory, Air Force Systems Command, United States Air Force, Hanscom, AFB, MA 01731

Colozza, A. & Dolce, J.L. (2005). High-Altitude, Long-Endurance Airships for Coastal Surveillance, *NASA Technical Report*, NASA/TM-2005-213427, 2005.

Dick, H.G. & Robinson, D.H. (1992). *The Golden Age of the Great Passenger Airships: Graf Zeppelin and Hindenburg*, Prentice Hall & IBD; ISBN-13: 978-1560982197, Nov 1992.

Donovan, D. & Van Lammeren A. (2001). Cloud effective particle size and water content profile retrievals using combined lidar and radar observations 1. Theory and examples. *Journal of Geophysical Research*. 106, 27425-27448.

El Amrani, A.; Mahrane, A.; Moussa, F. Y. & Boukennous, Y. (2007). Solar Module Fabrication. *International Journal of Photoenergy*, Vol. 2007, Article ID 27610

Glaser, P.E.; Maynard, O.E.; Mackovciak, J. & Ralph, E.L., Arthur D. Little, Inc., Feasibility study of a satellite solar power station, NASA CR-2357, NTIS N74-17784, Feb. 1974

Gueymard, C.A. (1995). SMARTS, A Simple Model of the Atmospheric Radiative Transfer of Sunshine: Algorithms and Performance Assessment. Technical Report No. FSEC-PF-270-95. Cocoa, FL: Florida Solar Energy Center.

Jackson, B. (1996). A software power model for a spin-stabilized LEO spacecraft utilizing V/T charge control. *Proceedings of the Aerospace Applications Conference IEEE*. 3. 219-227.

Khoury G.A. & Gillett J.D., (2004). *Airship Technology*, Cambridge Aerospace Series, Edited by Gabriel A. ISBN-10: 0521607531 2004.

Kirschner, E.J. (1986). *Aerospace Balloons: From Montgolfiere to Space*, pp. 21, TAB Books, Inc., Blue Ridge Summit, Pennsylvania, 1986.

Kurokawa, K. (2004). *Energy from the Desert: Feasibility of Very Large Scale Photovoltaic Power Generation (VLS-PV) Systems*, Earthscan Publications Ltd, ISBN-10: 1902916417, May 2004

Liou, K.N. (2002). *An Introduction to Atmospheric Radiation*, second ed. Elsevier Science

Mankins, J.C. (1997) A Fresh Look at Space Solar Power: New Architectures, Concepts and Technologies, *Proceedings of the 38th International Astronautical Congress*, IAF paper no IAF-97-R.2.03, 1997.

Markvart, T. & Castañer, L. (2003). *"Practical Handbook of Photovoltaics: Fundamentals and Applications"*, Elsevier, Oxford.

Mayer B. & Kylling A. (2005). Technical note: The libRadtran software package for radiative transfer calculations - description and examples of use. Atmos. Chem. Phys. 5, 1855-1877.

Miller J.I. & Nahon M. (2007). Analysis and Design of Robust Helium Aerostats. *AIAA Journal of Aircraft*, Vol. 44, No. 5, pp. 1447-1458, 2007

Redi S.; Aglietti, G.S.; Tatnall A.R.; Markvart T. (2009) An Evaluation of a High Altitude Solar Radiation Platform *ASME-Journal of Solar Energy Engineering* - in print.

Roberts, B.W.; Shepard, D.H.; Caldeira, K.; Cannon, M.E.; Eccles, D.G.; Grenier, A.J. & Freidin, J.F. (2007). Harnessing High-Altitude Wind Power, *IEEE Transactions On Energy Conversion*, Vol. 22, No. 1, March 2007.

Robinson, D.H. (1973). *Giants in the Sky: A History of the Rigid Airship*, University of Washington Press ISBN-13: 978-0295952499, 1973.

Williamson, C.H.K. & Govardhan, R.G. (1997). Dynamics And Forcing Of A Tethered Sphere In A Fluid Flow, *Journal of Fluids and Structures* (1997) 11 , 293 – 305

http://rosaerosystems.pbo.ru/english/products/puma.html, accessed 25/09/09

http://www.cloud-net.org/, accessed 25/09/09

http://www.emcore.com/assets/photovoltaics/btjm.solar.ds.pdf, accessed 25/09/09

http://www.magenn.com/news.php, accessed 25/09/09

http://www.tcomlp.com/aerostats.html, accessed 25/09/09

3

Solar Chimney Power Plants – Developments and Advancements

Marco Aurélio dos Santos Bernardes
Centro Federal de Educação Tecnológica de Minas Gerais – CEFET-MG
Brazil

1. Introduction

A wide range of existing power technologies can make use of the solar energy reaching Earth. Basically, all those ways can be divided into two basic categories: transformed for use elsewhere or utilized directly – direct – and involving more than one transformation to reach a usable form – indirect. The Solar Chimney Power Plant (SCPP) is part of the solar thermal group of indirect solar conversion technologies.

More specifically, a natural phenomenon concerning the utilization of the thermal solar energy involves the earth surface heating and consequently the adjacent air heating by the sun light. This warm air expands causing an upward buoyancy force promoting the flow of air that composes the earth atmosphere. The amount of energy available due to the upward buoyancy force associated with the planet revolution is so vast that can generate catastrophic tropical cyclones with disastrous consequences.

From another standpoint, such phenomenon can be enhanced and used in benefit of the human well-being. In this way, the SCPP is a device developed with the purpose to take advantage of such buoyancy streams converting them into electricity. For that, a greenhouse – the collector – is used to improve the air heating process, a tall tube – the chimney – promotes the connection between the warm air nearby the surface and the fresh air present in higher atmosphere layers and a system to convert the kinetic energy into electricity – the generator-turbine system (Fig. 1).

2. First steps and recent developments

One of the earliest descriptions of a solar chimney power station was written in 1903 by Isidoro Cabanyes, a Spanish artillery colonel. He made public the proposition "Proyecto de motor solar" (solar engine project) introducing an apparatus consisting of an air heater attached to a house with a chimney. In the house interior, a kind of wind propeller was placed with the purpose of electricity production, as shown in Fig. 2, (Cabanyes, 1903).

In 1926 Prof Engineer Bernard Dubos proposed to the French Academy of Sciences the construction of a Solar Aero-Electric Power Plant in North Africa with its solar chimney on the slope of the high height mountain, (Fig. 3., (Günther, 1931)). The author claims that an ascending air speed of 50 m/s can be reached in the chimney, whose enormous amount of energy can be extracted by wind turbines. Fig. 4 shows an solar chimney futurist representation presented by (Günther, 1931). Fig. 5 shows a simple experiment proposed by

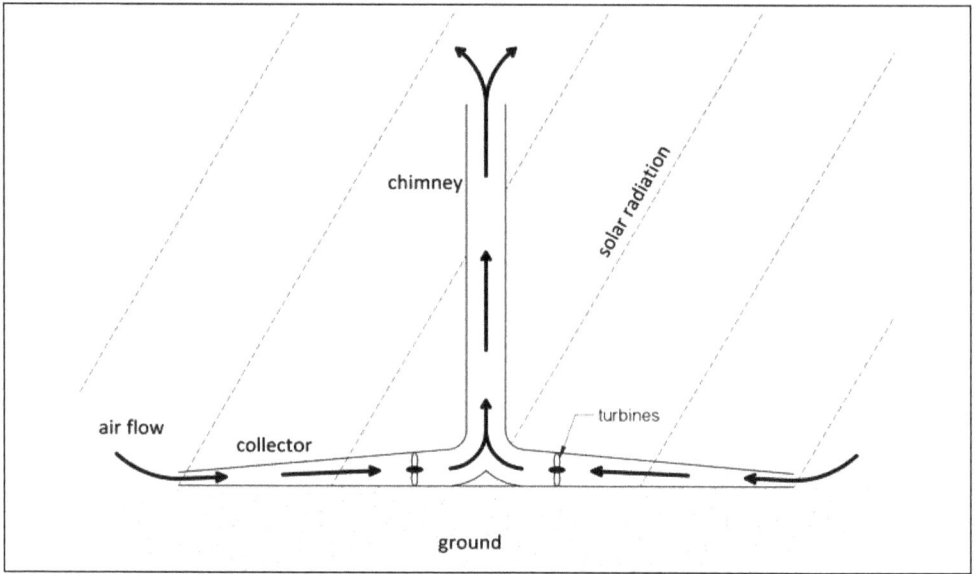

Fig. 1. SCPP components.

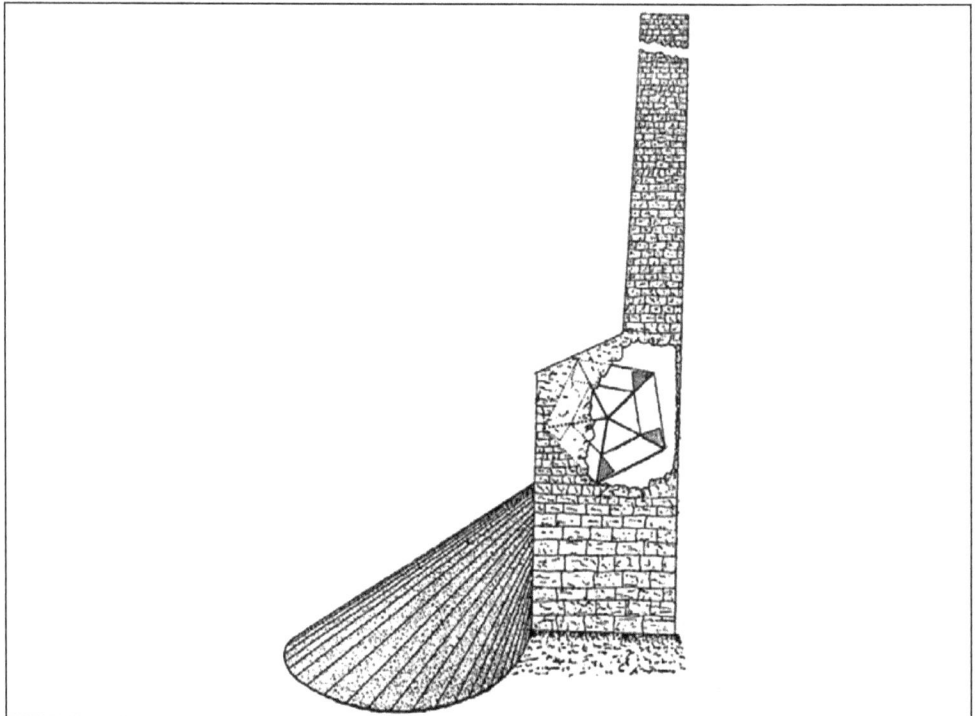

Fig. 2. Solar engine project proposed by Isidoro Cabanyes..

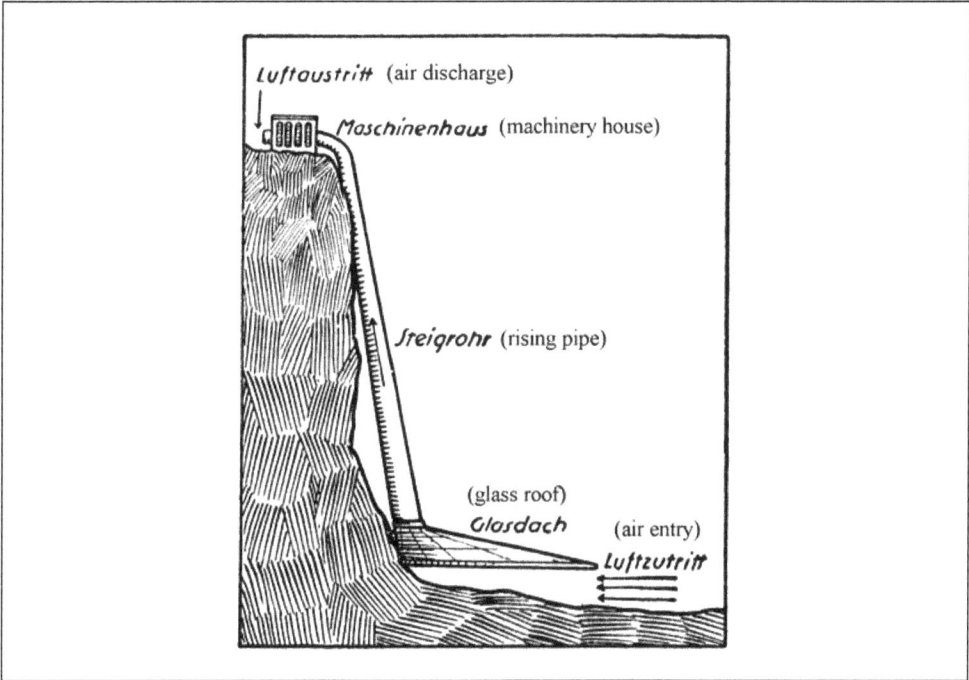

Fig. 3. Solar chimney proposal presented by (Günther, 1931).

Fig. 4. Solar chimney futurist representation presented by (Günther, 1931).

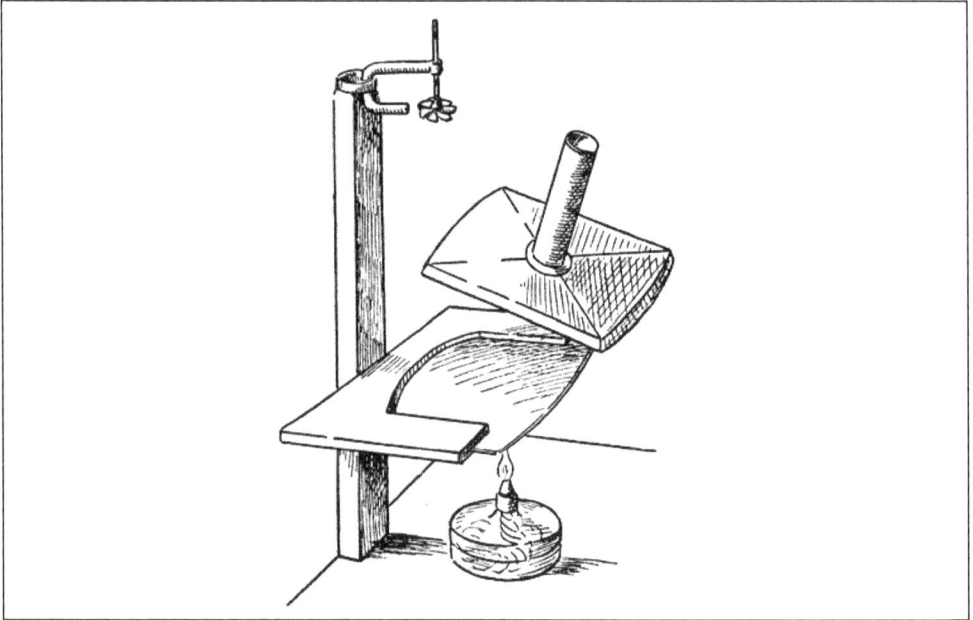

Fig. 5. Solar chimney proposal presented by (Günther, 1931).

Dubos confirming its concept. According to Günther (1931), the plate and the spirit lamp represent the Sahara desert and the solar heat, respectively. The small wind wheel placed on the top represents the wind turbines. If the spirit lamp is positioned under the plate, warm air flows concentrically through the plate reaching the tube. Consequently, the ascendant flow impels the wind wheel.

In the face of the original concepts, the first outstanding action for the SCPP development was the prototype erection in 1982 in Manzanares, Ciudad Real, 150 km south of Madrid, Spain. The chimney height was 195 m and its diameter 10 m. The collector area was 46,000 m² (about 11 acres, or 244 m diameter). Regardless of its dimensions, this prototype was considered as a small-scale experimental model. As the model was not intended for power generation, the peak power output was about 50 kW. Different glazing materials were tested, as well as, collector sections were used as an actual greenhouse, growing plants under the glass. The construction of the pilot plant were commissioned by the Minister of Research and Technology of the Federal Republic of Germany, (Haaf, *et al.*, 1983). The work was supervised by the energy research project management department of *Kernforschungsanalage (KFA) Jülich GmbH* (the Jülich Nuclear Energy Research Establishment) and fundamental principles of SCPP were found in the form of simple estimates. (Haaf, 1984) divulged preliminary test results including energy balances, collector efficiency values, pressure losses due to friction and losses in the turbine section. (Castillo, 1984) suggested a new "soft" structure approach to the chimney building instead of the conventional "rigid" one.

The SCPP has notable advantages in comparison with other power production technologies, namely - (Schlaich, 1995):

• the collector uses both direct und diffuse radiation;

- the ground provides a natural heat storage;
- the low number of rotating parts ensure its reliability;
- no cooling water is necessary for its operation;
- simple materials and known technologies are used in its construction;
- non OECD countries are able to implement such technology without costly technological efforts

After the prototype operation kick-off, several studies can be found about SCPP. They include transient and steady state fluid dynamic and thermal models, as well as, structural analysis for chimney, collector (including the ground natural heat storage capability) and turbine setups.

2.1 SCPP theoretical models

(Mullett, 1987) started the SCPP operational theoretical models development by deriving overall efficiency and relevant performance data. In his calculation, the overall efficiency is proportional the chimney height, returning about 1% for a height of 1000 m. He concluded that the solar chimney is essentially a power generator of large scale. The chimney efficiency is given by the equation (1).

$$\eta_t = \frac{gH}{c_p T_0} \tag{1}$$

Here, g is the gravity [m/s], H is the chimney height [m], c_p is the air heat capacity [J/kg·K] and T_0 is the ambient temperature [K]. For instance, with a chimney height of 1000 m and standard conditions for temperature and pressure, the chimney efficiency achieves the maximal value of 3 %. Considering a collector efficiency (η_c) of 60 % and a turbine efficiency (η_{tur}) of 80 %, the total system efficiency (η_{tot}) reaches 1.4%, as shown by equation (2).

$$\eta_{tot} = \eta_t \cdot \eta_c \cdot \eta_{tur} = 0.03 \cdot 0.6 \cdot 0.8 = 0.014 \tag{2}$$

Based on the data from the prototype of Manzanares, (Padki & Sherif, 1989) elaborated extrapolated SCPP models for medium-to-large scale power generation. (Yan, et al., 1991) described a more comprehensive analytical model for SCPP by using practical engineering correlations obtaining equations for air velocity, airflow rate, power output, and the thermo-fluid efficiency and (Padki & Sherif, 1992) also presented a mathematical model for SCPP. In the end of the 90's, (Pasumarthi & Sherif, 1998a) built a SCPP small-scale demonstration prototype to study the effect of various geometric parameters on the air temperature, air velocity, and power output of the solar chimney. Further studies conducted by (Pasumarthi & Sherif, 1998b) exploited the collector performance by extending the collector base and by introducing an intermediate absorber. According to them, both enhancements helped to increase the overall chimney power output. In addition, a brief economic assessment of the system costs is presented.

The first known attempt to solve by CFD (Computational Fluid Dynamics) the convective flow in a SCPP is showed by (Bernardes, et al., 1999). He presented a solution for Navier-Stokes and Energy Equations for the natural laminar convection in steady state, predicting its thermo-hydrodynamic behavior. The approach Finite Volumes Method in Generalized

Coordinates was employed allowing a detailed visualization of the effects of geometric of optimal geometric and operational characteristics.

(Kröger & Blaine, 1999) evaluated the influence of prevailing ambient conditions. Their work shows that the air moisture can enhance the driving potential and that under certain conditions condensation may occur. In the meanwhile, (Kröger & Buys, 1999) developed analytical relations for determining the pressure differential due to frictional effects and heat transfer correlations for developing radial flow between the roof and the collector.

A set of differential equations for SCPP is deducted and integrated by (Padki & Sherif, 1999). Expressions for the power generated and the efficiency were obtained by making simplifying assumptions.

(Gannon & von Backström, 2000) introduced a study including chimney friction, system, turbine and exit kinetic energy losses in the analysis. For that, a simple model of the solar collector is used to include the coupling of the mass flow and temperature rise in the solar collector. This work, verified by comparing the simulation of a small-scale plant with experimental data, is useful to predict the performance and operating range of a large-scale plant. A one-dimensional compressible flow approach for the calculation of all the thermodynamic variables as dependent on chimney height, wall friction, additional losses, internal drag and area change was developed by von Backström and Gannon (2000). They concluded that the pressure drop associated with the vertical acceleration of the air is about three times the pressure drop associated with wall friction. For a flared chimney (14%, in order to keep the through-flow Mach number constant virtually), the vertical acceleration pressure drop can be eliminated. (Kröger & Buys, 2001) developed relevant equations for a SCPP.

(Gannon & Von Backström, 2002a) proposed a turbine design based on the design requirements for a full-scale solar chimney power plant integrating the turbine with the chimney. In this way, the chimney base legs are offset radially and act as inlet guide vanes introducing pre-whirl before the rotor reducing the exit kinetic energy. By employing a three-step turbine design method and a free vortex analysis method, the major turbine dimensions are determined. The flow path through the inlet guide vanes and rotor is predicted by a matrix throughflow method. The blade profiles are optimized by using the scheme coupled to a surface vortex method to achieve blades of minimum chord and low drag. The authors stated that the proposed turbine design can extract over 80% of the power available in the flow. Additionally, an experimental investigation of a SCPP design was undertaken by (Gannon & Von Backström, 2002b) and (Gannon & Von Backström, 2003). Results of the experimental model turbine revealed a total-to-total efficiency of 85-90% and total-to-static efficiency of 77-80% over the design range.

(Bernardes, et al., 2003) developed a comprehensive SCPP analysis including analytical and numerical models to describe its performance, i.e., to estimate power output of solar chimneys as well as to examine the effect of various ambient conditions and structural dimensions on the power output. The mathematical model was validated with experimental results and the model was used to predict the performance characteristics of large-scale commercial solar chimneys. It turns out that the height of chimney, the factor of pressure drop at the turbine, the diameter and the optical properties of the collector are important parameters for the SCPP design.

(Schlaich, et al., 2003) describe the functional principle of the Solar Tower and give some results from designing, building and operating a small scale prototype in Mazanares. Future commercial Solar Tower systems are presented, as well as technical issues and basic economic data.

In his technical brief, (von Backström, 2003) develops calculation methods for the pressure drop in very tall chimneys. Equations for the vertical pressure and density distributions in terms of Mach number allowing density and flow area change with height, wall friction and internal bracing drag are presented. To solve the equations, two simplifications are presented, namely, the adiabatic pressure lapse ratio equation to include flow at small Mach numbers and extension of the hydrostatic relationship between pressure, density, and height to small Mach numbers. Accurate value of the average density in the chimney can be obtained by integration.

(Schlaich, et al., 2004) reintroduces the SCPP concept reinforcing its role as a sustainable option to produce solar electricity at low costs. (Koonsrisuk & Chitsomboon, 2004) analyzed the frictional effect on the flow in a SCPP.

The influence of the atmospheric winds on performance of SCPP was studied by (Serag-Eldin, 2004). By means of a computational model – governing partial differential equations expressing conservation of mass, energy and balance of momentum in addition to a two equation model of turbulence – the flow pattern in the neighborhood of a small-scale SCPP model was computed. The analysis results shown a total degradation of performance with strong winds and significant degradation for low speed winds, except for collector with low inlet height.

(Kirstein, et al., 2005) and (Kirstein & Von Backström, 2006) presented studies concerning the loss coefficient in the transition section between a SCPP turbine and the chimney as dependent on inlet guide vane stagger angle and collector roof height including scaled model experiments and commercial CFD simulations. The very good agreement between the experiments and the simulations permits predictions for a proposed full-scale geometry. Using the solar chimney prototype in Manzanares as a practical example, (Ming, et al., 2006) carried out numerical studies to explore the geometric modifications on the system performance, showing reasonable agreement with the analytical model. The SCPP performance was evaluated, in which the effects of various parameters on the relative static pressure, driving force, power output and efficiency were investigated.

(Pretorius & Kröger, 2006c) evaluated the performance of a large-scale SCPP. A particular reference plant under specified meteorological conditions at a reference location in South Africa was choose and developed convective heat transfer and momentum equations were employed. The authors claim that 24 hr plant power production is possible and that plant power production is a function of the collector roof shape and inlet height. Furthermore, more accurate turbine inlet loss coefficient, quality collector roof glass and various types of soil on the performance of a large scale SCPP are introduced ((Pretorius & Kröger, 2006b)). Results pointed out that the heat transfer correlations employed reduced the plant power output significantly. A more accurate turbine inlet loss coefficient has no noteworthy effect, at the same time as utilizing better quality glass enhances plant power production. Simulations employing Limestone and Sandstone soil gave virtually similar results to a Granite-based model.

(Von Backström & Fluri, 2006) investigated analytically the validity and applicability of the assumption that, for maximum fluid power, the optimum ratio of turbine pressure drop to pressure potential (available system pressure difference) is 2/3. They concluded that the constant pressure potential assumption may lead to overestimating the size of the flow passages in the plant, and designing a turbine with inadequate stall margin and excessive runaway speed margin.

The effects of the solar radiation on the flow of the SCPP were analyzed by (Huang, *et al.*, 2007). Boussinesq approximation and Discrete Ordinate radiation model (DO) were introduced in the model and simulations were carried out.

(Koonsrisuk & Chitsomboon, 2007) proposed the use of dimensionless variables to conduct the experimental study of flow in a small-scale SCPP for generating electricity. The similarity of the proposed dimensionless variables was confirmed by computational fluid dynamics.

(Ming, *et al.*, 2007) set up different mathematical models for heat transfer and flow for the collector, chimney and turbine. After the validation with the Manzanares prototype, they suggested that the output power of MW-graded can exceed 10 MW.

A new mathematical model based on the concept of relative static pressure was developed by (Peng, *et al.*, 2007). According to them, optimized local geometric dimensions between the collector outlet and the chimney inlet, the local velocity at this place ca increase 14% the SCPP performance, temperature profiles are more uniform, and the relative static pressure decreases about 50% improving the system energy conversion and reducing the energy losses.

(Pretorius & Kröger, 2007) presented a sensitivity analysis on the influence of the quality, thickness, reflectance, emissivity, shape, and insulation of the collector roof glass, the cross section of the collector roof supports, various ground types, ground surface roughness, absorptivity and emissivity, turbine and bracing wheel loss coefficients, and the ambient pressure and lapse rate on the performance of a large-scale (reference) solar chimney power plant. Computer simulation results point out that collector roof insulation, emissivity and reflectance, the ambient lapse rate, and ground absorptivity and emissivity all have a key effect on the power production of such a plant.

(Ming, *et al.*, 2008a) continued their work carrying out numerical simulations analyzing characteristics of heat transfer and air flow in the solar chimney power plant system with an energy storage layer including the solar radiation and the heat storage on the ground. They concluded that the ground heat storage depends on the solar radiation incidence. Higher temperature gradients also increase the energy loss from the ground ((Ming, *et al.*, 2008b)) and (Ming, *et al.*, 2008c)). Subsequently, (Tingzhen, *et al.*, 2008) included a 3-blade turbine in the model simulation and validated the model, turning out power out and turbine efficiency of 10 MW and 50%, respectively.

Due to the use of different heat transfer coefficients found in the literature, namely those of ((Bernardes, 2004b),(Pretorius & Kröger, 2006a)), (Bernardes, *et al.*, 2009) made a comparison of the methods used to calculate the heat fluxes in the collector and their effects on solar chimney performance. Notwithstanding the difference between the heat transfer coefficients, both approaches returned similar air temperature rises in the collector and therefore, comparable produced power.

(Koonsrisuk & Chitsomboon, 2009) went on with their previous study, namely (Koonsrisuk & Chitsomboon, 2007), trying to maintain dynamic similarity between the prototype and its theoretical model keeping the same solar heat flux. They showed that, for the same heat flux condition, all dimensional parameter, except the roof radius, must remain similar. They also revealed some engineering interpretations for the similarity variables.

SCPP energy and exergy balances were carried out by (Petela, 2009), turning out the energy distribution in the components for a input of 36.81 MW energy of solar radiation (or an equivalent input of 32.41 MW of radiation exergy), including also a sensibility analysis.

More recently (Bernardes and von Backström, 2010) performed a study regarding the performance of two schemes of power output control applicable to solar chimney power plants. It was revealed that the optimum ratio is not constant during the whole day and it is dependent of the heat transfer coefficients applied to the collector.

3. SCPP analysis for specific sites

(Dai, et al., 2003)   analyzed a solar chimney power plant to provide electric power for remote villages in northwestern China. Three counties in Ning Xia Hui Autonomous region, namely, Yinchuan, Pingluo, and Helan with good solar radiation availability were selected as sites to simulate a SCPP model. In according to the authors, a SCPP consisting of a 200 m height and 10 m diameter chimney, and the 500 m diameter solar collector can produce 110 ~ 190 kW electric power on a monthly average all year.

(Bilgen and Rheault, 2005) developed a mathematical model based on monthly average meteorological data and thermodynamic cycle to simulate the SCPP power production at high latitudes. Three locations in Canada, namely Ottawa, Winnipeg and Edmonton were choose to evaluate a 5 MW nominal power production plant. The authors also suggested the construction of a sloped collector field at suitable mountain hills in order to work as a collector. Next, for this proposition, a short vertical chimney is added to install the vertical axis air turbine. They also shown that such plant can produce as much as 85% of the same plants in southern locations with horizontal collector field and the overall thermal performance of these plants is a little less than 0.5%.

SCPP for rural villages was studied by (Onyango and Ochieng, 2006) emphasizing some features for power generating. They disclosed that for a temperature ratio = 2.9 (i.e., the difference between the collector surface temperature and the temperature at the turbine to the difference between the air mass temperature under the roof and the collector surface temperature) an 1000 W of electric power can be generated. The minimum dimension of a practical by a reliable SCPP to assist approximately fifty households in a typical rural setting has been determined to be chimney length = 150 m, height above the collector = chimney radius = 1.5 m.

(Zhou, et al., 2007b) developed mathematical model to investigate power generating performance of a SCPP prototype. The steady state simulation returned power outputs for different global solar radiation intensity, collector area and chimney height. The simulation results were validated with the measurements.

4. SCPP turbine developments

(Gannon & Von Backström, 2003), (Gannon & Von Backström, 2002b) and (Gannon & Von Backström, 2002b) were the first to develop an experimental investigation of the performance of a solar chimney turbine. The design consisted of a single rotor by using inlet guide vanes to introduce pre-whirl. Such strategy reduces the turbine exit kinetic energy at the diffuser inlet and assists the flow turning in the IGV-to-rotor duct. Total-to-total efficiencies of 85-90% and total-to-static of 77-80% over the design range were measured.

(Von Backström & Gannon, 2004) presented analytical equations in terms of turbine flow and load coefficient and degree of reaction, to express the influence of each coefficient on turbine efficiency. Analytical solutions for optimum degree of reaction, maximum turbine efficiency for required power and maximum efficiency for constrained turbine size were found. According to the authors, a peak turbine total-to-total efficiency of around 90% is achievable, but not necessarily over the full range of plant operating points.

Some structural aspects of classical wind energy turbines, like their high-cycle dynamic loading and reaction as well as their fatigue behavior were exposed by (Harte & Van Zijl, 2007). Structural challenges concerning wind action, eigenfrequencies, stiffening and shape optimization with special focus on the inlet guide vanes were discussed for SCPP's.

(Fluri & von Backström, 2008) analyzed many different layouts for the SCPP turbogenerator. Turbine layouts with single rotor and counter rotating turbines, both with or without inlet guide vanes were considered. They concluded that the single rotor layout without guide vanes performs very poorly; the efficiency of the other three layouts is much better and lies in a narrow band.

(Tingzhen, et al., 2008) performed numerical simulations on SCPP's coupled with turbine. The model was validated with the measurements from the Spanish prototype, obtaining a maximum power output higher than 50 kW. Subsequently, the authors presented the design and the simulation of a MW-graded solar chimney power plant system with a 5-blade turbine. The numerical simulation results show that the power output and turbine efficiency are 10 MW and 50%, respectively, which presents a reference to the design of large-scale SCPP's.

5. SCPP experimental analysis

(Pasumarthi & Sherif, 1998a; b): The solar chimney is a natural draft device which uses solar radiation to provide upward momentum to the in-flowing air, thereby converting the thermal energy into kinetic energy. A study was undertaken to evaluate the performance characteristics of solar chimneys both theoretically and experimentally. In this paper, a mathematical model which was developed to study the effect of various parameters on the air temperature, air velocity, and power output of the solar chimney, is presented. Tests were conducted on a demonstration model which was designed and built for that purpose. The mathematical model presented here, was verified against experimental test results and the overall results were encouraging. his paper describes details of the experimental program conducted to assess the viability of the solar chimney concept. A demonstration model was designed and built and its theoretical and experimental performance was examined. Two experimental modifications were tried on the collector: (1) extending the collector base and (2) introducing an intermediate absorber. The former modification helped in enhancing the air temperature, while the latter contributed to increasing the air temperature as well as the mass flow rate inside the chimney. Both enhancements helped to increase the overall chimney power output. Theoretical and experimental performance results of this demonstration model are presented in this paper, while the mathematical model developed in Part I was used to predict the performance of much larger systems. Mathematical model results were validated by comparing them to published data on the solar chimney system built in Manzanares, Spain. Also, an economic assessment of the system costs are presented.

(Gannon & Von Backström, 2002b): An experimental investigation of a solar chimney turbine design is undertaken. The aim of the program is to demonstrate and evaluate a proposed solar chimney turbine design. The measured results of an experimental model turbine are presented and the turbine efficiency calculated. The current turbine design has a total-to-total efficiency of 85-90% and total-to-static efficiency of 77-80% over the design range. Secondary objectives are to compare the measured and predicted results and through investigation of the experimental results suggest improvements to the turbine design.

(Zhou, et al., 2007a): A pilot experimental solar chimney power setup consisted of an air collector 10 m in diameter and an 8 m tall chimney has been built. The temperature distribution in the solar chimney power setup was measured. Temperature difference between the collector outlet and the ambient usually can reach 24.1 Â°C, which generates the driving force of airflow in the setup. This is the greenhouse effect produced in the solar collector. It is found that air temperature inversion appears in the latter chimney after sunrise both on a cool day and on a warm day. Air temperature inversion is formed by the increase of solar radiation from the minimum and clears up some time later when the absorber bed is heated to an enough high temperature to make airflow break through the temperature inversion layer and flow through the chimney outlet.

(Ferreira, et al., 2008): Solar dryers use free and renewable energy sources, reduce drying losses (as compared to sun drying) and show lower operational costs than the artificial drying, thus presenting an interesting alternative to conventional dryers. This work proposes to study the feasibility of a solar chimney to dry agricultural products. To assess the technical feasibility of this drying device, a prototype solar chimney, in which the air velocity, temperature and humidity parameters were monitored as a function of the solar incident radiation, was built. Drying tests of food, based on theoretical and experimental studies, assure the technical feasibility of solar chimneys used as solar dryers for agricultural products. The constructed chimney generates a hot airflow with a yearly average rise in temperature (compared to the ambient air temperature) of 13 ± 1 °C. In the prototype, the yearly average mass flow was found to be 1.40 ± 0.08 kg/s, which allowed a drying capacity of approximately 440 kg.

(Ming, et al., 2008c): A small scale solar chimney system model has been set up, and the temperature distribution of the system with time and space, together with the velocity variation inside the chimney with time, has been measured. The experimental results show that the temperature distributions inside the collector and the effects of seasons on the heat transfer and flow characteristic of system show great agreement with the analysis, while the temperature decrease significantly inside the chimney as the chimney is very thin which causes very high heat loss.

6. SCPP structural analysis

(Harte & Van Zijl, 2007) presented some structural aspects of classical wind energy turbines, like their high-cycle dynamic loading and reaction as well as their fatigue behaviour. Actual research results concerning pre-stressed concrete tower constructions for wind turbines will be focused on. For the solar chimney concept the structural challenges concerning wind action, eigenfrequencies, stiffening and shape optimization with special focus on the inlet guide vanes will be discussed.

7. SCPP ecological analysis

(Bernardes, 2004a) performed is a comprehensive evaluation of impacts caused by mass and energy flows of solar chimneys systems from its design through to production and then final disposal using the method of Life Cycle Assessment. The conventional Life Cycle Assessment method was improved by an additional sectoral analysis (input-output analysis), namely Hybrid Approach. The study was an important contribution for the integration of Life Cycle Assessment in the decision making process in the renewable energy sector and for an integrated evaluation of processes.

8. SCPP economical analysis

In their first approach (Haaf, *et al.*, 1983) concluded, from the relationships between the physical principles on the one hand and the scale and construction costs on the other, that economical power generation will be possible with large-scale plants designed for up to 400 MW/pk

(Pretorius & Kröger, 2008) undertook a study to stablish a thermoeconomically optimal plant configurations for a large-scale SCPP. For that, an approximated cost model was developed, giving the capacity for finding optimum plant dimensions for different cost structures. Thermoeconomically optimal plant configuration were obtained through multiple computer simulations and results comparison to the approximated cost of each specific plant.

A study developed by (Fluri, *et al.*, 2009) revealed that previous economical models may have underestimated the initial cost and levelised electricity cost of a large-scale solar chimney power plant. It also showed that carbon credits significantly can reduce the levelised electricity cost for such a plant.

9. Alternative concepts and applications

Probably (Ferreira, *et al.*, 2008) were the first to propose a solar chimney as a device to dry agricultural products. A small scale prototype solar chimney was built, in which the air velocity, temperature and humidity parameters were monitored as a function of the solar incident radiation. Based on theoretical and experimental studies Drying tests reveled the technical feasibility of solar chimneys used as solar dryers for agricultural products. A hot airflow with a yearly average rise in temperature (compared to the ambient air temperature) of 13 ± 1 °C could be achieved allowing a drying capacity of approximately 440 kg.

(Zhu, *et al.*, 2008) proposed different heat storage styles for SCPP. The experimental studies showed that the temperature difference in the sealed water system is the largest, while the open water system has the lowest one because of the latent heat consumed by water evaporation. The study also showed that there is the temperature distribution optimization of the system if the heat loss part of the collector to be avoided.

The concept for producing energy by integrating a solar collector with a mountain hollow is presented and described by (Zhou, *et al.*, 2009). As in a conventional SCPP, the hot air is forced by the pressure difference between it and the ambient air to move along the tilted segment and up the vertical segment of the 'chimney', driving the turbine generators to generate electricity. The author claimed that such concept provides safety and reduces a great amount of construction materials in the conventional chimney structure and the energy cost to a level less than that of a clean coal power plant.

The hypothesis of combining a salinity gradient solar pond with a chimney to produce power in salt affected areas is examined by (Akbarzadeh, *et al.*, 2009). The salinity in northern Victoria, Australia was analyzed and salinity mitigation schemes were presented. It was shown that a solar pond can be combined with a chimney integrating an air turbine for the production of power. A prototype of a solar pond of area 6 hectares and depth 3 m with a 200 m tall chimney of 10 m diameter was investigated and

10. Conclusion

The previous literature review about SCPP presents an outstanding technological development enlightening considerable advances in its construction, operation, including its technical economical and ecological relevant facets.

In contrast with other solar facilities, SCPPs can be used above and beyond power production. Very relevant byproducts are distilled water extracted from ocean water or ground water. Under certain conditions, agribusiness may be appropriate under the solar collector. It can involve fruits and vegetables, medicinal and aromatic essential oils from herbs and flowers, seaweeds and planktons, blue-green algae, ethanol and methane, biodiesel and all manner of vegetable and plant derivatives, etc. Besides, remaining biomass is useful creating additional heat during composting.

The insertion of SCPP in the power generation market requires scalability and base, shoulder and peak load electricity generation. Further developments should meet such localized requirements.

11. References

Akbarzadeh, A., et al. (2009), Examining potential benefits of combining a chimney with a salinity gradient solar pond for production of power in salt affected areas, *Solar Energy, 83*, 1345-1359.

Bernardes, M. A. d. S. (2004a), Life Cycle Assessment of solar Chimneys, paper presented at VIII World Renewable Energy Congress and Expo, Denver, USA, August 29-Sep 3.

Bernardes, M. A. d. S. (2004b), Technische, ökonomische und ökologische Analyse von Aufwindkraftwerken, PhD Thesis thesis, 230 pp, Universität Stuttgart, Stuttgart.

Bernardes, M. A. d. S., et al. (1999), Numerical Analysis of Natural Laminar Convection in a Radial Solar Heater, *International Journal of Thermal Sciences, 38*, 42-50.

Bernardes, M. A. d. S., et al. (2009), Analysis of some available heat transfer coefficients applicable to solar chimney power plant collectors, *Solar Energy, 83*, 264-275.

Bernardes, M. A. d. S., et al. (2003), Thermal and technical analyses of solar chimneys, *Solar Energy, 75*, 511-524.

Bilgen, E., and J. Rheault (2005), Solar chimney power plants for high latitudes, *Solar Energy, 79*, 449-458.

Cabanyes, I. (1903), Proyecto de Motor Solar, *La Energia Eléctrica - Revista General de Electricidad y sus Aplicaciones, 8*, 61-65.

Castillo, M. A. (1984), A New Solar Chimney Design to Harness Energy from the Atmosphere, in *Spirit of Enterprise - The Rolex 1984 Rolex Awards*, edited by M. Nagai and A. e. J. Heiniger.

Dai, Y. J., et al. (2003), Case study of solar chimney power plants in Northwestern regions of China, *Renewable Energy, 28*, 1295-1304.

Ferreira, A. G., et al. (2008), Technical feasibility assessment of a solar chimney for food drying, *Solar Energy, 82*, 198-205.

Fluri, T. P., et al. (2009), Cost analysis of solar chimney power plants, *Solar Energy*, *83*, 246-256.

Fluri, T. P., and T. W. von Backström (2008), Comparison of modelling approaches and layouts for solar chimney turbines, *Solar Energy*, *82*, 239-246.

Gannon, A. J., and T. W. von Backström (2000), Solar chimney cycle analysis with system loss and solar collector performance, *Journal of Solar Energy Engineering, Transactions of the ASME*, *122*, 133-137.

Gannon, A. J., and T. W. Von Backström (2002a), Solar chimney turbine part 1 of 2: Design, paper presented at International Solar Energy Conference, Reno, NV.

Gannon, A. J., and T. W. Von Backström (2002b), Solar chimney turbine part 2 of 2: Experimental results, paper presented at International Solar Energy Conference, Reno, NV.

Gannon, A. J., and T. W. Von Backström (2003), Solar chimney turbine performance, *Journal of Solar Energy Engineering, Transactions of the ASME*, *125*, 101-106.

Günther, H. (1931), *In hundert Jahren*, 78 pp., Kosmos - Gesellschaft der Naturfreunde, Stuttgart.

Haaf, W. (1984), Solar Chimneys - Part II: Preliminary Test Results from the Manzanares Pilot Plant, edited, pp. 141 - 161, Taylor & Francis.

Haaf, W., et al. (1983), Solar Chimneys - Part I: Principle and Construction of the Pilot Plant in Manzanares, edited, pp. 3 - 20, Taylor & Francis.

Harte, R., and G. P. A. G. Van Zijl (2007), Structural stability of concrete wind turbines and solar chimney towers exposed to dynamic wind action, *Journal of Wind Engineering and Industrial Aerodynamics*, *95*, 1079-1096.

Huang, H., et al. (2007), Simulation Calculation on Solar Chimney Power Plant System, in *Challenges of Power Engineering and Environment*, edited, pp. 1158-1161.

Kirstein, C. F., and T. W. Von Backström (2006), Flow through a solar chimney power plant collector-to-chimney transition section, *Journal of Solar Energy Engineering, Transactions of the ASME*, *128*, 312-317.

Kirstein, C. F., et al. (2005), Flow through a solar chimney power plant collector-to-chimney transition section, paper presented at International Solar Energy Conference, Orlando, FL.

Koonsrisuk, A., and T. Chitsomboon (2004), Frictional effect on the flow in a solar chimney paper presented at Proceedings of the 4th National Symposium on Graduate Research, Chiang Mai, Thailand.

Koonsrisuk, A., and T. Chitsomboon (2007), Dynamic similarity in solar chimney modeling, *Solar Energy*, *81*, 1439-1446.

Koonsrisuk, A., and T. Chitsomboon (2009), Partial geometric similarity for solar chimney power plant modeling, *Solar Energy*, *83*, 1611-1618.

Kröger, D. G., and D. Blaine (1999), Analysis of the Driving Potential of a Solar Chimney Power Plant, *South African Inst. of Mechanical Eng. R & D J.*, *15*, 85-94.

Kröger, D. G., and J. D. Buys (1999), Radial Flow Boundary Layer Development Analysis, *South African Inst. of Mechanical Eng. R & D J.*, *15*, 95-102.

Kröger, D. G., and J. D. Buys (2001), Performance Evaluation of a solar Chimney Power Plant, in *ISES 2001 Solar World Congress*, edited, Adelaide, south Australia.

Ming, T., et al. (2008a), Numerical analysis of flow and heat transfer characteristics in solar chimney power plants with energy storage layer, *Energy Conversion and Management, 49,* 2872-2879.

Ming, T., et al. (2008b), Numerical analysis of heat transfer and flow in the solar chimney power generation system, *Taiyangneng Xuebao/Acta Energiae Solaris Sinica, 29,* 433-439.

Ming, T., et al. (2006), Analytical and numerical investigation of the solar chimney power plant systems, *International Journal of Energy Research, 30,* 861-873.

Ming, T. Z., et al. (2008c), Experimental simulation of heat transfer and flow in the solar chimney system, *Kung Cheng Je Wu Li Hsueh Pao/Journal of Engineering Thermophysics, 29,* 681-684.

Ming, T. Z., et al. (2007), Numerical simulation of the solar chimney power plant systems with turbine, *Zhongguo Dianji Gongcheng Xuebao/Proceedings of the Chinese Society of Electrical Engineering, 27,* 84-89.

Mullett, L. B. (1987), Solar Chimney - Overall Efficiency, Design and Performance, *International Journal of Ambient Energy, 8,* 35-40.

Onyango, F. N., and R. M. Ochieng (2006), The potential of solar chimney for application in rural areas of developing countries, *Fuel, 85,* 2561-2566.

Padki, M. M., and S. A. Sherif (1989), Solar chimney for medium-to-large scale power generation, paper presented at Proceedings of the Manila International Symposium on the Development and Management of Energy Rexources, Manila, Philippines.

Padki, M. M., and S. A. Sherif (1992), A Mathematical Model for Solar Chimneys, paper presented at Proceedings of the 1992 International Renewable Energy Conference, in Renewable Energy: Research and Applications, University of Jordan, Faculty of Engineering and Technology, Amman, Jordan, June 22-26, 1992.

Padki, M. M., and S. A. Sherif (1999), On a simple analytical model for solar chimneys, *International Journal of Energy Research, 23,* 345-349.

Pasumarthi, N., and S. A. Sherif (1998a), Experimental and theoretical performance of a demonstration solar chimney model - Part I: Mathematical model development, *International Journal of Energy Research, 22,* 277-288.

Pasumarthi, N., and S. A. Sherif (1998b), Experimental and theoretical performance of a demonstration solar chimney model - Part II: Experimental and theoretical results and economic analysis, *International Journal of Energy Research, 22,* 443-461.

Peng, W., et al. (2007), Research of the optimization on the geometric dimensions of the solar chimney power plant systems, *Huazhong Keji Daxue Xuebao (Ziran Kexue Ban)/Journal of Huazhong University of Science and Technology (Natural Science Edition), 35,* 80-82.

Petela, R. (2009), Thermodynamic study of a simplified model of the solar chimney power plant, *Solar Energy, 83,* 94-107.

Pretorius, J. P., and D. G. Kröger (2006a), Critical evaluation of solar chimney power plant performance, *Solar Energy, 80,* 535-544.

Pretorius, J. P., and D. G. Kröger (2006b), Solar chimney power plant performance, *Journal of Solar Energy Engineering, Transactions of the ASME, 128,* 302-311.

Pretorius, J. P., and D. G. Kröger (2006c), Thermo-economic optimization of a solar chimney power plant, paper presented at CHISA 2006 - 17th International Congress of Chemical and Process Engineering, Prague.

Pretorius, J. P., and D. G. Kröger (2007), Sensitivity analysis of the operating and technical specifications of a solar chimney power plant, *Journal of Solar Energy Engineering, Transactions of the ASME, 129*, 171-178.

Pretorius, J. P., and D. G. Kröger (2008), Thermoeconomic optimization of a solar chimney power plant, *Journal of Solar Energy Engineering, Transactions of the ASME, 130*, 0210151-0210159.

Schlaich, J. (1995), *The Solar Chimney: Electricity from the Sun*, Edition Axel Menges, Stuttgart.

Schlaich, J., et al. (2003), Design of commercial solar tower systems - Utilization of solar induced convective flows for power generation, paper presented at International Solar Energy Conference, Kohala Coast, HI.

Schlaich, J., et al. (2004), Sustainable electricity generation with solar updraft towers, *Structural Engineering International: Journal of the International Association for Bridge and Structural Engineering (IABSE), 14*, 225-229.

Serag-Eldin, M. A. (2004), Computing flow in a solar chimney plant subject to atmospheric winds, paper presented at Proceedings of the ASME Heat Transfer/Fluids Engineering Summer Conference 2004, HT/FED 2004, Charlotte, NC.

Tingzhen, M., et al. (2008), Numerical simulation of the solar chimney power plant systems coupled with turbine, *Renewable Energy, 33*, 897-905.

von Backström, T. W. (2003), Calculation of pressure and density in solar power plant chimneys, *Journal of Solar Energy Engineering, Transactions of the ASME, 125*, 127-129.

von Backström, T. W., and T. P. Fluri (2006), Maximum fluid power condition in solar chimney power plants - An analytical approach, *Solar Energy, 80*, 1417-1423.

von Backström, T. W., and A. J. Gannon (2004), Solar chimney turbine characteristics, *Solar Energy, 76*, 235-241.

Yan, M. Q., et al. (1991), Thermo-fluid analysis of solar chimneys, paper presented at American Society of Mechanical Engineers, Fluids Engineering Division (Publication) FED, Publ by ASME, Atlanta, GA, USA.

Zhou, X., et al. (2009), Novel concept for producing energy integrating a solar collector with a man made mountain hollow, *Energy Conversion and Management, 50*, 847-854.

Zhou, X., et al. (2007a), Experimental study of temperature field in a solar chimney power setup, *Applied Thermal Engineering, 27*, 2044-2050.

Zhou, X., et al. (2007b), Simulation of a pilot solar chimney thermal power generating equipment, *Renewable Energy, 32*, 1637-1644.

Zhu, L., et al. (2008), Temperature rise performance in solar chimneyes with different heat storages, *Taiyangneng Xuebao/Acta Energiae Solaris Sinica, 29*, 290-294.

4

Photon Management in Dye Sensitized Solar Cells

Silvia Colodrero, Mauricio E. Calvo and Hernán Míguez
Instituto de Ciencia de Materiales de Sevilla
Consejo Superior de Investigaciones Científicas-Universidad de Sevilla
Spain

1. Introduction

Solar energy is nowadays one of the most promising future energy resources due to the depletion of fossil fuels, which supply the major part of all energy consumed worldwide. Among the different types of solar cell technologies, dye sensitization of mesoporous oxide based films has attracted a great deal of interest in the last few years because of the possibility it offers to achieve moderate efficiency devices at very low cost, being therefore an interesting alternative to conventional p-n junction solar cells. Dye sensitized solar cells (DSSC) consist of a nanocrystalline wide band gap semiconductor (usually TiO_2), which is deposited onto a transparent conductive substrate, and on whose surface a dye is adsorbed. The cell is completed with a counterelectrode, and both electrodes are put into electrical contact by infiltrating a liquid electrolyte in between them. Light is absorbed by the dye and charges are separated at the interface between the dye and the metal oxide it is anchored to. The optimization of the conversion efficiency, that is the fraction of light intensity that is converted into electrical power, is a key issue for this type of solar systems. In this way, different modifications of the originally proposed cell have been made in order to improve its performance, most of them based on the use of different semiconductors, dyes or ionic conductor. There is also an increasing interest in employing nanostructures to improve solar energy conversion devices (Kamat, 2007).

Another interesting route to enhance the cell efficiency is to modify its optical design in order to improve the light harvesting efficiency (LHE) or absorptance within the cell. The approach that has been explored the most has been the use of a diffuse scattering layer made of large TiO_2 colloids that are either deposited onto the nanocrystalline electrode or mixed with the nanocrystalline titania (nc-TiO_2) slurry. In both cases, increase of the optical light path within the absorbing layer rises the matter-radiation interaction time thus enhancing the probability of photon absorption. It has to be bear in mind that any structure introduced in the cell must permit the electrical contact between the electrolyte and the sensitized semiconductor slab, which forces it to have a porosity capable of sustaining the flow of charges. In recent times, different alternatives to light management in DSSC are being proposed and realized due to the development of novel porous periodic photonic nanostructures that can be easily integrated in these devices. The aim of this chapter is to give a brief review of the efforts performed to improve the light harvesting in DSSCs through the optimization of their optical design and provide a detailed description of the new emerging possibilities based on coupling dye-sensitized electrodes to different periodic photonic nanostructures. It will be shown that they have a great potential for the manipulation of light propagation. A detailed description of the integration processes used, of the different mechanisms of light harvesting enhancement that take place, and of actual examples showing the improvement of performance achieved will be presented in this chapter.

2. Brief description of DSSC

Photovoltaic devices have become a promising alternative energy source in the last decades. They are expected to increasingly and significantly contribute to overall energy production over the coming years. The photovoltaic field, dominated mainly by inorganic solid-state junction cells, is now being challenged by the emergence of new devices based on nanocrystalline and conducting polymer films, which offer a very low-cost fabrication and attractive features such as transparency, flexibility, etc. that might facilitate the market entry. Among all of them, dye sensitized solar cells (DSSC) are devices that have shown to reach moderate efficiencies, thus being feasible competitors to conventional cells.

DSSC combine the optical absorption and charge-separation processes by the association of a sensitizer as light-absorbing material with a wide band-gap semiconductor (usually titanium dioxide). As early as the 1970s, it was found that titanium dioxide (TiO_2) from photoelectrochemical cells could split water with a small bias voltage when exposed to light (Fujishima & Honda, 1972). However, due to the large band-gap for TiO_2, which makes it transparent for visible light, the conversion efficiency was low when using the sun as illumination source. Dye sensitization of semiconductor electrodes dates to the 1960s (Gerischer & Tributsch, 1968). This pioneering research involved an absorption range extension of the system into the visible region, as well as the verification of the operating mechanism by injection of electrons from photoexcited dye molecules into the conduction band of the n-type semiconductor. Since only a monolayer of adsorbed dye molecules was photoactive, light absorption was low and limited when flat surfaces of the semiconductor electrode were employed. This inconvenience was solved by introducing polycrystalline TiO_2 (anatase) films with a surface roughness factor of several hundreds (Desilvestro et al., 1985; Vlachopoulos et al., 1988). The amount of adsorbed dye was increased even further by using mesoporous electrodes, providing a huge active surface area thereby, and cells combining such electrodes and a redox electrolyte based on iodide/triiodide couple yielded 7% conversion efficiencies in 1991 (O´Regan & Gratzel, 1991). The current highest energy conversion efficiency is over 11% (Chiba et al., 2006), and further increase of the efficiency is possible by designing proper electrodes and sensitization dyes.

Figure 1 shows both a scheme and an energy level diagram of a liquid electrolyte dye sensitized solar cell. They usually consist of one electrode made of a layer of a few micrometers of titanium dioxide nanocrystals (average crystal size around 20 nm), that have been sintered together to allow electronic conduction to take place. A monolayer of a sensitizer dye, typically a ruthenium polypyridyl complex, is attached to the surface of the nanocrystalline electrode. This mesoporous film is deposited onto a conductive, transparent substrate, typically indium tin oxide (ITO) or fluorinated SnO_2 (FTO), and soaked with a redox electrolyte, typically containing I^-/I_3^- ion pairs. This electrolyte is also in contact with a colloidal platinum catalyst coated counter-electrode. Sunlight is harvested by the dye producing photo-excited electrons that are injected into the conduction band of the nanocrystalline semiconductor network, and then into the conducting substrate. At the same time the redox electrolyte reduces the oxidized dye and transports the electron acceptors species (I_3^-) to the counter-electrode, where the I_3^- is reduced back to I^- and the electrical circuit is completed via electron migration through the external load.

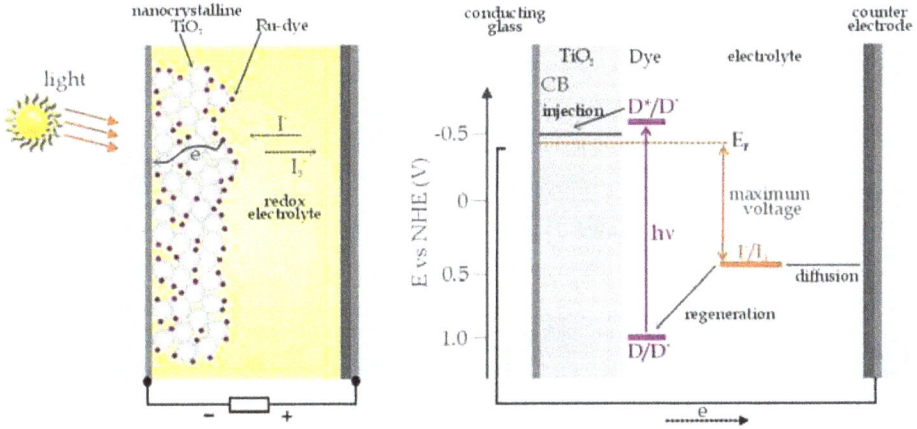

Fig. 1. Cross section view of the design of a dye sensitized solar cell under illumination conditions (left), and energy levels of the different components of the cell that represent the energetics of operation of such devices (right).

In contrast to silicon devices, charge separation is primarily driven by the oxidation/reduction potentials of the different species at the TiO_2/dye/electrolyte interface, being screened out any electric field gradient in the TiO_2 electrode due to the high concentration of mobile ions employed in the liquid electrolyte (Zaban et al., **1997**). Photoinduced charge separation takes place at the TiO_2/electrolyte interface. Thus, electron injection requires the dye excited state to be more reducing than the TiO_2 conduction band. In the same way, regeneration of the dye ground state by the redox couple requires the dye cation to be more oxidizing than the I^-/I_3^- redox couple (Mori & Yanagida, 2006). The voltage output of the device is approximately given by the splitting between the TiO_2 Fermi level and the chemical potential of the redox electrolyte, being the former related with the density of injected electrons and the density of charge traps in the band gap of TiO_2. Under illumination conditions, the density of electrons injected into the semiconductor conduction band increases, raising the Fermi level towards the conduction-band edge and generating a photovoltage in the external circuit.

Charge transport processes within the cell are considered to be diffusive (Södergren et al., 1994), (Cao et al., 1996), (Schwarzburg & Willig, 1999) and are driven by concentration gradients generated in the device, thus making electrons to go towards the working electrode and triiodide ions towards the counter electrode. During the diffusion process, photogenerated electrons can recombine with acceptors species, such us dye cations and triiodide ions. Another loss pathway includes decay of the dye excited state to ground (Huang et al., 1997), (Nelson et al., 2001). Kinetic competitions between the different forward and loss pathways are therefore critical to determine the quantum efficiencies of charge separation and collection, and so the conversion efficiency. A diagram showing the kinetics of a DSSC is presented in Figure 2. It should be noticed that not only energetics but also kinetics must be taken into account, and they constitute the key issues to achieve high energy conversion devices.

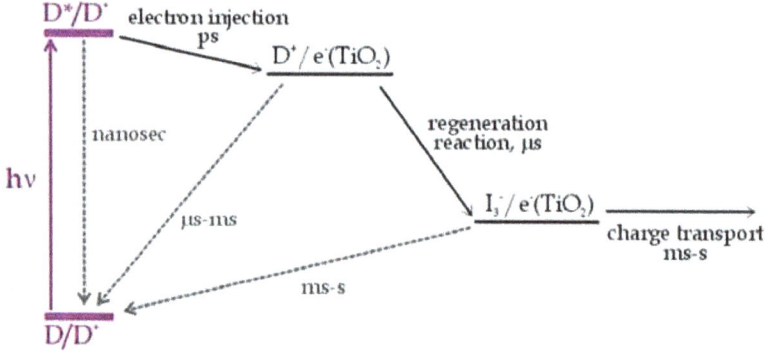

Fig. 2. Kinetics and competitive processes involved in the conversion of light to electric power in DSSC.

The overall conversion efficiency (η) of the dye-sensitized solar cell is determined by the photocurrent density measured at short circuit (J_{SC}), the open-circuit photovoltage (V_{oc}), the fill factor (ff), and the power of the incident light (P_{in}). These values can be extracted from the photocurrent density-voltage characteristics (IV curves) under AM 1.5 full sunlight (P_{in}=100 mWcm-2). The relation between J and V is determined by varying the resistance of the outer circuit, being J_{SC} obtained when the resistance of the outer circuit is zero (thus voltage is zero) and V_{oc} when the resistance is maximum (thus photocurrent is zero). The output power of the device equals the product of J and V, and the fill factor expresses the efficiency of the device compared to that of and ideal cell. P_{max} is commonly reported as the output power of the commercial device and corresponds to the maximum value that can reach the output power. The performance of DSSC can be therefore estimated using the following equations:

$$\eta(\%) = \frac{J_{SC} \cdot V_{OC} \cdot ff}{P_{in}} \tag{1}$$

Where,

$$ff = \frac{P_{max}}{J_{SC} \cdot V_{OC}} \tag{2}$$

A typical IV curve corresponding to a 7 µm thick dyed-TiO$_2$ electrode measured under 1 sun illumination is displayed in Figure 3 (left). The solar radiation and the ruthenium dye absorptance spectra are shown in Figure 3 (right).

For a detailed description of DSSCs, we refer the reader to M. Graetzel (Graetzel, 2000) and M. Graetzel and J. Durrant (Graetzel & Durrant, 2008).

In the next section, we analyze two different approaches that contribute to the enhancement of DSSC efficiencies through the control of photon absorption into the cell. We put special emphasis to describe the integration of new materials known as porous one-dimensional photonic crystal due to their ease of integration and demonstrated promising performance.

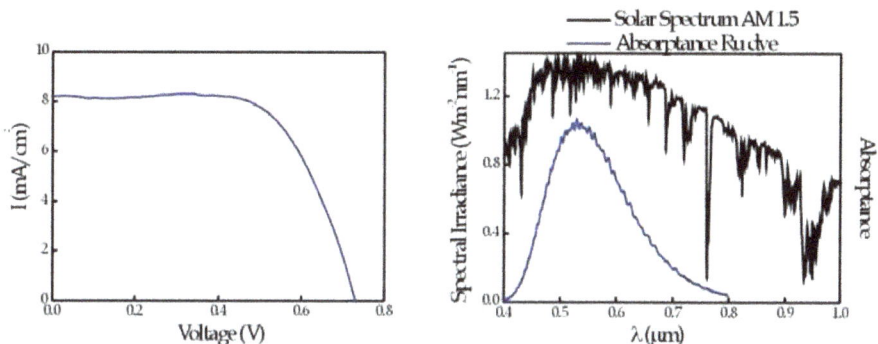

Fig. 3. IV curve for a 7 μm thick TiO$_2$ electrode, for which N719 as dye and a I$^-$/I$_3^-$ redox couple based liquid electrolyte have been employed (left), and mismatch between the dye absorption spectrum and that for the AM 1.5 solar spectrum (right).

3. Approaches to light management in DSSC

In the last years, many attempts using different modifications of the originally proposed cell have been made in order to improve its performance, most of them based on the use of different semiconductors (Tennakone et al., 1999), dyes (Wang et al., 2005) or ionic conductors (Wang et al., 2004). However, the extremely delicate sensibility of the charge transport and recombination dynamics to any alteration of the nature of the interfaces present in the cell should be considered (Haque et al., 2005). For instance, some of the most important routes of research have focused on the molecular engineering of suitable dyes having broader absorption spectra that show a better matching to the solar spectrum and higher molar extinction coefficients (Wang et al., 2005), thus yielding higher short circuit currents. However, further improvements in terms of cell stability and durability should be done. On this respect, the performance of cells using solid state hole conductor based DSSC (Bach et al., 1998) to increase the long-term stability still remains far from that achievable when liquid organic electrolytes are employed.

The quantification of the electrical kinetic parameters of the cell has attracted the attention of many research groups and a great effort has been made in this direction. However, less interest has been paid to the study and development of optical elements that could be introduced in the cell for boosting the optical path of light, thus increasing the probability for the photons to be absorbed. Although it was well-known that by optical means the output power of the cell can be enhanced through a higher photogenerated current, since it depends on the number of photons collected by the dye (Tachibana et al., 2002), it was also clear that the introduction of optical elements that can enhance light harvesting in DSSC was not straightforward. First, they are typically made of dense materials, which would block the flow of charged species in solution. Second, the standard fabrication and integration processes usually employed to make optical materials did not seem to be compatible with the colloidal chemistry approaches normally taken to prepare a DSSC. The first and most successful approach to light management in DSSC was based on the introduction of a diffuse scattering layer, as described below, which largely enhances the photon path length through the working electrode, thus increasing the probability of optical absorption to take place. More recently, approaches based on periodic structures are also being investigated with promising results. Apart from the large enhancements of efficiency these latter structures gives rise to, they present the added advantage of allowing for the precise selection of the spectral range at which optical absorption is amplified, leading to both control over the aspect and the semi-transparency of the cell.

3.1 Effect of increase optical absorptance on the efficiency of the cell

Optical approaches to raise J_{SC}, and therefore efficiencies, are based on the increase of optical absorption caused by either an enlargement of the photon path length through the working electrode or light trapping effects occurring within the TiO_2 electrodes. J_{SC} can be attained by integrating the product of the ratio between the solar spectral irradiance and the photon energy density, $F(\lambda)$, and the photon-to-current conversion efficiency, IPCE, of the cell over the wavelength of the incident light:

$$J_{SC} = \int q\xi(\lambda)F(\lambda)IPCE(\lambda)d\lambda \qquad (3)$$

Here q is the electron charge and $\xi(\lambda)$ is a factor that accounts for the losses at the air-substrate interface. IPCE can be expressed as the product of light harvesting efficiency (LHE) and the electron-transfer yield $\Phi(\lambda)$, that is the product of the electron injection yield and the charge collection efficiency.

$$IPCE(\lambda) = LHE(\lambda)\Phi(\lambda) \qquad (4)$$

The LHE or optical absorptance, A, at a certain wavelength is defined as the fraction of incident photons that are absorbed by the dyed electrode: $A=I_A/I_0$, where I_0 is the incident intensity and I_A the intensity absorbed. Therefore the relationship between LHE and J_{SC} is given by the expression:

$$J_{SC} = \int q\xi(\lambda)F(\lambda)LHE(\lambda)\Phi(\lambda)d\lambda \qquad (5)$$

In a first approximation, absorptance for a standard dyed electrode is related with the extinction coefficient and the concentration of the dye. Since only a monolayer of dye molecules is attached to the surface of the TiO_2 nanoparticles, its total amount is directly related to the oxide layer thickness. For example, a 7 μm thick dye sensitized (N719) mesoscopic electrode can absorb nearly the 80% of incident photons at the maximum absorption wavelength. However, photons still in the visible range but of lower energy are weakly absorbed. Looking for a dye absorption enhancement, devices having thicker electrodes (around 10 μm) have been previously reported (Ito et al, 2006). Nevertheless, the thickness of TiO_2 layer cannot be increased at will without affecting its mechanical properties, reaching mass transport limitations in the electrolyte or/and reducing the photovoltage of the cell. In addition, electrons injected in the conduction band must travel a longer distance to reach the back contact, increasing the probability of recombination at grain boundaries and diminishing both current and voltage of the cell, as experimentally demonstrated (Ito et al., 2008). Finally, another disadvantage to scale-up the device would appear due to the high cost of sensitizer dyes (approx. 1000€/g). The thicker the electrodes, the higher the dye loads required, therefore raising the final cost of the DSSC. All these nondesirable features make it preferable to increase the absorption of the cell for a given dye and film thickness, modifying the optical path length within the film and improving the spectral response of the photoelectrode. Keeping in mind that a standard dye sensitized layer of around 7-8 μm thick will not absorb light strongly, these achievements can be obtained by reflecting light back to the dyed electrode. In what follows, we provide an overview of the different approaches taken towards the integration of optical passive components in order to increase the power conversion efficiency of DSSCs.

3.2 Diffuse scattering layer

TiO_2 working electrodes used in DSSC are composed of 20 nm size crystallites. These electrodes are essentially transparent since visible light is not scattered for titania particles of sizes on the order of the few tens of nanometers. In fact, the incident photons that are not absorbed by the dye sensitized electrode are either lost through the counter electrode or partially absorbed by the electrolyte solution. From a photo-chemical point of view, this implies that part of the reagent (light) is wasted. The first attempt to collect these escaping photons were based on the use of polydisperse packings of sub-micron size spheres as highly diffusive reflecting layers (see figure 4). The intensity of the scattering effect depends on the size and refractive index of the particles as well as on the refractive index of the medium surrounding them. Hence, particle sizes between 300 nm and 1000 nm made of transition metal oxides with high refractive index, such as TiO_2 (rutile n=2.8 or anatase, n=2.4) or ZrO_2 (n=2.1) can be used as efficient scatterers. The introduction of a reflecting layer in DSSC to scatter photons and re-inject them into the electrode was proposed in a theoretical work by Usami (Usami, 1997). In other attempt, Ferber and Luther simulated the scattering process for a mixture of small and large particles, concluding that an enhancement in photon absorption was produced (Ferber & Luther, 1998). Subsequent simulations using different approaches were made (Rothenberger et al., 1999), for which a 6% of increase of the photon flux was predicted.

The integration of scattering centre particles in DSSC can be experimentally done under different architectures. In one approach, they are jointly included with the TiO_2 nanocrystallites that form the electrode (Tachibana et al., 2002). In another, they are deposited as a second layer on top of the dye sensitized electrode in a well-known configuration referred to as "double-layer" (Hore et al., 2006). The latter configuration is normally the most employed. Other ways of integrating scattering layers made of submicron size disordered particles have also been reported (Wang et al., 2004), (Zhang et al., 2007). In general, the scattering layer is deposited using methods similar to those used for the TiO_2 electrode deposition, such as doctor blade, screen printing, etc. A porous network connection between both layers is needed to allow the dye load and a proper diffusion of the charge carriers.

Improvements of efficiency around 20% in average are reported using the arrangements before described for nc-TiO_2 layers with thicknesses between 5 and 7 microns. In fact, the efficiency record attained for a DSSC corresponds to a cell that incorporates a highly scattering layer (Chiba et al., 2006). The boost in efficiency is mainly a consequence of the increase of J_{SC} up to 20%. On the other hand, IPCE values can be incremented between 20% and 50% depending on the spectral region considered. The highest improvement is obtained in the red part of the spectrum, where the dye extinction coefficient is small. Additionally, this light-scattering layer has been shown to act not only as a photon-trapping system but also to be an equally active layer in photovoltaic generation (Zhang et al., 2007).

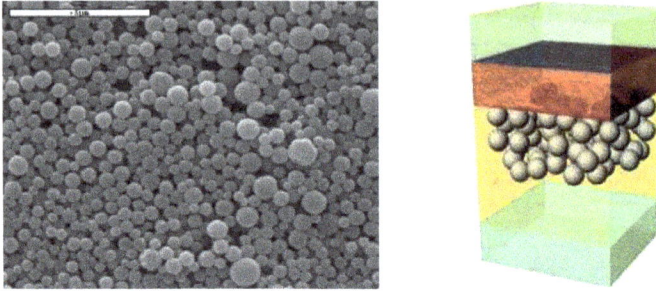

Fig. 4. Left: Scanning electron microcospy image of a slab made of polydisperse sub-micrometer TiO$_2$ particles that can be used as diffuse scattering layer in DSSC. Right: Scheme representing a DSSC with a "double layer" architecture.

The disadvantage of employing diffuse light scattering layers or mixed light scattering particles in DSSC is the loss of transparency of the cell. Unfortunately, the cells turn opaque, leaving them useless as window modules or any other application where transparency, one of the added values of these cells, was required. Also, the thickness of the electrode largely increases, particularly in the case of the double layer configuration, which might cause an increase of the resistance of the cell and a reduction of the voltage.

3.3 Periodic structures

An alternative to the use of disordered structures to enhance diffuse light scattering is the introduction of porous materials in which a periodic variation of the refractive index has been built up. As it will be shown next, periodic structures allows to achieve high reflectance within the cell at targeted and well-defined wavelength ranges, which may prevent the drawback of the loss of transparency. Also, in some cases, highly reflecting structures can be only a few hundreds of nanometres thick, which reduces the potential problems of increase of resistance and reduction of the photovoltage. Depending on the spatial dimensions where the modulation of refractive index is found, we will refer these structures as one, two, or three dimensional photonic crystals (1DPC, 2DPC, or 3DPC, respectively) (Joannopoulos et al. 1995). The interference effects associated with these periodic dielectrics give rise to the opening of a photonic band gap whose effect is detected as a maximum in the specular reflectance spectrum of the structure. In the following section, the effect of integrating both 3DPC and 1DPC in DSSC is described.

3.3.1 Three dimensional ordered structures

In 2003, Mallouk and co-workers proposed the use of a novel type of optical elements to improve the efficiency of DSSC. Their approach was based on the coupling of a particular type of 3DPC to dye-sensitized nc-TiO$_2$ films (Nishimura et al., 2003). This approach consisted of integrating a 3 µm layer of a TiO$_2$ ordered porous structure, known as inverse opal (see figure 5). By doing so, IPCE was shown to increase with respect to that of a standard cell used as reference (Figure 6a). Although the origin of this enhancement was first attributed to the reduction of the group velocity of photons near the edge of a stop band or photonic pseudogap, which implies the increase of the probability of absorption, it was later found that the absorption enhancement effect was partly due to the diffuse scattering caused by the imperfections present in the opal film (Halaoui et al., 2005) and, mainly, to the coupling between a standard electrode and the ordered structure, as explained next.,

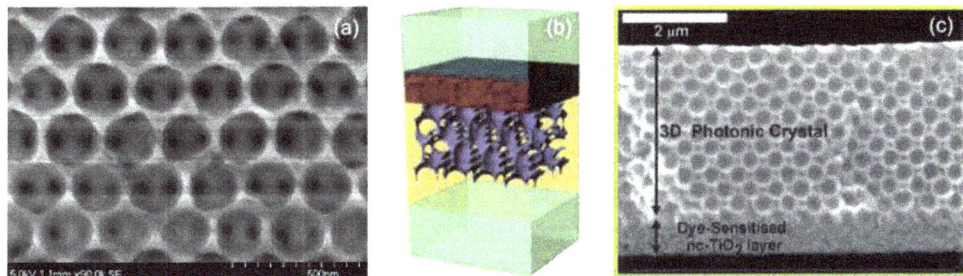

Fig. 5. (a) SEM cross section image of an inverse TiO_2 opal. 300 nm latex spheres were used as template, and then removed by thermal treatment. (b) Scheme and (c) SEM cross section image of a TiO_2 3D inverse opal deposited on top of a TiO_2 nanocrystalline electrode.

The theoretical analysis of the phenomena (Mihi & Míguez, 2005) demonstrated that the enhancement of the photocurrent in bilayer structures formed by a dye sensitized nanocrystalline TiO_2 film coupled to an inverse titania opal is mainly due to the surface resonant modes confined within the overlayer (figure 6b). This is a consequence of the mirror effect of the photonic crystal. The theoretical results identified an increase in the absorptance spectrum of the modelled bilayer and an estimated photocurrent enhancement factor very similar to that reported in Mallouk experiments (Figure 6c). Later experimental results confirmed convincingly the model proposed by Mihi & Míguez (Lee et al., 2008; Mihi et al., 2008) and as they predicted, the improvement was found for frequencies comprised within the photonic pseudogap of the inverse opal structure. In that way, the dependence of the increased IPCE with the position of the forbidden interval range can be employed to enhance absorption at desired spectral ranges. The selection of this region is made through the lattice parameter of the photonic crystal which is controlled as well by the size of the templating spheres and the infiltration degree. In these specific experiments, DSSC required an opal with a lattice parameter that gives red photocurrent enhancement.

Additionally it is possible to couple two or more 3DPC with different lattice parameters to enhance photocurrent in a wider spectral range (Mihi et al., 2006). Since this effect was confirmed theorically, we can mention that the disadvantage of these materials is the several steps involved to achieve these types of structures. This leads to defects within the 3D photonic crystal, and therefore a lower reflectivity is obtained.

3.3.2 DSSC coupled to 1DPC

Although the coupling of inverse opals to dye sensitized electrodes demonstrated an increased IPCE with respect to that of a reference cell (Nishimura et al., 2003), the main drawback of these 3D structures is the difficult assembly process to achieve reasonable reflecting periodic materials, which leads usually to thick structures (between 5-10 micron thick). This might have a deleterious effect on charge transport and recombination through the cell. Very recently, new types of one-dimensional photonic crystals (1DPC) have been prepared by alternate deposition of either mesoporous (Choi et al., 2006); (Fuertes et al., 2007) or nanoparticles (Wu et al., 2007); (Colodrero et al., 2008) based films. These structures

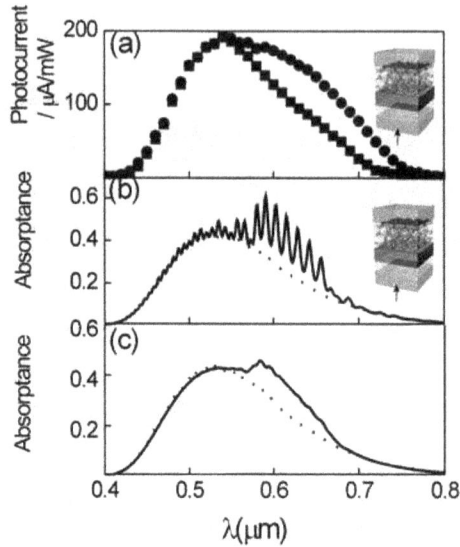

Fig. 6. (a) Circles: Photogenerated current observed for a bilayer DSSC like the one shown in the inset when illuminated from the rear side. Squares: Photocurrent corresponding to a non periodically structured, standard DSSC. These data have been extracted from Nishimura et al., 2004. The curve in (b) shows the calculated absorptance (or LHE) for the structure shown in the inset under rear illumination. (c) Averaged absorptance of bilayer DSSCs formed by nc-TiO$_2$ inverse opals of different width (from 3 to 17 sphere monolayers), each one of them having in turn different nc-TiO$_2$ layer thickness on top (from 6.5 μm to 7.5 μm). Dotted lines in (b) and (c) are the calculated absorptance spectra of standard DSSC having the same amount of absorbing material than in the bilayer system. The insets show schemes of the modelled structure. The corresponding illumination direction is indicated by an arrow. (*Extracted with permission from Mihi & Míguez, 2005*)

are usually easier to build and integrate than those abovementioned of higher dimensionality and present attractive features, such as very intense and wide Bragg reflections and reduced thickness (less than a micron versus the several micron thickness of opals). Furthermore, the advantage of such lattices lies on the wide range of materials available to be deposited as multilayers, which implies accurate control over the optical properties of the periodic ensemble, and on the high structural and optical quality attainable. These nanostructures could therefore be a potentially interesting alternative to other type of light scattering layers used within the solar cell field, having created high expectations due to the large improvement of the performance achieved for this type of devices.

In this section, we will focus on mesostructured Bragg reflectors in which the building blocks are nanoparticles of different sort (Colodrero et al., 2008) that can be easily coupled to DSSC to enhance the optical absorption. The novelty of these nanostructures is mainly the large and highly accessible interconnected mesoscopic porosity that they can present, which makes them suitable for this type of solar devices. In fact, some of the most successful approaches developed to improve the LHE in silicon photovoltaic systems are based on the implementation of coherent scattering devices such as highly reflecting distributed Bragg reflectors (Johnson et al., 2005), surface gratings (Llopis & Tobias, 2005), or a combination of both (Zeng et al., 2006). However, the implementation of such structures in DSSC had been no possible due to the need for porous back reflectors that allowed a proper flow of the electrolyte through the cell and, at the same time, due to the complicated deposition process of solid layers from colloidal suspensions.

The fabrication of DSSC containing nanoparticle based 1DPC involves two basic steps: first, the deposition of the nanocrystalline TiO_2 layer that acts as electrode onto a transparent conducting substrate, and second, the stacking of layers made of nanoparticles of different kind deposited alternately by spin-coating onto the sintered electrode. In this case, silica and titanium dioxide suspensions are employed because of the very high refractive index contrast they present, which allows achieving broad and intense Bragg reflections. The nanoparticle multilayer is periodic with a period of around a hundred nanometers, the thickness of these layers being controlled through either the concentration of the precursor suspensions or the rotation speed of the substrate during the spin coating process. Figure 7 shows a scheme of the described 1DPC based cell, as well as FESEM images corresponding to a cross section of both the nanocrystalline-TiO_2 electrode and the periodic structure deposited onto the former. The uniformity in the thickness of both types of layers composing the 1DPC, and even the different morphology of the particles employed, can be clearly distinguished in the picture below. The total thickness of the photonic crystal can vary between 0.5 to less than 2 microns, depending mainly on the lattice parameter of the structure and the number of layers that compose the periodic stack.

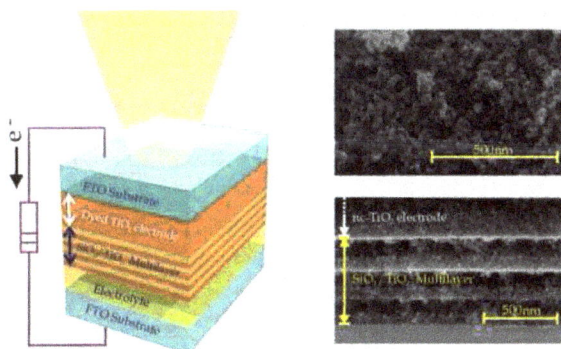

Fig. 7. Left: Design of a DSSC coupled to a nanoparticle based 1DPC. Right: FESEM images showing the TiO_2 nanocrystals forming the solar cell electrode (top), on which the porous periodic stack made of nanoparticles of different kind is deposited (bottom). In this case, a six layer photonic crystal has been implemented.

The procedure that follows to complete the solar cell is the same than the one usually employed for standard DSSC. It should be noticed that the nanoparticle multilayer integrated into the solar cell in this way behaves as a distributed Bragg reflector, providing the cell with a brilliant metallic reflection whose colour can be tuned by varying the thickness of the layers forming the periodic nanostructure. This can be readily seen in the photographs shown in Figure 8, in which the appearance of a reference cell and the same cell including two different 1DPC under perpendicular illumination are shown. Another remarkable issue from these systems is that the multilayer implemented like that does not alter significantly the cell semi-transparency, contrary to what happens when other scattering layers made of large titania nanoparticles are employed in DSSC to increase the photogenerated current. When these diffuse scattering layers are used, the solar cell becomes almost completely opaque as a consequence of the lack of spectral selectiveness of the incoherent scattering by slurries with a wide particle-size distribution. The comparison between the optical transmission spectra for the case of a standard reference cell (7.5 micron thick) and those corresponding to solar cells possessing the same TiO_2 electrode thickness but coupled to different 1DPC and to a diffuse scattering layer are also included in Figure 8.

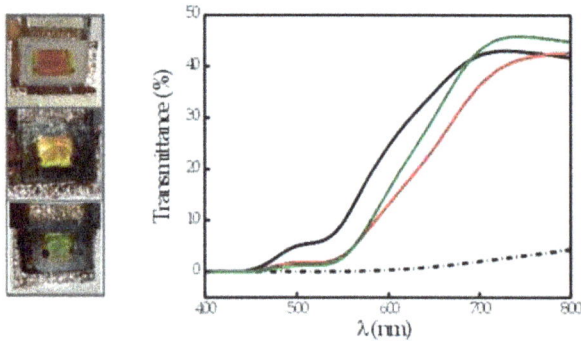

Fig. 8. Left: Images of a reference cell and different photonic crystal based DSSC. The brilliant colours displayed by the cell (bottom) arise from the periodic structures with different lattice parameter coupled to the dyed electrode (top image). Right: Transmittance spectra of a DSSC composed of a 7.5 micron thick electrode (black curve) and of the same electrode coupled to periodic structures with different lattice parameters (green and red curves). For comparison, the transmittance spectrum of a DSSC with the same electrode thickness but coupled to a 7.5 micron thick porous diffuse scattering layer is also plotted (black dashed line). (*Extracted with permission from Colodrero et al., 2009 [b]*)

As explained in section 3.3.1, enhancement of optical absorption is primarily due to the partial localization of photons of certain narrow frequency ranges within the dyed TiO_2 electrode (that acts as absorbing layer) as a result of its coupling to the photonic crystal, which acts as a porous low-loss dielectric mirror (Mihi & Míguez, 2005). These optical modes could, in principle, be recognized as narrow dips in the reflectance spectra at frequencies located within the photonic band gap, the enhancement range being determined by the spectral width of the photonic band gap (Mihi et al., 2005). The first experimental demonstration of the mechanism of light harvesting enhancement that takes place in DSSC coupled to photonic crystals has been recently reported using the nanostructures under the scope of this section (Colodrero et al., 2009[a]). The effect of well defined optical absorption resonances was detected both in optical spectroscopy and photogenerated current experiments of very thin and uniform dye-sensitized TiO_2 electrodes coupled to high quality porous 1DPC, an unambiguous correspondence between them being established. This study demonstrated that light trapping within absorbing electrodes is responsible for the absorption enhancement that had previously been reported. Figure 9 shows the spectral response of the IPCE for three DSSC having increasing electrode thicknesses range from 350 nm thick to 1.5 micron thick but the same 1DPC implemented. In each case, an enhancement factor γ was calculated as the ratio between the IPCE of the 1DPC based cell and that of the reference one. The spectral behaviour of γ for each cell is compared to its corresponding optical reflectance measured under front-side illumination. It can be clearly seen that peaks of photocurrent correspond to the dips in reflectance are obtained, which are the fingerprint of optical resonant modes localized in a film coupled to a photonic crystal.

For these modes, matter-radiation interaction times are much longer; thus, the probability of optical absorption, and therefore the photogenerated current, is enhanced. As the thickness of the dye-sensitized electrode increases, the number of localized modes rises and so does the number of peaks in the γ curve. The presence of a photonic crystal not only enhances the photogenerated current but also allows one to vary the spectral photoelectric response of thin electrodes in a controlled manner. For instance, in the example shown in Figure 9a, the largest current is attained at λ = 470 nm instead of at λ = 515 nm, where the dye absorption curve reaches its maximum. Thus, the photonic crystal allows tailoring to measure the enhanced absorption window of the dye, and thus, its overlap with the solar spectrum.

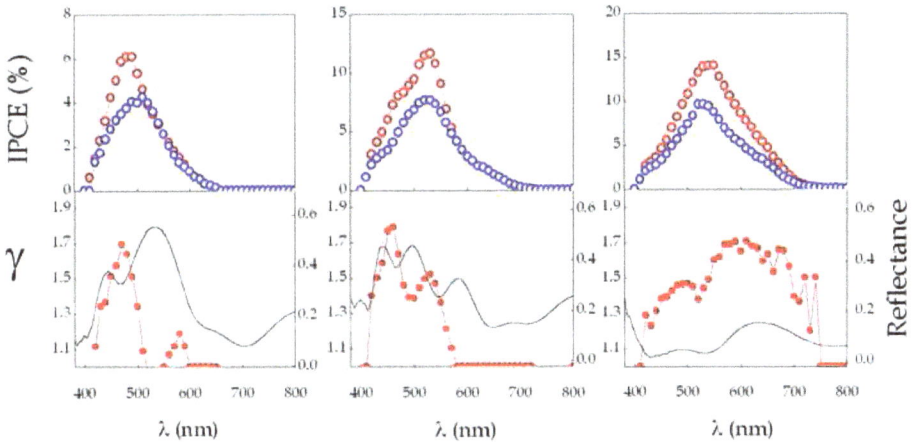

Fig. 9. Top: IPCE versus wavelength for cells containing the same 6 layer-1DPC coupled to dyed electrodes of increasing thicknesses in each case (from left to right). It is also plotted the IPCE for reference cells of the same electrode thickness without photonic crystal (blue circles). Bottom: Reflectance spectra of the 1DPC based solar cells (solid line) and the corresponding photocurrent enhancement factor (red circles). (*Extracted with permission from Colodrero et al., 2009 [a]*)

On the other hand, besides the experimental demonstration and confirmation of the light harvesting enhancement mechanism achieved using 1DPC based solar cells, great improvements in power conversion efficiency (η) have also been observed in this type of solar devices coupled to highly reflecting nanostructures (Colodrero et al., 2009 [a]), (Colodrero et al., 2009 [b]). After analyzing the photocurrent density-voltage (J-V) curves under 1 sun illumination of DSSC, on which photonic crystals reflecting different ranges of wavelengths were coupled, it was found that the photocurrent was largely improved while leaving the open-circuit voltage almost unaltered. The magnitude of this effect depends mainly on two factors: first, the spectral width and position of the photonic band gap relative to the absorption band of the ruthenium dye; second, the degree of optical coupling to the dye-sensitized electrode, which depends in turn on the thickness of that electrode. The magnitude of the photocurrent enhancement effect caused by the coupling to the 1DPC is therefore expected to be lower as the thickness of the electrode increases, since more photons are absorbed by the dyed nc-TiO₂ layer when they first pass through it. For this case, red reflecting 1DPCs might perform better, since the ruthenium dye captures less effectively solar radiation precisely for $\lambda > 600$ nm. Results on the power conversion efficiency (η) for DSSC with a 7.5 micron thick electrode coupled to 1DPC reflecting different ranges of wavelengths (green and red) are shown in figure 10. An enhancement in the efficiency close to 20% with respect to that of the reference cell was obtained for the 1DPC based solar cell showing a better matching with the absorption spectrum of the dye.

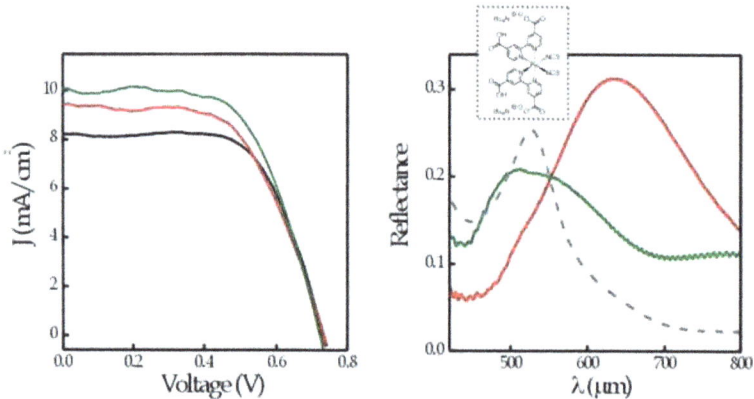

Fig. 10. Left: IV curves of a 7.5 micron thick electrode coupled to different 1DPC under 1 sun illumination. The corresponding IV curve for a reference cell is also plotted (black line). Right: Reflectance spectra measured under frontal illumination conditions of the PC-based solar cells together with the absorption spectrum of the Ru-dye (arbitrary units).

The photocurrent enhancement reported using PC based solar cells could be even larger at lower incident radiation intensities, reaching up to 30% of the reference value under 0.1 sun for the samples above described. This is mainly due to the decrease of density of carriers when so does the incident light intensity, which has a positive effect on electron transport and recombination through the cell. Besides, any resistance potentially introduced by the photonic crystal will have a minor effect at lower illumination conditions, since its effect increases with the number of carriers. In order to illustrate this effect, values of efficiency, photogenerated current and open-circuit voltage obtained for DSSC having 7.5 micron thick electrodes coupled to different 1DPC are presented in Figure 11. The variation of J_{sc} and V_{oc} with intensity of the incident radiation confirms that the presence of the PC enhances the photocurrent significantly, but has a minor effect on the photovoltage. The linear and logarithmic dependence observed for Jsc and Voc, respectively, versus incident radiation intensity are in good agreement with theoretical predictions (Nazeerudin et al., 1993); (Södergren et al., 1994).

To conclude this section and in order to prove the performance of these nanoparticle based structures as light harvesters, DSSC based on both 1DPC and diffuse scattering layers were evaluated and compared. For this purpose, a 7.5 micron diffuse scattering layer made of titania spheres 130 nm in diameter mixed with a paste similar to that employed to prepare the nanocrystalline titania layer was deposited onto a 7 micron thick reference electrode. A similar electrode was coupled to a 700 nm thick highly reflecting 1DPC. In order to perform a comparison of the effect on light harvesting these different architectures have, the 7 micron thick diffuse scattering layer was electrically isolated from the dye-sensitized electrode by introducing a thin layer of silica spheres 30 nm in diameter between them. By doing so, no contribution to the photocurrent from the different scattering layers employed is measured, since the 1DPC is also based on alternate layers of SiO_2 and TiO_2 nanoparticles, the first layer deposited onto the electrode being insulating. The effect of the PC on the short circuit photocurrent is observed to be similar and comparable to that of a diffuse scattering layer, provided that a suitable PC is chosen, as displayed in figure 12. Furthermore, the open circuit voltage is slightly higher in the case of the PC, which might be due to its much smaller width. It should be reminded that the enhancement in the case of PC is based on the partial confinement of light of a selected frequency range within the absorbing electrode, whereas in the case of diffuse scattering layers, the increase in efficiency is based on the random and non-selective scattering of visible light in all directions.

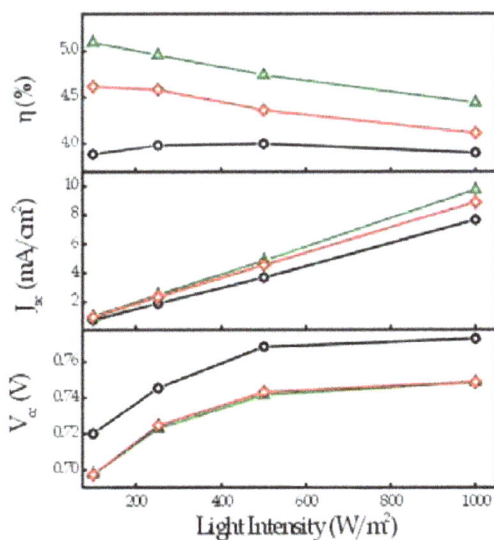

Fig. 11. Efficiency (η), short-circuit current density (J_{sc}) and open-circuit voltage (V_{oc}) for a reference cell (open black circles) and for those PC based cells having the same electrode thickness coupled to 1DPC with different lattice parameters (green and red symbols) under illumination at different light intensities. (*Extracted with permission from Colodrero et al., 2009 [b]*)

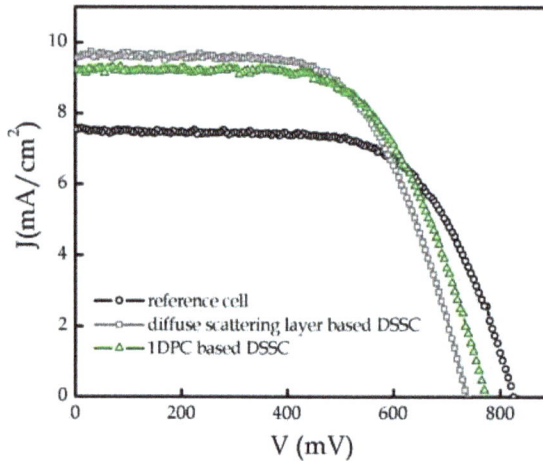

Fig. 12. Comparison between the efficiencies for a DSSC made of a 7 micron thick electrode and those corresponding to the same electrode thickness coupled to a diffuse scattering layer and a 1DPC. The thicknesses employed for the diffuse scattering layer and the periodic structure are 7.5 micron and 700 nm, respectively. (*Extracted with permission from Colodrero et al., 2009 [b]*)

4. Conclusions

Colloidal chemistry approaches are suitable for implementing optical devices of high quality in DSSC in order to improve their performance. This opens the door for the conscious optimization of the photonic design of DSSC, as is commonly done for their silicon counterparts. This may also open the way to amplifying the absorption of other dyes with low extinction coefficients that cover other regions of the visible and near-IR solar spectrum. In this respect, a thorough analysis in terms of the interplay between the effect of the electrode thickness, the dye absorption spectrum, and the characteristics of the Bragg reflection, such as intensity, spectral position and width, is needed for designing 1DPC based DSSC of optimized performance. On the other hand, the greater enhancement of efficiency attained for thin electrodes coupled to these photonic structures highlight the potential that they might have in other cells using very thin absorbing layers, in which the main source of loss of efficiency is frequently the low amount of light absorbed.

5. References

Bach, U.; Lupo, D.; Comte, P.; Moser, J.E.; Weissortel, F.; Salbeck, J.; Spreitzer, H. & Graetzel, M. (1998). Solid-state dye-sensitized mesoporous TiO2 solar cells with high photon-to-electron conversion efficiencies. *Nature*, 395, 6702 (October 1998) 583-585, 0028-0836

Baruch, P. (1985). A 2-level system as a model for a photovoltaic solar cells. *Journal of Applied Physics*, 57, 4, (February 1985) 1347-1355, 0021-8979

Cao, F.; Oskam, G.; Meyer, G.J. & Searson, P.C. (1996). Electron transport in porous nanocrystalline TiO2 photoelectrochemical cells. *Journal of Physical Chemistry* 100, 42, (October 1996) 17021-17027, 0022-3654

Colodrero, S.; Ocaña, M. & Míguez, H.(2008) Nanoparticle-based one-dimensional photonic crystals. *Langmuir* 24, 9, (May 2008) 4430-4434, 0743-7463

Colodrero, S.; Mihi, A.; Anta, J.A.; Ocaña, M. & Míguez, H. (2009) [a]. Experimental Demonstration of the Mechanism of Light Harvesting Enhancement in Photonic-Crystal-Based Dye-Sensitized Solar Cells. *Journal of Physical Chemistry C, 113, 4, (January 2009) 1150-1154*, 1932-7447

Colodrero, S.; Mihi, A.; Anta, J.A.; Ocaña, M. & Míguez, H. (2009) [b]. Porous One-Dimensional Photonic Crystals Improve the Power-Conversion Efficiency of Dye-Sensitized Solar Cells. *Advanced Materials*, 21,7, (February 2009) 764-768, 0935-9648

Chiba, Y.; Islam, A.; Komiya, R.; Koide, N. & Han, L. (2006). Conversion efficiency of 10.8% by a dye-sensitized solar cell using a TiO_2 electrode with high haze. *Applied Physics Letters*, 88, 22 (May 2006) 223505, 0003-6951

Choi, S. Y.; Mamak, M.; Freymann von, G.; Chopra, N. & Ozin, G. A. (2006). Mesoporous Bragg stack color tunable sensors. *Nano Letters* 6, 11 (November 2006) 2456-2461, 1530-6984

Desilvestro, J.; Gratzel, M.; Kavan, L.; Moser, J. & Augustynski, J. (1985). Highly Efficient Sensitization of Titanium Dioxide. *Journal of the American Chemical Society*, 107, 10, (May 1985) 2988-2990, 0002-7863

Ferber, J. & Luther, J. (1998). Computer simulations of light scattering and absorption in dye-sensitized solar cells. *Solar Energy Materials and Solar Cells*, 54, 1-4, (August 1998) 265-275, 0927-0248

Fuertes, M. C.; López-Alcaraz, F. J.; Marchi, M. C.; Troiani, H. E.; Míguez, H. & Soler Illia, G. J. A. A. (2007). Photonic crystals from ordered mesoporous thin-film functional building blocks. *Advanced Functional Materials, 17*, 8, (May 2007) 1247-1254, 1616-301X

Fujishima, A. & Honda, K. (1972). Electrochemical photolysis of water at a semiconductor electrode. *Nature*, 38, 5358, (1972) 37, 0028-0836

Gerisher H. & Tributsch, H. (1968). Electrochemistry of ZnO monocrystal spectral sensivity. *Berichte der Bunsen-gesellschaft fur physikalische chemie*, 72, 3, (March 1968) 437

Graetzel, M. (2000). Perspectives for dye-sensitized nanocrystalline solar cells. *Progress in Photovoltaics: Research and Applications*, 8, 1, (February 2000), 171-185, 1062-7995

Graetzel, M & Durrant, J. (2008) Dye Sensitized Mesoscopic Solar Cells, In: *Nanostructured and Photoelectrochemical Systems for Solar Photon Conversion*, Mary D Archer & Arthur J Nozik, (Ed.), (503-536), Imperial College Press., 978-1-86094-255-6, England

Halaoui, L.I.; Abrams, N.M. & Mallouk, T. (2005) Increasing the conversion efficiency of dye-sensitized TiO2 photoelectrochemical cells by coupling to photonic crystals. *Journal of Physical Chemistry B*, 109, 13 (April 2005), 6334-6342, 1520-6106

Haque, S.A.; Palomares, E.; Cho, B.M.; Green, A.N.M; Hirata, N.; Klug, D.R. & Durrant, J.R. (2005). Charge separation versus recombination in dye-sensitized nanocrystalline solar cells: the minimization of kinetic redundancy. *Journal of the American Chemical Society*, 127, 10, (March 2005) 3456-3462, 0002-7863

Hore, S.; Vetter, C.; Kern, R.; Smit, H. & Hinsch, A. (2006). Influence of scattering layers on efficiency of dye-sensitized solar cells. *Solar Energy Materials and Solar Cells*, 90, 9, (May 2006) 1176-1188, 0927-0248

Huang, S.Y.; Schlichthorl, G.; Nozik, A.J.; Graetzel, M. & Frank, A.J. (1997). Charge recombination in dye-sensitized nanocrystalline TiO2 solar cells. *Journal of Physical Chemistry B*, 101, 41, (April 1997) 2576-2582, 1089-5647

Ito, S.; Nazeeruddin, M.K.; Liska, P.; Comte, P.; Charvet, R.; Pechy, P.; Jirousek, M.; Kay, A.; Zakeeruddin, S.M. & Graetzel, M. (2006).Photovoltaic Characterization of Dye-sensitized Solar Cells: Effect of Device Masking on Conversion Efficiency. *Progress in photovoltaics*, 14, 7, (November 2006) 589-601, 1062-7995

Ito, S.; Murakami, T.N.; Comte, P.; Liska, P.; Graetzel, C.; Nazeerudin, M.K. & Graetzel, M. (2008). Fabrication of thin film dye sensitized solar cells with solar to electric power conversion efficiency over 10%. *Thin solid films*, 516, 14, (May 2008) 4613-4619, 0040-6090

Joannopoulos, J. D.; Meade, R. D. & Winn, J. N. (1995). *Photonic Crystals: Molding the Flow of Light*; Princeton University Press, 0691-03744-2, Princeton, NJ, USA.

Johnson, D. C.; Ballard, I.; Barnham, K. W. J.; Bishnell, D. B.; Connolly, J. P.; Lynch, M. C.; Tibbits, T. N. D.; Ekins-Daukes, N. J.; Mazzer, M.; Airey, R.; Hill, G. &. Roberts, J.S.(2005). Advances in Bragg stack quantum well solar cells. *Solar Energy Materials and Solar Cells*, 87, 1-4, (May 2005) 169-176, 0927-0248

Kamat, P. (2007). Meeting the clean energy demand: Nanostructure architectures for solar energy conversion. *Journal of Physical Chemistry C*, 111, 7, (February 2007) 2834-2860, 1932-7447

Lee, S.H.A.; Abrams, A.; Hoertz, P.G.; Barber, G.D.; Halaoui, L.H. & Mallouk, T.H. (2008). Coupling of Titania Inverse Opals to Nanocrystalline Titania Layers in Dye-Sensitized Solar Cells. *Journal of Physical Chemistry B*, 112, 46, (November 2008) 14415-14421, 1520-6106

Llopis, F. & Tobias, I. (2005). The role of rear surface in thin silicon solar cells. *Solar Energy Materials and Solar Cells*, 87, 1-4, (May 2005) 481-492, 0927-0248

Mihi, A. & Miguez, H. (2005). Origin of light-harvesting enhancement in colloidal-photonic-crystal-based dye-sensitized solar cells. *Journal of Physical Chemistry B*, 109, 33, (August 2005) 15968-15976, 1520-6106

Mihi, A.; Míguez, H.; Rodríguez, I.; Rubio, S.; Meseguer, F. (2005). Surface resonant modes in colloidal photonic crystals. *Physical Review B*, 71, 12 (March 2005) 125131, 1098-0121

Mihi, A; López-Alcaraz, F.J.; Míguez, H. (2006). Full spectrum enhancement of the light harvesting efficiency of dye sensitized solar cells by including colloidal photonic crystal multilayers. *Applied Physics Letters*, 88, 19 (May 2006), 193110, , 0003-6951

Mihi, A.; Calvo, M. E.; Anta, J. A. & Míguez, H. (2008). Spectral response of opal-based dye-sensitized solar cells. *Journal of Physical Chemistry C*, 112, 1, (January 2008) 13-17, 1932-7447

Mori, S. and Yanagida, S. (2006) Dye Sensitized Solar Cells, In: *Nanostructured Materials for Solar Energy Conversion*, Tatsuo Soga, (Ed.), (193-225), Springer B.V., 0-444-52844-X, England

Nazeerudin, M. K.; Kay, A.; Rodicio, I.; Humphry-Baker, R.; Müller, E.; Liska, P.; Vlachopoulos, N. & Graetzel, M. (1993). Conversion of light to electricity by cis-X2bis(2,2'-bipyridyl-4,4'-dicarboxylate)ruthenium(II) charge-transfer sensitizers (X = Cl-, Br-, I-, CN-, and SCN-) on nanocrystalline titanium dioxide electrodes *Journal.of the American Chemical Society* 115, 14, (July 2003) 6382-6390, 0002-7863

Nelson, J.; Haque, S.A.; Klug, D.R. & Durrant, J.R. (2001). Trap-limited recombination in dye-sensitized nanocrystalline metal oxide electrodes. *Physical Review B*, 63, 20 (May 2001) 205321, 0163-1829

Nishimura, S.; Abrams, N.; Lewis, B.; Halaoui, L.I.; Mallouk, T.E.; Benkstein, K.D.; Van de Lagemaat, J. & Frank, A.J. (2003). Standing wave enhancement of red absorbance and photocurrent in dye-sensitized titanium dioxide photoelectrodes coupled to photonic crystals. *Journal.of the American Chemical Society* 125, 20, (May 2003) 6306-6310, 0002-7863

Rothenberger, G.; Comte, P. & Graetzel, M. (1999). A contribution to the optical design of dye-sensitized nanocrystalline solar cells. *Solar Energy Materials and Solar Cells* 58, 3, (July 1999) 321-336, 0927-0248

Schwarzburg, K. & Willig, F. (1999). Origin of photovoltage and photocurrent in the nanoporous dye-sensitized electrochemical solar cell. *Journal of Physical Chemistry B* 103, 28, (July 1999) 5743-5746, 1089-5647

Södergren, S.; Hagfeldt, A.; Olsson, J. & Lindquist, E. (1994). Theoretical Models for the Action Spectrum and the Current-Voltage Characteristics of Microporous Semiconductor Films in Photoelectrochemical Cells. *Journal of Physical Chemistry* 98, 21, (May 1994) 5552-5556, 0022-3654

Tachibana, Y.; Hara, K.; Sayama, K. & Arakawa, H. (2002). Quantitative analysis of light-harvesting efficiency and electron-transfer yield in ruthenium-dye-sensitized nanocrystalline TiO2 solar cells. *Chemistry of Materials* 14, 6, (June 2002) 2527-2537, 0897-4756

Tenakkone, K.; Kumara, G.R.R.A.; Kottegoda, I.R.M. & Perera, V.P.S. (1999). An efficient dye-sensitized photoelectrochemical solar cell made from oxides of tin and zinc. *Chemical Communications*, 1, (January 1999) 15-16, 1359-7345

Usami, A. (1997). Theoretical study of application of multiple scattering of light to a dye-sensitized nanocrystalline photoelectrochemical cell. *Chemical Physics Letters* 277, 1-3, (October 1997) 105-108, 0009-2614

Vlachopoulos, N.; Liska, P.; Augustynski, J. & Gratzel, M. (1988). Very Efficient Visible Light Energy Harvesting and Conversion by Spectral Sensitization of High Surface Area Polycrystalline Titanium Dioxide Films. *Journal of the American Chemical Society* 110, 4, (February 1988) 1216-1220, 0002-7863

Wang, P.; Zakeeruddin, S.M; Moser, J.E.; Humphry-Baker, E. & Graetzel, M. (2004). A solvent-free, SeCN-/(SeCN)(3)(-) based ionic liquid electrolyte for high-efficiency dye-sensitized nanocrystalline solar cells. *Journal of the American Chemical Society,* 126, 23 (June 2004) 7164-7165, 0002-7863

Wang, P.; Klein, C.; Humphry-Baker, R.; Zakeeruddin, S.M. & Graetzel, M. (2005). A high molar extinction coefficient sensitizer for stable dye-sensitized solar cells. *Journal of the American Chemical Society,* 127,3, (January 2005) 808-809, 0002-7863

Wu, Z.; Lee, D.; Rubner, M.F. & Cohen, R.E. (2007) Structural color in porous, superhydrophilic, and self-cleaning SiO_2/TiO_2 Bragg stacks. *Small* 3, 8 (August 2007) 1445-1451, 1613-6810

Zaban, A.; Meier, A. & Gregg, B.A. (1997). Electric potential distribution and short-range screening in nanoporous TiO2 electrodes. *Journal of Physical Chemistry B* 101, 40, (October 1997) 7985-7990, 1089-5647

Zeng, L.; Yi, Y.; Hong, C.; Liu, J.; Feng, N.; Duan, X.; Kimerling, L.C. & Alamariu, B.A. (2006). Efficiency enhancement in Si solar cells by textured photonic crystal back reflector. *Applied Physics Letters,* 89, 11, (September 2006) 111111, 0003-6951

Zhang, Z. et al. (2007). The electronic role of the TiO_2 light-scattering layer in dye-sensitized solar cells. *Zeitschrift für physikalische chemie* 221, 3, (March 2007) 319-327, 0942-9352

5

Contact Definition in Industrial Silicon Solar Cells

Dr. Luis Jaime Caballero
Isofoton S.A.
Spain

1. Introduction

The incredible development that industrial silicon based photovoltaic devices have followed for the last decades has been related to the consecution of a simple, easy and economically feasible way to define the electrical contacts of the devices. Due to the size of silicon photovoltaic device in relation to its substrate wafer size (one device per wafer), traditional microelectronic means to define contact (using metal evaporation in vacuum and photolithographic processes) are not economically appropriate for the mass production devices. Screen-printing technique has represented the perfect means to allow the production cost reduction, and the strong introduction of photovoltaic devices for terrestrial applications in the global market.

Although similar primitive printing techniques have been known by the mankind thousand of years ago, the industrial introduction of the screen-printing technique had to wait till the end of the XIX century, beginning of the XX century. Its application in the textile industry started the beginning of its intensive use. It was in the second half of the XX century when electronic industry started to employ this technique in the field of the hybrid circuits for the deposition of dielectric and conductive layers; but its first applications in the photovoltaic industry dates from 1975-6 (Ralph, 1975); (Haigh, 1976). Since then, improvements of the screen-printing techniques and metal pastes for the creation of contacts have been crucial for the development and improvement that industrial produced silicon solar cells have followed, becoming the heart of its fabrication processes.

But the technical characteristics of the screen-printed contacts are far from the ones obtained when a metal is deposited on the silicon surface with a microelectronic process in vacuum conditions, introducing limitations to the final energy conversion efficiency that devices can reach. The strong interest in increasing the conversion efficiency of industrial solar cells is encouraging the appearance of new research focused on overcoming this limitations. For this reason, advances and new developments in other techniques seem to be the future way to get improved contacts in mass production device fabrication. The application of these could mean changes in the nowadays standard processing technology that would be introduced by the industry in coming years.

This chapter aims to present the curent contact definition technique, reviewing the screen-printing technology, and the ways followed to optimize its results, getting higher conversion efficiencies in devices; some design topics for the front grid patterns will be discussed, finishing with a quick introductory view on the future of the industrial silicon solar cell with special attention to its contact definition.

2. Screen printing of metallic contacts basic fundamentals

The screen printing process consists in the transference of an ink or paste (with a specific viscosity) through a screen that allows its pass in a defined pattern, thanks to the pressure applied by a squeegee fixed in a moving part. Basic parts of a system for a solar cell screen-printing definition process are:

- The screen, comprising a frame that holds a stretched fabric with a photo stencil attached to the mesh with the required design of the grid pattern.
- A squeegee, comprising a holder with a fixed, flexible, resilient blade.
- A metallic conductive paste that is transferred to the device surface,
- And a silicon substrate located in a chuck aligned with the pattern to be transferred.

After a flooding of the screen with paste, the moving of the squeegee produces the deposition of the paste in the wafer surface as can be seen in the Fig. 1 where all the steps in a printing cycle are shown.

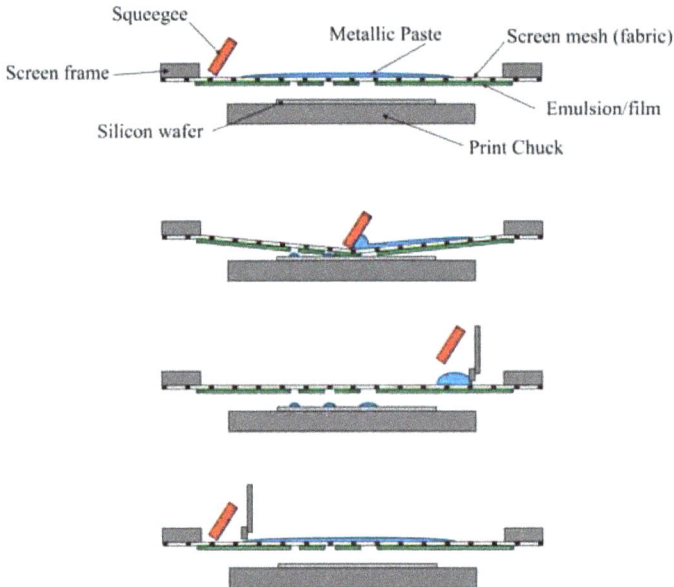

Fig. 1. Printing cycle steps, main parts of screen-printing machine are detailed along with the process sequence followed to define the contact

The result of this printing process that affects the final solar cell efficiency is highly dependent on a big number of different parameters, not only related to the raw materials involved in this production step, such as the properties of metallic pastes (viscosity and rheology) or wafers characteristics such as homogeneity in thickness or flatness, but also related to the used screens, squeegees, and the set of processing parameters related to the printing process. All of these parameters are related and require a practical optimization in order to get an improved result of the printing definition.

It can be said that the more homogeneous the device structure (in the whole surface area), the more efficient the final solar cell. Therefore a homogeneous definition of the front contact is important in order to get an optimal industrial solar cell.

Among the processing parameter are the pressure applied by the squeegee, the speed of its printing movement and the related position of the wafer surface with respect to the screen position. While among the parameter related with the screens that define the pattern, it can be found the material of the fabric (typically stain-steel due to its good response for mass production), its constituent wires' diameter and mesh density (wires per centimetre), the thickness of the fabric, its attached emulsion or film, and the stretching of the mesh of the screen.

All this set of parameters and its effects (studied early in the literature of the thick film technology field (Holmes & Loasby, 1976)) makes the printing a complex process to be optimised, which must be analysed on the industrial environment with the help of statistical tools based in the final device performance, and the optical and electrical characterization of the final contact definition of the produced solar cells; always inside specific designs of experiments where parameter are changed in defined working ranges (using a DOE analysis tool).

3. Composition of metallic pastes

3.1 Front paste composition

Metallic contact nowadays employed by the industry are based on the use of silver as a conductive metal due to its good contact properties with n type silicon, its good conductivity and its excellent solderability (needed for the later interconnection of cells). To make the deposition of this metallic contact with a screenprinting technique possible, with the goal of reaching a final correct mechanical adhesion to the surfaces and good electrical properties, special metallic pastes have been developed for the solar cell industry.

Silver powder represents the 70-85% in weight of the commercial pastes with a mixture of different shaped particles, of different sizes (as spherical powder grains or flakes), that are responsible for the paste conductivity and final cohesion of the contact.

The rest of the paste components are extremely sensitive to the device surface that want to be contacted (with different dielectrics layers of different thickness that pastes must go through), and to the specific processing of contact creations such as temperature profiles and processing times. These are:

- Powder of glass frits.
  Glass frits are metal oxides that play the most important role in the formation of the contact because its function is to melt the dielectric layers (by forming eutectic alloys of lower melting point), that are deposited or grown on the silicon emitter, allowing the metal particles to reach the silicon surface. Additionally its content determines the adhesion of the paste to the silicon substrate.
  An example of the different glass frits contain of two pastes is shown in Table 1 (Firor & Hogan, 1981)

- Organic compounds used as a vehicle to transport the suspended silver and glass particles, allowing its disposal with the screen printing techniques. Among these organic compounds are:
  Organic solvent to allow the mixture to be used as a paint, and

Organic binders to maintain the particles joined once the solvents have been evaporated (cellulosic resins) after transferring of pattern.

- Other additives to modify the rheological properties of the mixture and its interaction with the substrate surfaces (wetting agents).

Composition of the glass frits contain (% weight)		
Glass	Paste 1 Phosphate glass based	Paste 2 Borosilicate glass based (the most important type)
Al_2O_3	11.6	14.6
B_2O_3	--	2.1
BaO	0.2	--
CdO	--	0.6
CaO	8.6	0.2
CuO	7.7	0.6
Fe_2O_3	1.4	--
Na_2O	4.1	--
P_2O_5	65.8	4.4
PbO	--	51.8
SiO_2	0.61	25
ZnO	--	0.8

Table 1. Different glass frit contain, two examples of different pastes are evaluated

Although the exact formulations of metallization pastes are kept as industrial secrets it is possible to summarise the general components of a typical front contact paste as it is shown in table 2

Components	(Wt.%)
Silver	70 - 85
Glass Frits	0 - 5
Cellulosic Resin	3 - 15
Solvent (Pine Oil or Glycol Ethers)	3 - 15
Additives (Rheological Modifiers and Surfactants)	0 - 2

Table 2. Components of industrial screen-printing pastes for the definition of the front contact

All the components are dispersed and intimately mixed by using agitators, three roll mills or any similar equipment.

Front Pastes are typically designed to find a compromise between a situation that obtains a good final diode quality after processing (reached with higher metal particle sizes and frits with a higher melting temperature points), and a situation that makes easy the electrical contact with the emitter (reached with lower metal particle sizes and frits with a lower melting temperature points).

3.2 Rear paste composition

As for the front contact, the rear contact is also deposited by screen-printing techniques in nowadays industrial solar cells. In this case, as the type of silicon that must be contacted is p, the metal that constitutes the paste for the rear contact is Aluminium instead of Silver. But due to the problems of solderability that this material presents, with the formation of its surface oxide, industrial definition of the rear contact is carried out with a double step printing that defines:

- First the area that will be used, later on, to bond the connection ribbons (tabs), using a paste based on silver or silver and aluminium mixtures;
- And a second printing process that must be aligned with the first, with a paste based on aluminium particles to create the rear structure of the industrial solar cells knows as *bsf* (back surface field), due to the electrical field that aluminium doping generates in the silicon bulk near the surface.

Apart from the different metal particles of this pastes, the rest of the components that constitutes the final mixture are more or less the same because frits to improve adhesion and to remove residual dielectric layers, that can appear due to processing, are also needed, although in a lesser extend.

3.3 Supplied pastes properties

Among the parameters typically reported by the paste suppliers it is possible to find:

- The 'Viscosity', whose characterization should report the measured value along with equipment used to measure viscosity (spindle or cone/plate) and temperature.
- Possible 'Thinners' that are listed in the event of needed viscosity adjustments.
- The 'Solids Content' that is reported for combined inorganic content (in absence of organic) and 'Metal Content'.
- The 'Fineness of Grind' (FOG) that is reported for the first scratch ($\leq$).
- Some after processing parameters such as: the 'Dried Thickness' and 'Fired Thickness' that are given as a range reported in µm, and the 'Resistivity' (for a fired thickness given in mΩ/square).
- Some processing parameters such as the 'Drying Profile' giving applicable ranges and necessary durations, the 'Peak temperature' that gives the optimal firing range, the 'Peak Duration' reporting the optimal value or a range and the 'Firing Atmosphere' that is typically air.
- And other supplied parameters such as the 'Shelf Life' of the paste that is listed for a properly sealed container.

4. Contact formation, the co-firing process

As both contacts of solar cells are nowadays deposited using the same technique, pastes are designed to follow a common thermal final treatment that creates the contact in both faces of the cell. This process is called 'co-firing' and industrially is carried out with in-line belt furnaces in a (typically) air atmosphere during processing. The sequence to produce the final cell structure is:

- Printing of the front (silver) contact
- Drying of the front contact
- Printing of the back contact (aluminium contact with two printing processes as was introduced in previous section, with a drying process before the second printing process)

- Drying of the Back contact
- Co-firing process, to create the contact, and a
- Laser isolation process (if wafers have not been previously isolated chemically)

It is important to keep the defined order in these steps to avoid any possible aluminium contamination of the front side of devices that would destroy the performance of the final cells (by creating a serious shunting problem of the p-n junction).

During the co-firing process wafers follow a fast process with a defined temperature profile similar to the one that is shown in Fig. 2, resulting in chemical and structural changes inside the printed pastes and substrate surfaces.

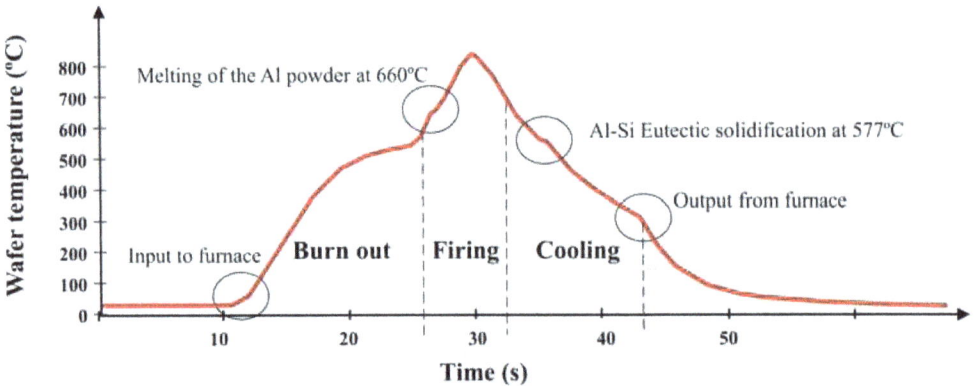

Fig. 2. Temperature profile of the furnace for the co-firing process, and time of processing

The organic components that keep the dried metallic pastes attached to the surfaces are burn out at the beginning of the process at temperatures below $600°$ C, step that is also used to produce the diffusion of passivating hydrogen from the antireflective layer to the silicon bulk. After this step the formation of the contact takes place at higher temperatures during the firing peak. At this point glass frits ($PbO-BO_3-SiO_2$) melt the silicon nitride while silver particles are transported through this melted mixture and suffer a sintering process that creates a conductive film.

Liquid Pb generated during the chemical reaction etches the silicon surface in places where, during cooling, silver uses to recrystallize creating silver crystallites that are responsible for the ohmic contact with the emitter.

After these steps the silver and glass frits mixture solidify creating a conductive contact that combines a direct contact between the sintered silver and the silver crystallites, a tunnelling contact through frits, and regions without electrical contact.

On the other hand, at the rear surface aluminium paste follow the same burn out of organic components as a first step. When temperature are above 660°C the aluminium melt and can go through the oxide that covers the particles (its oxide). The melted aluminium reaches the silicon surface, and silicon is solved following the phase diagram of the Si-Al alloy. During cooling down, silicon is rejected from the melt and recrystallize as an epitaxial Al-doped layer creating a *bsf*.

When temperature are below 577°C, the melted alloy solidify with the eutectic composition creating a layer whose contact with the silicon produces the typical bow of this aluminium *bsf* device structure.

Controlling the bowing produced at the creation of the contact is one of the most important issues for the industry nowadays. The most extended technique to keep controlled the bowing is to control the weight of aluminium deposited through the weighting of the deposited paste (between 6 and 10 mg/cm^2 of dried paste).

5. Practical industrial optimization of the metallization process

Industrial optimization of the firing process are carried out by means of a simplified DOE (Design of Experiment) that tries to find the optimal firing peak temperature for a specific thickness of substrates with a defined weight of rear aluminium paste deposited, once the wafers are right printed, and pastes are dried (in-line furnaces must be previously adjusted to ensure a perfect drying that eliminates many in other way existing problems).

Fig. 3 shows the typical shape of the optimal metallization process working window as a function of the processing speed (speed of the furnace belt) and the furnace firing zone temperature; the effect of the substrate thickness on the right working window and its orientation inside the furnace during processing is also shown.

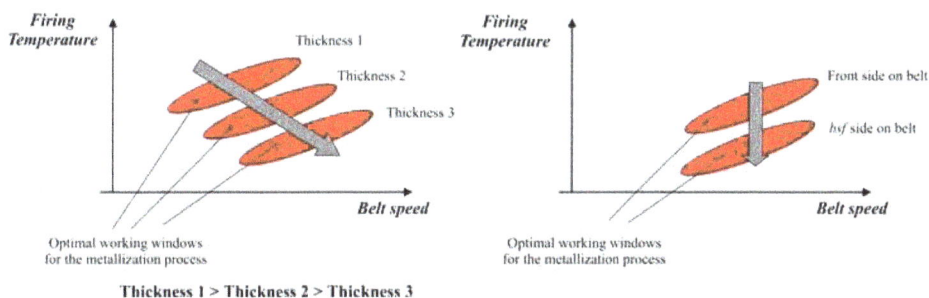

Thickness 1 > Thickness 2 > Thickness 3

Fig. 3. Optimal Firing working window variation, picture on the left shows the dependency of the thickness of substrates, while picture on the right shows that when the front side is put on the belt, higher firing peak temperatures are needed

A set of wafers with the same characteristics are split in groups that will be processed at a different firing temperatures, always with the same chain speed that is previously fixed. Taking the electrical results of the processed cells (Maximum Power and Fill Factor) it can be established the position of the optimal processing temperature in each case. Fig. 4 shows a practical example of contact optimisation.

Other characterisation techniques such as Corescan (Van der Heide et al., A 2002); (Van der Heide et al., B 2002) can also be used with optimization purposes, its measurements allow to determining local problems related to the existence of cool spots during firing, contamination problems, and processing defects in other steps of the production that can generate problems during contact creation.

6. Series resistance analytical modelling and optimal design for the front grid of industrial solar cells

The continuous rises in current that industrial photovoltaic devices have been undergoing due to the continuous improvement of materials, processing, device design and over all, substrate wafer size increase, make its associated device power losses also increase. Thus,

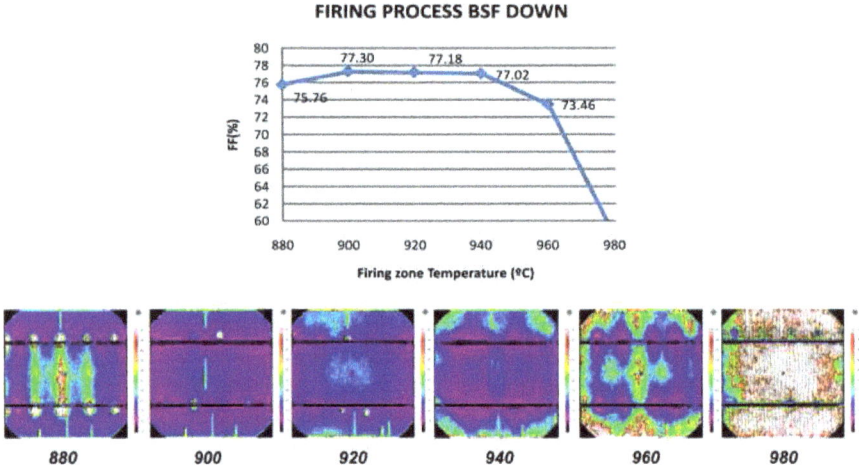

Fig. 4. Firing process optimization, example of a firing optimization attending to the final Fill Factor of resulting cells processed with its *bsf* rear contact on the furnace belt during firing. Corescan analysis for each temperature is also shown

series resistance of cells that introduces this power loss constitutes a key factor that must be taken into account in order to reach higher energy conversion efficiencies. Series resistance components characterization and its associated power losses have a great importance in the cell and module fabrication field (Luque et al., 1986); (Roberts et al., 2000). It has a potential application in grid design, metallization paste research and development, solder or conductive adhesives contact resistance requirements, and number of solder points (Caballero et al., 2006) and interconnection ribbon thickness selection.

Impact of the grid pattern design in the final series resistance of an industrial screen-printed silicon solar cell is known to be one of the strongest, so the correct design of the typical H pattern industrial front grids (a set of fingers orthogonal to two main collecting bus-bars) can improve the final device performance.

This section will deal with this task, showing how it is possible to optimise the design of front grid through the use of an analytical modelling derived from the theoretical expressions of the power loss for each part of the device. The presented modelling will take into account only the contributions that compose the series resistance of a solar cell with total aluminium rear *bsf*, leaving out of this text the cell interconnection inside a module and the analysis of all the additional contributions to the module series resistance that can appear, just in order to get a simple reference tool for further future analysis.

6.1 Cell series resistance components
When a cross-section of a simple cell is analysed it can be easily distinguished all the layers responsible for the different components of the series resistance, each one produces a power loss associated to the pass of the current through its volume. The total series resistance is the addition of all these components as it is plotted in the equivalent circuit of Fig. 5.

The proposed analytical model calculate the series resistance of a square cell as the result of the parallel association of multiple unitary basic cells (neglecting the interconnection resistance between them) in which all the components of the series resistance are calculated.

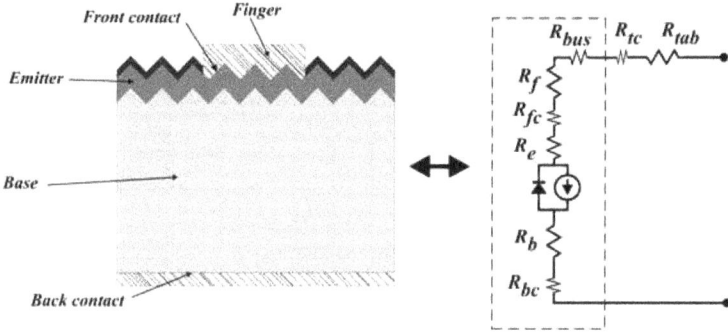

Fig. 5. Components of the cell's series resistance, equivalent circuit for an Aluminium *bsf* solar cell are shown inside the dashed box

Considering that current is extracted from the soldered points that join the bus bar with the interconnection conductive ribbons (tabs), and these points are homogeneously distributed along the bus bar, it is possible to define a basic unitary cell and the number of them that are needed to get the total series resistance (being each soldered point in contact with four unitary cells), as it is shown in Fig. 6.

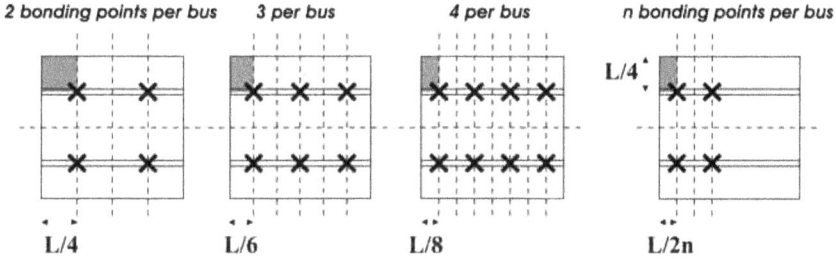

Fig. 6. Basic calculus cell for two bus-bar, it is shown the basic unitary cell size for the case of having two collecting bus-bars with a variable number of bonding points per bus-bar in the solar cell

The higher the number of soldered points per bus-bar the more accurate the result of this simple modelling, because in that way, the real current flow in an interconnected solar cell produces a natural distribution of the device in regions that are more similar to the basic cell proposed by this modelling.

Components of the unitary basic cell's series resistance are detailed in Table 3; expressions such as the emitter, finger and bus components are deduced from matching the power loss integrated along the current path (increasing the current along this path) with the power loss associated to an equivalent resistance crossed by the total current ($I_{Total}^2 R_{equivalent}$).

It must be noted that series resistance associated to the BSF layer has been neglected due to its low value.

Equations that are summarized in Table 3 include characteristic parameters from the different materials, such as ρ_{base} (bulk resistivity of the wafer), ρ_{metal}, $R_{FrontPaste}$ and $R_{BackPaste}$ (resistivity of the metallic grid in Ωcm and semiconductor-metal contact specific resistivities in Ωcm^2); characteristic parameters from the process and wafer such as R_e (Emitter layer resistance in Ω/square), L(wafer side) and w_{base} (base width); and grid design parameters such as

Component	Expression
Emitter	$R_{Emitter} = \dfrac{n \cdot s^2}{3L} \dfrac{R_e}{\left(L/2 - w_{bus}\right)}$
Base	$R_{base} = \rho_{base} \dfrac{8n \cdot w_{base}}{L^2}$
Metallic finger	$R_{finger} = \dfrac{n \cdot s}{3L} \cdot \dfrac{\rho_{metal}}{w_f h_f} \left(\dfrac{L}{2} - w_{bus}\right)$
Bus bar	$R_{bus} = \dfrac{\rho_{metal}}{3n} \cdot \dfrac{L}{w_{bus} h_{bus}}$
Rear contact resistance	$R_{fc} = \dfrac{8n \cdot s \cdot R_{FrontPaste}}{L \cdot \left(w_f \cdot L + 2w_{bus} \cdot \left(s - w_f\right)\right)}$
Front contact resistance	$R_{bc} = \dfrac{8n \cdot R_{BackPaste}}{L^2}$

Table 3. Analytical expressions of the series resistance components for a two bus-bar cell

$w_{bus, f}$ (width of buses or fingers), $h_{bus, f}$ (high of buses or fingers) and s (separation between fingers). All the expressions include the number of bonding points per busbar referred as n. As the total series resistance for the solar cell is the parallel of all the unitary basic cells, it can be expressed as:

$$R_{Series\,Total} = \frac{\sum R_{Components}}{8n}$$

With a set of values typical from the industrial environment as could be:

L	156 mm
w_{base}	200-240 µm
ρ_{base}	1 Ωcm
w_{bus}	1.8-2 mm
w_f	100-150 µm
h_{bus}	25 µm
h_f	12 µm
s	1.8-2.5 mm
R_e	45 Ω/sqr
$R_{FrontPaste}$ (Recart,2001)	~10 mΩcm²
$R_{BackPaste}$ (Recart,2001)	~10 mΩcm²
ρ_{metal} (Recart,2001)	2-3 ·10⁻⁶ Ωcm

Table 4. Typical technological values for the industrial solar cell

The modelling of the series resistance is complete and can be used to include its effect in the simulation of a device performance.

With these parameters it is possible to evaluate the effect of the number of solder spots (contact points with the interconnection tab) in the series resistance as it is shown in Fig. 7. This analysis shows that a number of contact points higher than eight points per bus in industrial solar cells produces an increment in series resistance of devices that can be neglected.

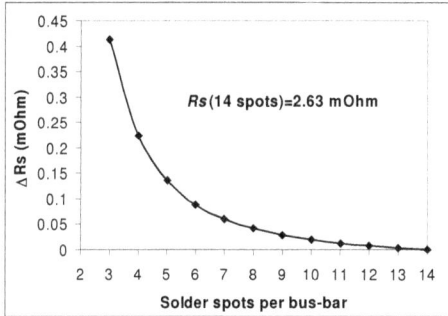

Fig. 7. Reduction of the series resistance with the increment of bonding points per bus-bar, result of the modelling are presented as an increment with respect to the situation of having 14 bonding point.

Joining this series resistance parameters and the response of an ideal diode, inside a modelling of solar cell, allows to optimise the design of front grids for each specific case, knowing for example the optimal finger separation that must be selected to get the highest conversion efficiency. So using the mathematical expressions proposed elsewhere (Luque et al., 1986), it is possible to have a modelling whose input parameters are just the photo-generated current density (J_{SC}) without front grid, the open circuit voltage (V_{OC}) reached by the device and the parameters needed to model the series resistance. An example of the results of such a modelling is shown in Fig. 8 where the effect of finger separation on efficiency is studied for the case of having a $J_{SC}=36.5$ mA/cm2, $V_{OC}= 616$ mV and the set of defining parameters presented previously in this section.

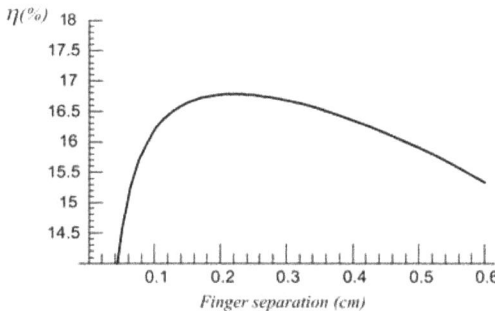

Fig. 8. Efficiency as a function of finger separation, the resulting efficiency of the mathematical modelling is plotted as a function of the finger separation when the set of parameters of Table 4 is used (with a 100 µm fingers width)

6.2 Three buses vs. two buses design for an industrial solar cell

The modelling of the previous subsection can be used as a easy tool to evaluate changes in the industrial development of solar cells, not only the improvement in material properties, but also changes of the design of the metallic contact.

The increase in size solar cells have followed last years has motivated the appearance of new designs with three bus-bars to collect the generated current instead of two, but, what is the effect of this change of the front grid design in the performance of the solar cells?. This section will discuss on this topic. A comparison is carried out based on the analytical modelling results of the two alternative designs that nowadays it is possible to find in the industry (two and three bus-bars). The proposed model needs to be adapted to introduce the calculus of the resistance of the three bus-bars cell. Table 5 summarises the expressions of the components of the resistance, while Fig. 9 presents the efficiencies of both cases as a function of the shadowing factor for the optimal finger separation in each point. In both cases fingers of the grid have 100 μm width, so to get a specific shadowing factor, just the bus-bar width can be modify (as it is plotted in Fig. 9). The rest of design parameters are the same presented in Table 4 with a w_{base} of 200 μm and a ρ_{metal} of 3 $\cdot 10^{-6}$ Ωcm.

Component	Expression
Emitter	$R_{Emitter} = \dfrac{n \cdot s^2}{3L} \dfrac{R_e}{\left(\dfrac{L}{3} - w_{bus}\right)}$
Base	$R_{base} = \rho_{base} \dfrac{12 \cdot n \cdot w_{base}}{L^2}$
Metallic finger	$R_{finger} = \dfrac{n \cdot s}{3L} \cdot \dfrac{\rho_{metal}}{w_f h_f} \left(\dfrac{L}{3} - w_{bus}\right)$
Bus bar	$R_{bus} = \dfrac{\rho_{metal}}{3n} \cdot \dfrac{L}{w_{bus} h_{bus}}$
Rear contact resistance	$R_{fc} = \dfrac{12 \cdot n \cdot s \cdot R_{FrontPaste}}{L \cdot \left(w_f \cdot L + 3 \cdot w_{bus} \cdot (s - w_f)\right)}$
Front contact resistance	$R_{bc} = \dfrac{12 \cdot n \cdot R_{BackPaste}}{L^2}$

Table 5. Analytical expressions of the series resistance components for a three bus-bar cell

With a total series resistance that is:

$$R_{Series\,Total} = \frac{\sum R_{Components}}{12n}$$

From Fig. 9 it is clear that three bus-bars design presents an improved performance that produce an increase of 0.1 points in efficiency. This efficiency increase comes from the reduction of the finger resistivity as can be seen in the resistance distribution carried out for a fixed shadowing factor of 7.34% that is shown in Fig. 10, this is why it is important to notice that an improvement in material parameters, such as conductivity or contact resistance of pastes, can reduce the difference between the results of the two patterns (for example a reduction of paste resistivity to $1 \cdot 10^{-6}$ Ωcm and a contact resistance of metal-semiconductor of 1 mΩcm² would result in a difference of just 0.05 points in efficiency).

In addition to the efficiency increase the three bus-bars design has another advantage that is the reduction of interconnection power loss in the final module, although the width of the bus-bar is reduced in the three buses design to have the same metal covering factor (when finger width is kept as a constant), as can be seen in Fig. 9, it must be taken into account that the total interconnection width (addition of the width of all the bus-bars) is higher, reducing the resistance associated to the tabbing interconnection (that have the same width than the bus-bar) and thus the final module power loss. For example, for a shadowing factor of 7.34% in the presented case of Fig. 9, the relative reduction of power loss in the tab interconnection can be estimated in 12.4% for a series association of cells independently of the tab thickness used (when the same tab thickness is used in both two and three buses case).

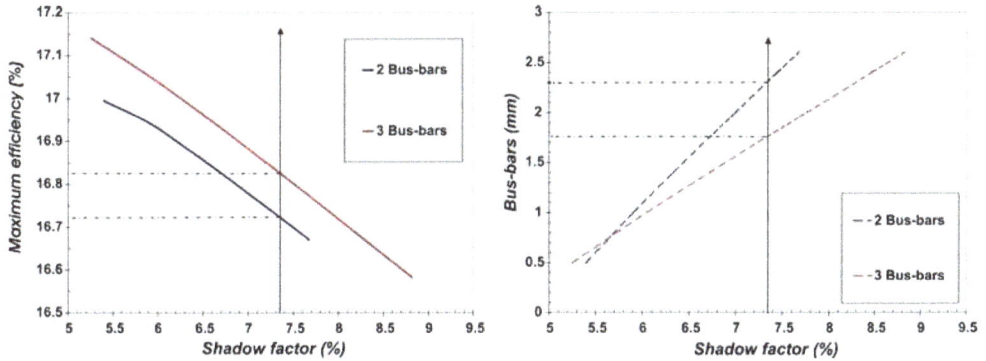

Fig. 9. Comparison between two and three bus-bar cell behaviour for the case of having fingers of 100μm width (fixed), picture on the left shows the efficiency as a function of shadowing factor of the front grid, while picture on the right shows the bus-bar width needed to reach an specific shadowing factor

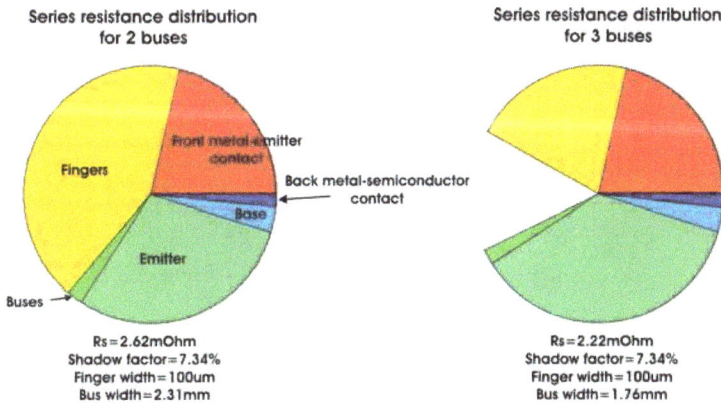

Fig. 10. Series resistance component distribution for cells with two and three bus-bars, cells with the same shadowing factor of the grid and finger width (different bus-bar width) are studied for the optimal finger separation of each case. The reduction of total resistance is mainly due to a reduction of the component associated to fingers

6.3 Multi-bus bar solar cell concept
Taking the results shown in the previous section, it is clear that solar cell performance is improved with the addition of one bus more to its front contact grid, but what happens

when a bigger number of buses are added to the front grid design? Can we expect further improvement?. To answer these questions we need to modify the mathematical expressions of the analytical model to evaluate the new situation that is shown in Fig. 11.

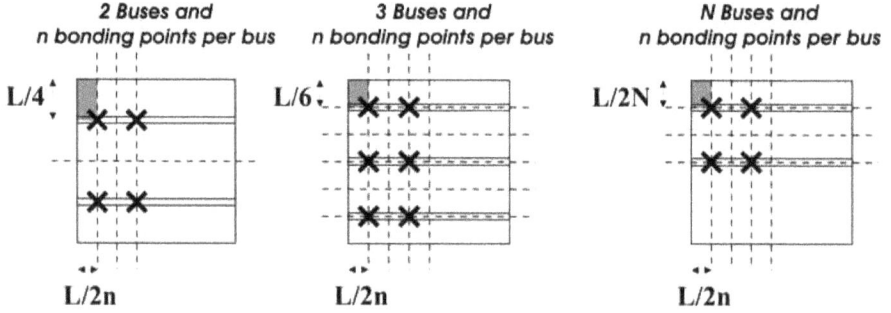

Fig. 11. Basic calculus cell for a variable number of bus-bar, it is shown the basic unitary cell size for the case of having N collecting bus-bars with n bonding points per bus-bar in the solar cell

With this configuration, expressions of the different components of the series resistance will change as it is shown in Table 6, where N is the number of buses and the rest of symbols represent the same parameter as in Table 3.

Component	Expression
Emitter	$R_{Emitter} = \dfrac{n \cdot s^2}{3L} \dfrac{R_e}{\left(\dfrac{L}{N} - w_{bus}\right)}$
Base	$R_{base} = \rho_{base} \dfrac{4N \cdot n \cdot w_{base}}{L^2}$
Metallic finger	$R_{finger} = \dfrac{n \cdot s}{3L} \cdot \dfrac{\rho_{metal}}{w_f h_f} \left(\dfrac{L}{N} - w_{bus}\right)$
Bus bar	$R_{bus} = \dfrac{\rho_{metal}}{3n} \cdot \dfrac{L}{w_{bus} h_{bus}}$
Rear contact resistance	$R_{fc} = \dfrac{4N \cdot n \cdot s \cdot R_{FrontPaste}}{L \cdot \left(w_f \cdot L + N \cdot w_{bus} \cdot (s - w_f)\right)}$
Front contact resistance	$R_{bc} = \dfrac{4N \cdot n \cdot R_{BackPaste}}{L^2}$

Table 6. Analytical expressions of the series resistance components for a multi bus-bar cell (with a number of bus-bars equal to N)

Total solar cell series resistance corresponds in this case to the expression:

$$R_{Series\,Total} = \frac{\sum R_{Components}}{4n \cdot N}$$

With these new analytical modelling and the set of values for each technological parameter presented in the previous section, it is possible to extend the study of the performance of cells with two and three buses to a bigger number of buses, as it is plotted in Fig. 12, where the maximum possible efficiencies (reached with the optimal design of grid) are shown as a function of the covering factor for a grid with 100 microns finger width.

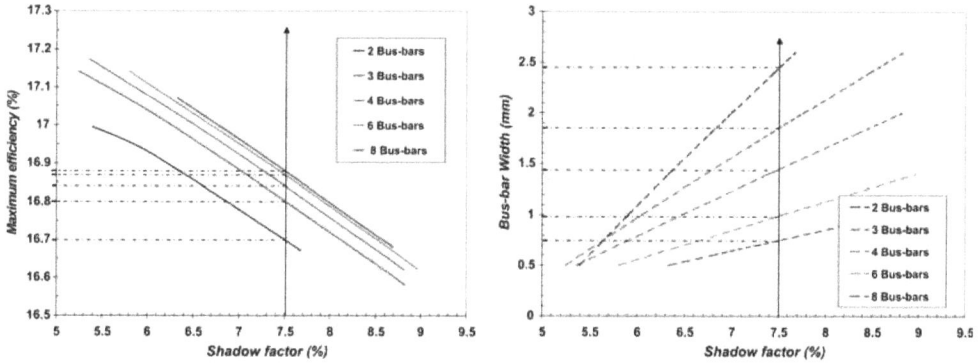

Fig. 12. Comparison between behaviour of cells with a different number of bus-bars for the case of having fingers of 100µm width (fixed), picture on the left shows the efficiency as a function of shadowing factor of the front grid, while picture on the right shows the bus-bar needed to reach an specific shadowing factor

When the number of bus-bars increases an improvement in efficiency is reached, but this improvement, due to the addition of one new bus-bar, makes lower when the total number of buses increase as can be seen in Fig. 12.

As in the case of using three bus-bars, when the number of bus-bars is increased a further reduction of the interconnection module power losses is obtained. It can be calculated from the example referred in Fig. 12 (for a shadowing factor of 7.5%) that a relative power loss reduction of 11.7% (with respect to the two bus-bars configuration) would be reach with three bus- bars, while it would be 15.5%, 18.3%, and 18.3% with four, six and eight bus-bars respectively.

Nowadays technological parameters make possible the improvement of solar cells with the addition of more buses mainly thanks to the reduction of the finger resistance. The improvement is not so high for more than three or four buses, and it must be taken into account that a future improvement in conductivity of metal pastes or its resulting finger cross section's aspect ratio would produce a lower improvement, due to the addition of more buses.

6.4 Modelling the effects of future improvements in metallization of solar cells

The proposed modelling of the series resistance constitutes a powerful tool to know how industrial solar cells can evolve thanks to the improvement of metallic pastes, or new passivation processing for bulk and surface. With this purpose the modelling will be used to evaluate the efficiency of a solar cell (with an optimal H-pattern grid design) when it happens a change in the technological parameters related to the metallic contact (resistivity of metal and contact resistance) or in the Voc (parameter that is related to the passivation of volume and surfaces). All the rest of parameters of the modelling have been kept constant for this analysis with the values referred in Table 7.

L	156 mm
Jsc(without grid shadow)	36.5 mA/cm²
Number of bus-bars	2 and 3
Number of contacts per bus-bar	8
w_{base}	200 μm
ρ_{base}	1 Ωcm
w_{bus}	2.45 mm (2 bus-bars) 1.85 mm (3 bus-bars)
w_f	100 μm
h_{bus}	25 μm
h_f	12 μm
R_e	45 Ω/sqr

Table 7. Model parameters used in the analysis of the evolution of the cell behaviour when the technological parameters associated to the screen-printing technology are modified

Results of the modelling are plotted in Fig. 13 for two different situations, with a Voc of 616 mV and for a Voc of 630 mV.

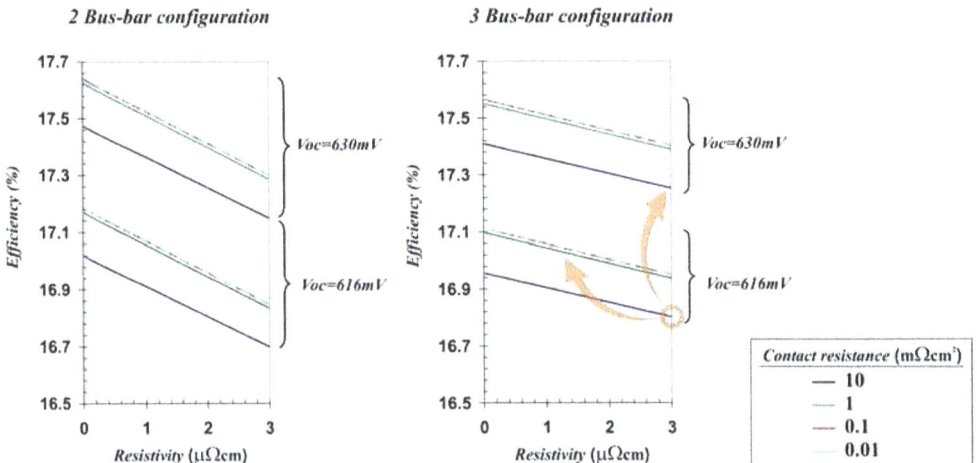

Fig. 13. Modelling results for the efficiency as a function of the finger resistivity and contact resistance for cells of two and three bus-bars with the optimal design of grid in each case, when generated current are kept constant and two different Voc are considered. Rest of parameters for the modelling are taken from Table 7

As it is shown in the figure a jump in *Voc* of 14mV has an stronger impact in cell efficiencies than any possible improvement in the material of the front metallic grid; even when it is use the resistivity of the pure silver (1.63 μΩcm (Lide, 1974)) for the optimal grid.

When an increase in generated current is studied, as it is shown in the Fig. 14, a similar result is obtained, showing that although it is possible to improve the solar cell performance thanks to the pattern optimization and enhancement of the materials properties for the front grid contact, the improvement of the device design and materials under the front contact have a bigger importance in the road to the industrial solar cells improvement.

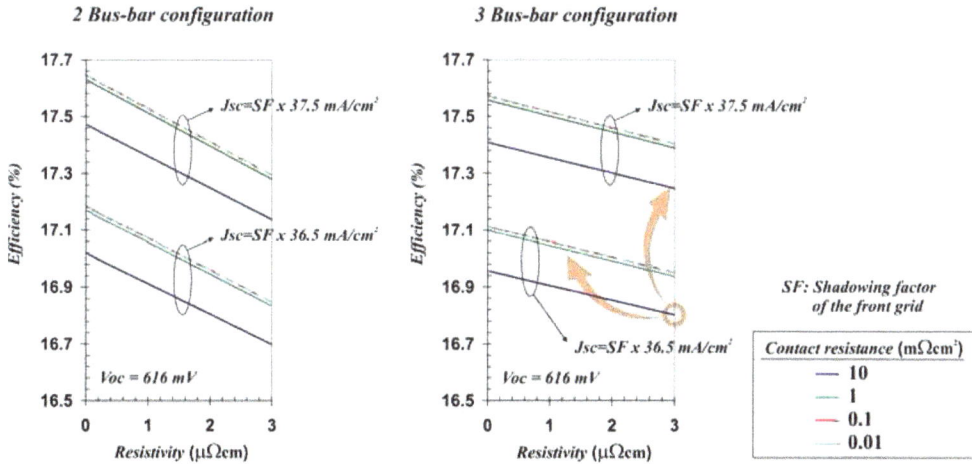

Fig. 14. Modelling results for the efficiency as a function of the finger resistivity and contact resistance for cells of two and three bus-bars with the optimal design of grid in each case, when *Voc* are kept constant and two different generated current are considered (36.5 and 37.5 mA/cm² without grid shadowing). Rest of parameters for the modelling are taken from Table 7

7. Coming future for the industrial front face cell definition

7.1 Changes of the device structure

As can be extracted from the previous section, the coming future for the solar cells front face definition, to get an effective improvement on the energy conversion efficiency, is not only related to the use of an optimal metal grid definition, but is also related to the introduction of different device's structures that will improve the performance of cells.

The Different working lines that can be followed by industry are related to the introduction in production lines of high efficiency concepts that are being successfully tested by research institutes and universities, among these lines it can be found several approach such as:

- To change the silicon surface topology introducing new more efficient texturization processing (Zhao et al., 1999), (Kumaravelu et al., 2002).
- And improving the optical performance of these with and optimised design of the anti-reflective coatings and rear reflector in order to get higher generated current (Nilsen et al., 2005); (Glunz et al., 2007).

- Improving the front, bulk and rear surface electrical passivation in order to enhance the final open circuit voltage of the devices, by means of better aluminium rear pastes, using gettering steps or introducing more radical changes in the cell design as a rear local contact with an improved electrical passivation. (Glunz et al., 2007).

Among all the high efficiency concepts, the idea of a 'selective emitter' in the devices is prone to be one of the first concepts introduced by the industry in its production lines without the need of radical changes in the production processes. This is why a special attention must be devoted to this concept.

The concept of 'selective emitter' consist in rising the surface doping level in the emitter area where metal grid will be deposited to improve later on its electrical contact (as it is plotted in Fig. 15), and keeping a low surface doping level in the rest of the front side of the cell (area that in that way will be later electrically better passivated) (Green, 1995). This concept not only improves the generated current due to the improvement of the low wavelength response of the resulting cell emitters, but also makes the open circuit voltage to rise due to the improvement in the front surface passivation.

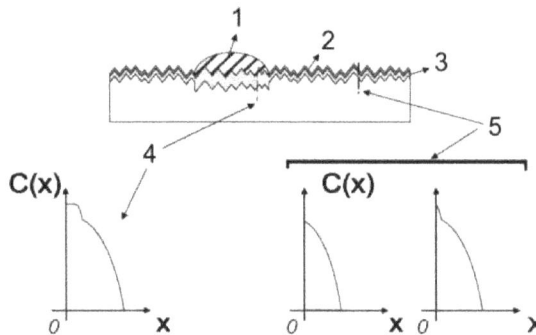

Fig. 15. Selective emitter concept, picture shows a cross section of the upper part of a solar cell where it can be observe the metal fingers (1); antireflective coating (2) and diffused layer (3). A different Phosphorus dopant concentration as a function of depth must be appreciated under the metallic grid (4) and in the non-contacted area (5)

But technical difficulty in obtaining this structure has made the selective emitter concept typical from fabrication processes of research labs, or implemented in solar cells with more complex structures and complicated fabrication processes due to the alignment requirements (as the LGBC or Laser Grove Buried Contact solar cell (Wenham & Green, 1993)), leaving this concept out of simple processing production due to its high implementation cost.

Apart from the typical microelectronic way to get a selective emitter with several diffusion steps and using diffusion barriers deposited on the (later on) non contacted areas (with an expensive photolithographic process due to the restrictive alignment between patterns and the high number of processing steps), new innovative techniques to get a selective emitter with screen-printing contacts are appearing recently due to the great industrial interest on improving photovoltaic device efficiencies (Raabe et al., 2007). All these new techniques can be classified in two big groups according to the way the selective emitter structure is obtained:

A. Selective emitter developed without masking processes, carrying out local phosphorus diffusion on the wafer surface.
    Several alternatives for this kind of processes has been studied, among these, it could be emphasized:
    1. The local deposition of phosphorus sources on the wafer surface by using a screen-printing process previous to the high temperature diffusion step, generating most highly doped areas in a self-aligned process that only requires the right location of the contact grid during the metallization step. Different examples of these processes can be found in (Horzel et al., A 1997); (Horzel et al., B 1997); (Salami et al., 2004).
    2. The use of lasers to create, after a soft phosphorus diffusion step in all the cell area, a higher doped contact area (Besi-Vetrella et al., 1997).
    3. Or the use of special metallization pastes that include doping material in its composition, thereby these pastes are used as a source of dopants for the contact area during the firing step of the metallization, creating a selective emitter structure when they are deposited on wafers with a low doped emitter in a self-aligned fabrication process. Examples of these processes can be found in (Rohatgi et al., 2001); (Porter et al., 2002); (Hilali et al., 2002).
B. Selective emitter developed using masking processes to protect, with barriers, part of the front area from the diffusion step (creating zones with a softer diffusion due to these barriers) (Bultman et al., 2000); (Bultman et al., 2001). Or masking to protect the surface from a selective etching. As it is refer in (Ruby et al., 1997); (Zerga, A. et al., 2006); (Haverkamp et al., 2008).

The alternatives exposed in the first group produce a correct selective emitter structure but have a drawback, after the phosphorus diffusion and the gettering step it introduces, when it is carried out in super-saturation conditions, the impurities concentrations in the silicon bulk keep constant, because impurities remain mainly trapped in the 'dead layer' that appears near the surface, without been effectively removed from the device, reducing the potential impact of the improvement this step could have.

Among the different alternatives of the second group, however, it exists processes that can carry out an effective reduction of impurities when these include a surface etching, what result in a better device performance; but also present some drawbacks related to the needed mask treatments and processing (such as deposition, curing and removing steps), giving slightly more complex fabrication process.

All the developed alternatives in both groups (with exception of the use of self-doped metallization pastes) present a fundamental problem for the selective emitter structure, that is the need of an alignment with the next processing steps for the contact definition, complicating the fabrication routes. This problem gets worse when it is taken into account the random deformation screen-printing technique presents for the transferred patterns with the increase in the number of prints (deformation that is associated to the relaxation of the fabrics, that compose the screens, and gets a maximum value in the mass production environment). But appearance of new alignment relaxed device structures concepts (Caballero, 2009) can help to develop new and easier industrial fabrication processes that could finally result in the implantation of the selective emitter as a common part of the typical industrial solar cell device structures. The introduction of the selective emitter structure would force the re-design of the front contact grid that could present a different optimal finger separation (with closer fingers) due to the increase in the emitter resistance of the new cells, but no other additional important changes.

7.2 The future of the metallic contact definition

Apart from the improvement of the general parameters of the device (*Voc* and *Jsc*), as previous sections has shown, improving the front contact of solar cells is possible and it can produce an increase in the final efficiency of the industrial solar cells. In this section it will be reviewed the strategies that research centres and industry are following for a future improvement of the front grid.

The approach for improving the front grid is based on increasing the aspect ratio of the cross-section of fingers as it is shown in Fig. 16, reducing the seepages that increase the shadowing factor of the grid without reducing the grid resistance, increasing the finger height, and reducing its width in order to produce a lower grid shadowing.

Fig. 16. Future improvement of the metal finger cross-section

Several strategies can be followed or will be followed by the industry with this purpose; from the introduction of slight modifications or changes of the nowadays production technology, such as:

1. Optimising the paste composition with different combinations of the silver particles with different shapes and sizes, in order to maximise the finger heights after printing.
2. Optimising the printing process, modifying the fabrics and emulsions characteristics in the screens, and the processing parameters to reduce the paste seepages during printing.
3. Introducing a heated chuck inside the standard printer units in order to produce an increase in resolution by heating the wafer substrates during printing process. Reaching finger width as lower as 50 microns (Erath et al., 2009).
4. Substituting the standard pastes by the Hotmelt technology pastes (Williams et al., 2002) that produce an improved aspect ratio of the final metal fingers. This technology changes the traditional solvent of the paste by a long Chain alcohol with a melting point in the range of 40 to 90°C, so final paste needs a heated screen to be disposed.

To the introduction of additional processing steps, that needs the addition of new machines in the production lines, such as:

5. Growing pure silver over the screen-printed contact in an electrolytic bath, reducing the resistance of the fingers and improving the contact resistance in its edges, due to the silver filling of the empty space of the fingers volume.

    Several approaches can be found in the market, based on the classical electrolytic growing, that needs a current contact with the front grid, or based on the LIP technique (Pysch et al., 2008); (Glunz et al., 2008) that doesn't need any contact. In both cases grid fingers width increases after processing.

Or the complete change of the technology for the contact definition using new techniques nowadays under development, in research projects of several companies and institutions, such as:

6. Growing the complete finger in an electrolytic processing, but defining previously a grid which is used as a seed for the electrolytic grown with narrow lines to avoid an excessive increase of the final finger width.

This seed for growing could be produced with an electroless nickel plating over the bare emitter silicon (Glunz et al., 2008) (processing that would introduce previously, a masking of the front silicon nitride layer using an inkjet printing system to deposit the mask with the needed definition of lines; and an etching of the masked structure to open the nitride layer), or defined with new promising techniques such as inkjet printing (Mills & Branning, 2009), aerosol jet printing (technique able to reach finger definition of 40 microns) (Hörteis et al., 2008), or laser direct-printing (able to reach a finger definition below 20 microns) (Shin et al., 2008); (Arnold et al., 2004) over the silicon nitride layer (with its later firing through previous to the electrolytic growing) or also over the bare silicon emitter (processing that would need also a previously masked etchings of the nitride).

The introduction of the plating of new materials to grow the contact grid can reduce additionally the final cost of the metallization step. Thus, copper with a good conductive properties (97.61% of the silver conductivity (Brady et al., 2002)), is stirring up the industrial interest on the creation of new multiple metal layer contact such as:

Seed of Nickel/ Copper/ Tin
Seed of Silver/ Nickel/ Cooper/ Tin or
Directly Copper without seed/ Tin

On the other hand, situation for the back aluminium contact of cells is different because it is not related with the improvement of resolution, but with the need of an improvement in Voc to increase the final cell efficiencies. In this case the possible paths for industry to follow are:

*   Improving the characteristics of the Aluminium pastes for the total bsf device structure, designing new formulations of the used pastes and removing the limitation that nowadays the aluminium contact has, related with the bowing of devices during the firing process, by means of, for example, adding special thermal treatments processing (Huster, 2005).
*   Changing the total rear contact by a local rear contact where non contacted areas must be passivated with a new deposited layer with improved properties of passivation for p type silicon. Laser techniques for the creation of the local contacts (Schneiderlöchner et al., 2002); (Tucci et al., 2008) are taking a good position to be industrialised for the creation of this kind of local contacted structures.

8. Conclusions

This chapter has presented the technique used by the mass production industry to define the contacts of the silicon solar cell, its basic principles and factors that have an influence in its result with the main aim of giving an introductory view of a technique responsible for the development and expansion of the nowadays photovoltaic terrestrial market.

It has been shown how through a simple analytical modelling the performance of different designs for the front contact in the commercial solar cells can be optimised and compared. And a quick review of the coming changes in the device structure design, and the future techniques that are under research to substitute the screen-printing technology have been done in order to give an idea of how the industry can evolve in the coming years.

9. References

Arnold, C. B.; Sutto, T. E.; Kim, H.; Piqué, A., 'Direct-write laser processing creates tiny electrochemical systems', *Laser Focus World*, May 2004.

Besi-Vetrella, U. et al., 'Large area, screen printed Silicon solar cells with selective emitter made by laser overdoping and RTA spin-on glasses', U. Besi-Vetrella et al., 26th IEEE PVSC, Anaheim, CA, 1997.

Brady, G. S.; Clauser, H. R.; Vaccari, J. A., *Materials handbook 15th Edition*, Ed. Mc Graw Hill, pg 1106, 2002

Bultman, J. H. et al., 'Single step selective emitter using diffusion barriers', *Proceedings of the 16th EPVSEC*, Glasgow, 2000.

Bultman, J. H. et al., 'Ideal single diffusion step selective emitters: a comparison between theory and practice', *Proceedings of the 17th EPVSEC*, Munich, 2001.

Caballero, F. J., *Patent pending concept*, 'Structure of selective emitter in solar cell of easy alignment between patterns for its industrial mass production, with possible local phosphorus gettering process integration and able for in-line processing', 2009.

Caballero, L.J.; Sánchez-Friera, P.; Lalaguna, B.; Alonso, J.; Vázquez, M.A., 'Series Resistance Modelling Of Industrial Screen-Printed Monocrystalline Silicon Solar Cells And Modules Including The Effect Of Spot Soldering', *Proceedings of the 4th WCPVSC*, Hawaii, 2006.

Erath, D.; Filipovic, A.; Retzlaff, M.; Goetz, A. K.; Clement, F.; Biro, D.; Preu, R., 'Advanced screen printing technique for high definition front side metallization of crystalline silicon solar cells', *Solar Energy Materials & Solar Cells*,V5,Issue 18, 2009.

Firor, K.; Hogan, S., 'Effects of processing parameters on thick film inks used for solar cell front metallization', *Solar Cells*, vol. 5, Dec. 1981, p. 87-100.

Glunz, S. W. et al., 'High-Efficiency Crystalline Silicon Solar Cells', Advances in OptoElectronics, Volume 2007, Article ID 97370, Hindawi Publishing Corporation, 2007

Glunz, S. W. et al., 'Progress in advanced metallization technology at Fraunhofer ISE', *Proceedings of the 33rd IEEE PVSC*, San Diego, 2008.

Green , M. A., *'Silicon solar cells, Advanced principles & Practice'*, Centre for photovoltaic devices and systems UNSW, Sydney, 1995.

Haigh, A. D. (Ferranti Ltd.), 'Developments in Polycrystalline Silicon Solar Cells And A Novel Form Of Printed Contact'. *Proceedings of the International Conference on Solar Electricity*. France, 1976.

Haverkamp, H. et al., 'Minimizing the electrical losses on the front side: development of a selective emitter process from a single diffusion', *Proceedings of the 33rd IEEE PVSC*, San Diego, CA, 2008.

Hilali, M. et al., 'Optimization of self-doping Ag Paste Firing to achieve high Fill Factors on screen-printed Silicon solar cells with 100Ω/sq. Emitter', *Proceedings of the 29th IEEE PVSC*, New Orleans, 2002.

Holmes, P. J.; Loasby, R. G., *Handbook of Thick Film Technology*, Electrochemical Publications Limited, 1976.

Horzel, J. et al., A, 'Novel method to form selective emitters in one diffusion step without etching or masking', *Proceedings of the 14th EPVSEC*, Barcelona, 1997.

Horzel, J. et al., B, 'A simple processing sequence for selective emitters', *Proceedings of the 26th IEEE PVSC*, Anaheim, CA, 1997.

Hörteis, M.; Richter, P. L.; Glunz, S. W., 'Improved front side metallization by aerosol jet printing of Hotmelt inks', *Proceedings of the 23rd EPVSEC*, Valencia, 2008.

Huster,F., Aluminium-Back Surface Field: Bow investigation and Elimination, *Proceedings of the 20th EPVSEC*, Barcelona, pp. 635-638, 2005.

Kumaravelu, G.; Alkaisi, M. M.; Bittar, A., 'Surface texturing for silicon solar cells using reactive ion etching technique', *Proceedings of the 29th IEEE Photovoltaic specialists conference*, New Orleans, Louisiana, USA, 2002.

Lide, D. R., *Handbook of Chemistry and Physics*, CRC Press 1974.

Luque A. et al. *Solar Cells and Optics for Photovoltaic Concentration*, chapter 4, Adam Hilger series on optics and optoelectronics, 1989.

Mills, R. N.; Branning, P., 'Inkjet systems for use in photovoltaic production', *Photovoltaic World*, September 2009.

Nilsen , D.; Stensrud, E. and Holt , A., Double layer anti-reflective coating for silicon solar cells, Proceedings of the 31st IEEE PVSC, Orlando, Florida, USA, pp.1237-1240, 2005.

Porter L. M.. et al., 'Phosphorus-Doped, Silver-Based Pastes for Self-Doping Ohmic Contacts for Crystalline Silicon Solar Cells', *Sol. Energ. Mat. and Sol. Cells* 73 (2), 209-219, 2002.

Pysch, D.; Mette, A.; Filipovic, A.; Glunz, S. W., 'Comprehensive analysis of advanced solar cell contacts consisting of printed fine-line seed layers thickened by silver plating', *Progress in Photovoltaics*, V17, Issue 2, pp. 101-114, 2008.

Raabe, B. et al. 'Monocrystalline Silicion-Future Cell Concepts', *Proceedings of the 22nd EPVSEC*, Milan, 2007.

Ralph, E. L. (Spectrolab Inc.), 'Recent advancements in low-cost solar cell processing', *Proceedings of the 11th IEEE PVSC*, 1975.

Recart, F.,*Evaluación de la serigrafía como técnica de metalización para células solares eficientes´*,.PhD Thesis, Universidad del Pais Vasco, 2001.

Roberts, S.; Heasman, K. C.; Bruton T. M., 'The reduction of module power losses by optimisation of the tabbing ribbon', *Proceedings of the 16th EUPVSEC*, Glasgow, 2000.

Rohatgi, A. et al., 'Self-aligned self-doping selective emitter for screen-printed silicon solar cells', *Proceedings of the 17th EPVSEC*, Munich, Germany, 2001.

Ruby, D. S. et al., 'Recent progress on the self-aligned, selective-emitter Silicon solar cell', *Proceedings of the 26th IEEE PVSC*, Anaheim, CA, 1997.

Salami, J. et al., 'Characterization of screen printed phosphorus diffusion paste for Silicon solar cells', *Proceedings of the PVSEC-14*, Bangkok, Thailand, 2004.

Shin, H.; Lee, H.; Sung, J. and Lee, M., 'Parallel laser printing of nanoparticulate silver thin film patterns for electronics', *Appl. Phys. Lett.* 92, 233107 (2008).

Schneiderlöchner, E.; Preu, R.; Lüdemann, R.; Glunz, S. W., Laser-fired rear contacts for crystalline silicon solar cells, *Progress in Photovoltaics*, V. 10 Issue 1, Pages 29 - 34, 2002

Tucci M.; Talgorn, E.; Serenelli L.; Salza, E.; Izzi, M. and Mangiapane, P., 'Laser fired back contact for silicon solar cells', Thin Solid Films, Volume 516, Issue 20, pp. 6767-6770, 2008.

Van der Heide A. S. H., Bultman J. H., Hoornstra J., et al. 'Locating losses due to contact resistance, shunts and recombination by potential mapping with the Corescan'.

Proceedings of the 12th NREL Workshop on Crystalline Silicon Solar Cells, Materials and Processes, Breckenrige (CO), USA, A, 2002

Van der Heide A. S. H., Bultman J. H., Hoornstra J., et al., 'Optimizing the front side metallization process using corescan', *Proceedings of the 29th IEEE PVSC*, New Orleans, USA, B, 2002.

Williams, T.; McVicker, K.; Shaikh, A.; Koval, T.; Shea, S.; Kinsey, B. and Hetzer, D., 'Hot Melt Ink Technology for Crystalline Silicon Solar Cells', 29th IEEE PVSC, New Orleans, 2002.

Wenham, S. R. and Green, M.A. Australian Patent 570309, 1993.

Zerga, A. et al., 'Selective emitter formation for large-scale industrially MC-Si solar cells by hydrogen plasma and wet etching', *Proceedings of the 21st EPVSEC*, Dresden, 2006.

Zhao, J.; Wang, A.; Campbell, P.; and Green, M. A., 'A 19.8% Efficient Honeycomb Multicrystalline Silicon Solar Cell with Improved Light Trapping', *IEEE Transactions On Electron Devices*, Vol. 46, No. 10, pp 1978-1983, 1999.

6

Potential of the Solar Energy on Mars

Dragos Ronald Rugescu[1] and Radu Dan Rugescu[2]
[1]*University of California at Davis,*
[2]*University Politehnica of Bucharest*
[1]*U.S.A.,*
[2]*Romania E.U.*

1. Introduction

The problem of energy accessibility and production on Mars is one of the three main challenges for the upcoming colonisation of the red planet. The energetic potential on its turn is mainly dependent on the astrophysical characteristics of the planet. A short insight into the Mars environment is thus the compulsory introduction to the problem of energy on Mars. The present knowledge of the Martian environment is the result of more than two centuries of attentive observation on its astronomical appearance and, more recently, on its on-site astrophysical features. Recent surface measurements of Martian geology, meteorology and climate had fixed the sometime-unexpected image of a completely desert planet. Mars is one of the most visible of the seven planets within the solar system and thusfor its discovery cannot be dated, still the interest for Mars is old. It was easily observed from the ancient times by necked eye and the peculiar reddish glance of the planet had induced the common connection of the poor planet with the concept of war. The god of war and the planet that inherited his name had provoked, still from antiquity, curiosity and disputes on the most diverse themes. These disputes are at a maximum right now regarding the habitability of Mars. The red planet owes his color to still unexplained causes, where a yet undisclosed chemistry of iron oxides seems to be the main actor. The visit card of Mars is fast increasing in the quantity of data and is now quite well known (Bizony, 1998), as we observe from the description that follows.

1.1 Mars as seen before the space age

As far as the knowledge of the solar system has gradually extended, from optical, ground-based observations to the present astrophysical research on site, Mars appears as the fourth planet as starting from the Sun. The reddish planet of the skies, nicely visible by necked eyes, has attracted the most numerous comments during the time regarding the presence of life on an extraterrestrial planet. With all other eight planets, except for Pluto-Charon doublet, Mars aligns to a strange rule by orbiting the Sun at a distance that approximates a multiple of $\sqrt{2}$ from that of the Earth. This means that the rough 149.6 mil km of the Earthsemi-major axis is followed by a rough 212 mil km for Mars. In fact there are 227.92 mil km at mean from the center of Sun. The power rule of Titius-Bode[1], modified several times, but originally described as $a = (4 + 3 \times \mathrm{sgn}\, n \times 2^{n-1}) / 10 \,|\, n = \overline{0,9}$ gives a better distribution,

Planet	n	Titius-Bode rule	Actual semi-major axis
Mercury	0	0.4	0.39
Venus	1	0.7	0.72
Earth	2	1.0	1.00
Mars	3	1.6	1.52
Asteroids	4	2.8	2.80
Jupiter	5	5.2	5.20
Saturn	6	10.0	9.54
Uranus	7	19.6	19.20
Neptune/Pluto	8	38.8	30.10/39.20
Sedna	9	77.2	75.00

Table 1. Mars within Titius-Bode's rule (astronomical units)

It is immediately seen that the primary solar radiation flux is roughly two times smaller for Mars than it is for Earth. More precisely, this ratio is equal to 2.32. This observation for long has suggested that the climate on Mars is much colder than the one on Earth. This has not removed however the belief that the red planet could be inhabited by a superior civilization. Nevertheless, beginning with some over-optimistic allegations of Nicolas Camille Flammarion (*Flamarion*, 1862) and other disciples of the 19-th century, the planet Mars was for a century considered as presenting a sort of life, at least microbial if not superior at all. The rumor of Mars channels is still impressing human imagination. When estimates begun to appear regarding the Martian atmosphere and figures like 50 mbar or 20 mbar for the air pressure on Martian ground were advanced (Jones 2008), a reluctant wave of disapproval has been produced. It was like everybody was hoping that Mars is a habitable planet, that we have brothers on other celestial bodies and the humankind is no more alone in the Universe. As more data were accumulating from spectroscopic observations, any line of emission or absorption on Mars surface was immediately related to possible existence of biological effects. Even during the middle 20-th century the same manner was still preserving. In their book on "Life in the Universe" Oparin and Fesenkov are describing Mars in 1956 as still a potential place for biological manifestations (*Oparin & Fesenkov*, 1956). The following two excerpts from that book are relevant, regarding the claimed channels and biological life on Mars: "...up to present no unanimous opinion about their nature is formed, although nobody questions that they represent real formations on the planet (Mars)..." and at the end of the book "On Mars, the necessary conditions for the appearance and the development of life were always harsher than on Earth. It is out of question that on this planet no type of superior form of vegetal or animal life could exist. However, it is possible for life, in inferior forms, to exist there, although it does not manifest at a cosmic scale."

---

[1] In 1768, Johann Elert Bode (1747-1826), director of Berlin Astronomical Observatory, published his popular book, "Anleitung zur Kenntnis des gestirnten Himmels" (*Instruction for the Knowledge of the Starry Heavens*), printed in a number of editions. He stressed an empirical law on planetary distances, originally found by J.D. Titius (1729-96), now called "Titius-Bode Law".

The era of great *Mars Expectations*, regarding extraterrestrial life, took in fact its apogee in 1938, when the radio broadcast of Howard Koch, pretending to imaginary fly the coverage of Martian invasion of Earth, had produced a well-known shock around US, with cases of extreme desperation among ordinary people. Still soon thereafter, this sufficed to induce a reversed tendency, towards a gradual diminution of the belief into extraterrestrial intelligence and into a Martian one in particular. This tendency was powered by the fact that no proofs were added in support of any biological evidence for Mars, despite the continuous progress of distant investigations. Still every of the 36 great oppositions of Mars, since the "canali" were considered by Giovanni Schiaparelli in 1877, prior to the space age in 1957, like the series in 1901, 1911, 1941, 1956 was only adding subjective dissemination of channels reports, with no other support for the idea that on Mars any form of biological life exists. After Schiaparelli and Flammarion, the subjective belief in a Martian life is successively claimed by the well known astronomers Antoniadi, Percival Lowel, A. Dollfus, G. A. Tihov and others.

Any documentation of a hostile environment on Mars was received with adversity. Despite later spectroscopic and radiometric measurements from Earth, which were revealing a very thin atmosphere and extreme low temperatures, still in the immediate down of the space age the pressure on the Martian soil was yet evaluated at 87 mbar (*Oparin & Fesenkov*, 1956), overrating by more than ten times the actual value, irrefutably found after 1964. It is a pregnant evidence of how subjective the world could be in administrating even the most credible scientific data in this delicate subject. A piece of this perception is surviving even today.

1.2 Mars during the space age

With the availability of a huge space carrier, in fact a German concept of Görtrupp, the Soviets started to built a Mars spacecraft during 1959, along to the manned spacecraft Vostok. The launch took place as early as in October 1960, mere 3 years after Sputnik-1, but proved that only mechanical support is of no much use. The restart of the accelerator stage from orbit was a failure that repeated a few weeks later with similar non-results. The boost towards Mars commenced again with Mars-1 and a companion during the 1962 window, ending in failure again. It followed this way that a much smaller but smarter US device, called Mariner-4, despite of the nose-hood miss-separation of its companion Mariner-3, had marked the history of Mars in 1964 with a shaking fly-by of the planet and several crucial observations. These stroke like a thunder: the radio occultation experiment was suddenly revealing an unexpectedly low atmospheric pressure on Mars of approximately 5 mbar, much far below the most of the previous expectations. The life on Mars was bluntly threatened to became a childish story. Primitive but breathtaking images from Mariner-4 were also showing a surface more similar to the Moon's one than any previous expectation could predict. A large number of craters were mixed with dunes and shallow crevasses to form a desert landscape and this appearance produced a change in the perception of the formation of the solar system at large. No channel was revealed and none of the previously mentioned geometrical marks on the Martian surface. The wave of disappointment grew rapidly into a much productive bolster for deepening these investigations, so incredible those results were. Mars's exploration proceeded with Mariner-6 and 7 that performed additional Martian fly-byes in 1967 only to confirm the portrait of a fully deserted planet.

The present hostile environment on Mars and the absence of life seem for now entirely proven, but it still remains to understand when this transform took place, whether there was sometime a more favorable environment on Mars and all issues regarding the past of Mars. We say *for now* because our human nature bolsters us towards an ideal which we only dream of, but which, as we shall prove, is almost accomplishable to the level of present technologies. Anyhow, the nephews of our nephews will perhaps take it up to the end. We are speaking of Mars's colonization, process that could only take place after the complete transformation of the surface and of the atmosphere to closely resemble those on Earth, process we call terraforming. Whether planet Mars worth more than its terraforming, then it deserves all the money spent with this process.

For the moment however, the Martian environment is far of being earth-like, as the compared data from the next table reveal (*Almanac* 2010):

Parameter	Mars	Earth	units.
Orbital eccentricity	$\varepsilon = 0{,}09341233$	0.01669518	-
Semi-major axis	$a = 227{,}920{,}000$	149,597,871	km
Focal semi-chord	$p = 225{,}931{,}201$	149,556,174	km
Perihelion	$r_P = 206{,}629{,}462$	147,100,308	km
Aphelion	$r_A = 249{,}210{,}538$	152,095,434	km
Inclination to ecliptic	$i = 1.84964°$	0°	°
Ascending node	$\Omega = 49.6296364°$	0°	°
Perihelion argument	$\omega = 336.04084°$	102.94719°	°
Sideral revolution	$T = 59355072$	31558118	s
Equatorial radius	$R_e = 3{,}397{,}515.0*$	6,378,136.6	m
Flattening factor	$f = 1/154.321$	1/298.25642	-
Rotational velocity	$w = 7.088244 \cdot 10^{-05}$	$7.292700 \cdot 10^{-05}$	rad/s
Obliquity	$v = 25.19°$	23.45°	°
Gravitational field**	$K = 4.28283149 \cdot 10^4$	$3.986004391 \cdot 10^5$	km³/s²
Solar irradiance	$q = 589.2$	1367.6	W/m²
Ground pressure***	$p_0 = 6.36$	1013.25	mbar
Ground temperature	$T_0 = 210.0$	283.0	°K .

* (JPL 1998)
** Earth without Moon
*** Mean for Martian latitude and year

Table 2. Comparative Mars and Earth astrophysical data

The data in table 2 are based on the mean eccentricities of the year 2009 as given in the reference. Accordingly, a value of $g = 6.67428 \cdot 10^{-11}$ is used for the universal constant of gravitation. Second is the time unit used to define the sidereal periods of revolution around the Sun and is derived, on its turn, from the solar conventional day of 24 hours in January 1, 1900. The atmosphere of Mars is presently very well known and consists, in order, of Carbon dioxide (CO_2) 95.32%, Nitrogen (N_2) 2.7%, Argon (Ar) 1.6%, Oxygen (O_2) 0.13%, Carbon monoxide (CO) 0.07%, Water vapor (H_2O) 0.03%, Nitric oxide (NO) 0.013%, Neon (Ne) 2.5 ppm, Krypton (Kr) 300 ppb, Formaldehyde (H_2CO) 130 ppb [1], Xenon (Xe) 80 ppb, Ozone (O_3) 30 ppb, Methane (CH_4) 10.5 ppb and other negligible components. This composition is further used to assess the effect of solar radiation upon the dissociation and los of the upper Mars atmosphere and upon potential greenhouse gases.

The present atmosphere of Mars is extremely tenuous, below 1% of the Earth one and seemingly unstable. Seasonal worming and cooling around the poles produce variations of up to 50% in atmospheric pressure due to carbon dioxide condensation in winter. The values in table 18.2 are rough means along an entire year, as measured by Viking-1 and Viking-2 lenders, Pathfinder and Phoenix station recently. The greenhouse effect of carbon dioxide is considered responsible for 5° increment of atmospheric temperature, very low however, with only occasional and local worming above 0°C on privileged equatorial slopes. The chart of the present Martian atmosphere (*Allison & McEwen*, 2000) is given in figure 18.1. The exponential constant for pressure on Mars is H=11,000 m.

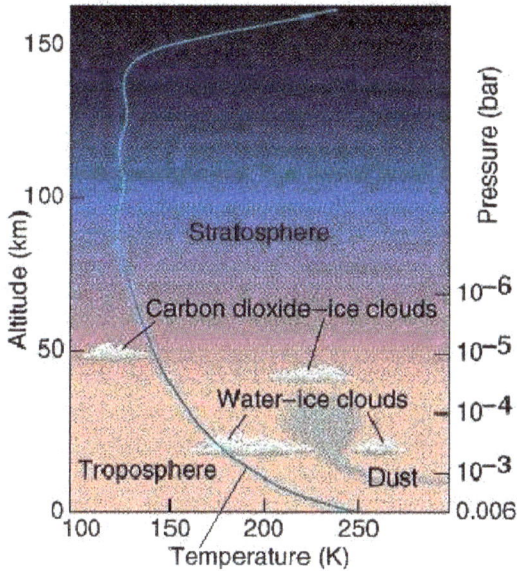

Fig. 1. Mars's atmosphere profile.

These atmospheric characteristics stand as the database in evaluating the efficiency of the solar-gravitational draught, closed-circuit powerplant, which we propose to be employed as an energy source on Mars.

1.3 Mars presumptive past

It is generally considered that the planetary atmospheres, including that of Mars, went through major transformations at the beginning of formation of the solar system (Brain 2008). Present models show a gradual and fast depletion of Martian atmosphere, with essential implications for designing the Mars's terraforming. Even today, the depletion of Mars's atmosphere continues on a visible scale. The observations, recently made by Mars Express and Venus Express, have allowed scientists to draw similar conclusions for both Mars and Venus, through direct comparisons between the two planets. The phenomenon proves this way as nonsingular and systematic (*Brain*, 2008). The results have shown that both planets release beams of electrically charged particles to flow out of their atmospheres. The particles are being accelerated away by interactions with the solar wind released by the Sun. This phenomenon was observed by the two spacecrafts while probing directly into the magnetic regions behind the planets, which are the predominant channels through which electrically-charged particles escape. The findings show that the rate of escape rose by ten times on Mars when a solar storm struck in December 2006. Despite the differences among the two atmospheres, the magnetometer instruments have discovered that the structure of the magnetic fields of both planets is alike. This is due to the fact that the density of the ionosphere at 250 km altitude is surprisingly similar, as shows the Venus Express magnetometer instrument. By observing the current rates of loss of the two atmospheres, planetary scientists try to turn back the clock and understand what they were like in the past. To build a model of Mars's atmosphere past we are only basing our presumptions on the in-situ observed tracks impressed by that past evolution. A very detailed computer modeling of the charged particles erosion and escape may be found in a recent paper from Astrobiology (*Terada et al.*, 2009), with the tools of magneto-fluid-dynamics and today's knowledge of the Martian upper atmosphere and of the solar radiation intensity.

First, recent orbital observations of networks of valleys in high lands and cratered areas of the Southern Hemisphere, especially those by Mars Express and Mars suggest that Mars exhibited a significant hydrologic activity during the first Gyr (Giga-year) of the planet lifetime. Some suggest that an ancient water ocean equivalent to a global layer with the depth of about 150 m is needed for the explanation of the observed surface features (*Bullock & Moore*, 2007). These authors suppose that the evolution of the Martian atmosphere and its water inventory since the end of the magnetic dynamo (*Schubert et al.*, 2000) at the late Noachian period about 3.5-3.7 Gyr ago was dominated by non-thermal atmospheric loss processes. This includes solar wind erosion and sputtering of oxygen ions and thermal escape of hydrogen (*Zhang et al.*, 1993). Recent studies (*Wood at al.*, 2005) show that these processes could completely remove a global Martian ocean with a thickness of about 10–20 m, indicating that the planet should have lost the majority of its water during the first 500 Myr. As far as Mars was exposed before the Noachian period to asteroid impacts, as the entire solar planets did, their effect on atmospheric erosion was simulated. Furthermore, the study uses multi-wavelength observations by the ASCA, ROSAT, EUVE, FUSE and IUE satellites of Sun-like stars at various ages for the investigation of how high X-ray and EUV fluxes of the young Sun have influenced the evolution of the early Martian atmosphere. Terada et al. apply for the first time a diffusive-photochemical model and investigate the heating of the Martian thermosphere by photo-dissociation and ionization processes and by exothermic chemical reactions, as well as by cooling due to CO_2 IR-radiation loss. The model used yields high exospheric temperatures during the first 100–500 Myr, which result in blow off for hydrogen and even high loss rates for atomic oxygen and carbon. By applying a hydrodynamical model for the estimation of the atmospheric loss rates, results were obtained, which indicate that the early Martian atmosphere was strongly evaporated by the young Sun and lost most of its water during the first 100 – 500 Myrs after the planets origin. The efficiency of the impact erosion and hydrodynamic blow off are compared, with the conclusion that both processes present the same rating.

It is a common believe now that during the early, very active Sun's lifetime, the solar wind velocity was faster then the one recorded today and the solar wind mass flux was higher during this early active solar period (Wood *et al.*, 2002, 2005; Lundin *et al.*, 2007).

As the solar past is not directly accessible, comparisons to neighboring stars of the same G and K main-sequence, as observed e. g. By the Hubble Space Telescope's high-resolution spectroscopic camera of the H Lyman-α feature, revealed neutral hydrogen absorption associated with the interaction between the stars' fully ionized coronal winds and the partially ionized interstellar medium (Wood *et al.* 2002, 2005). These observations concluded in finding that stars with ages younger than that of the Sun present mass loss rates that increase with the activity of the star. A power law relationship was drawn for the correlation between mass loss and X-ray surface flux, which suggests an average solar wind density that was up to 1000 times higher than it is today (see Fig. 2 in Lammer *et al.*, 2003a).

This statement is considered as valid especially for the first 100 million years after the Sun reached the ZAMS (Kulikov *et al.* 2006, 2007), but not all observations agree with this model. For example, observations by Wood *et al.* (2005) of the absorption characteristic of the solar-type G star τ-Boo, estimated as being 500-million-year-old, indicate that its mass loss is about 20 times less than the loss rate given by the mass loss power law, for a star with a similar age. Young stars of similar G- and K-type, with surface fluxes of more than 10^6 erg/cm^2/s require more observations to ascertain exactly what is happening to stellar winds during high coronal activity periods. Terada *et al.* (2009) are using a lower value for the mass loss, which is about 300 times larger than the average proton density at the present-day Martian orbit. These high uncertainties regarding the solar wind density during the Sun's youth prevent from attracting a confident determination of how fast the process of atmospheric depletion took place. To overcome this uncertainty, Terada *et al.* (2009) has applied a 3-D multispecies MHD model based on the Total Variation Diminishing scheme of Tanaka (1998), which can self-consistently calculate a planetary obstacle in front of the oncoming solar wind and the related ion loss rates from the upper atmosphere of a planet like Mars.

Recently, Kulikov *et al.* (2006, 2007) applied a similar numerical model to the early atmospheres of Venus and Mars. Their loss estimates of ion pickup through numerical modeling strongly depended on the chosen altitude of the planetary obstacle, as for a closer planet to the star more neutral gas from the outer atmosphere of the planet can be picked up by the solar plasma flow. We shall give here a more developed model of the rarefied atmosphere sweeping by including the combined effect of electrically charged particles and the heterogeneous mixture of gas and fine dust powder as encountered in the Martian atmosphere along the entire altitude scale.

Data were sought from the geology of some relevant features of the Mars surface like the Olympus Mons (Phillips *et al.*, 2001; Bibring *et al.*, 2006), the most prominent mountain in the solar system, to derive an understanding of the past hydrology on the red planet and the timing of geo-hydrological transformations of the surface. Olympus Mons is an unusual structure based on stratification of powdered soils. To explain the features, computer models with different frictional properties were built (McGovern and Morgan, 2009). In general, models with low friction coefficient reproduce two properties of Olympus Mons' flanks: slopes well below the angle of repose and exposure of old, stratigraphically low materials at the basal scarp (Morris and Tanaka, 1994) and distal flanks (Bleacher et al., 2007) of the edifice. The authors show that such a model with a 0.6° basal slope produces asymmetries in averaged flank slope and width (shallower and wider downslope), as actually seen at Olympus Mons. However, the distal margins are notably steep, generally creating planar to convex-upward topography.

This is in contrast to the concave shape of the northwest and southeast flanks and the low-slope distal benches (next figure) of Olympus Mons. Incremental deformation is concentrated in the deposition zone in the center of the edifice.

Outside of this zone, deformation occurs as relatively uniform outward spreading, with little internal deformation, as indicated by the lack of discrete slip surfaces. This finding contradicts the evidence for extension and localized faulting in the northwest flank of Olympus Mons. The wedgelike deformation and convex morphology seen in the figure appear to be characteristic of models with constant basal friction. McGovern and Morgan (2009) found that basal slopes alone are insufficient to produce the observed concave-upward slopes and asymmetries in flank extent and deformation style that are observed at Olympus Mons; instead, lateral variations in basal friction are required.

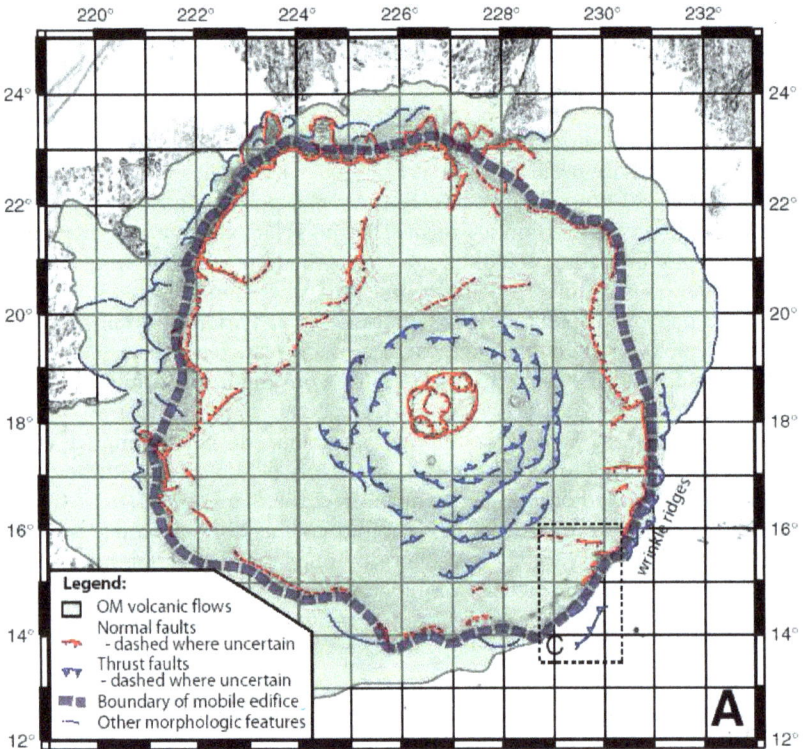

Fig. 2. Geology and interpretation of the Olympus Mons.

The conclusion is that these variations are most likely related to the presence of sediments, transported and preferentially accumulated downslope from the Tharsis rise. Such sediments likely correspond to ancient phyllosilicates (clays) recently discovered by the Mars Express mission. This could only suggest that liquid water was involved into the formation of ancient Olympus Mons, but neither of the conclusions of this work is strong enough. Up to the present, those findings are some of the hardest arguments for ancient water on Mars for a geologically long period.

The importance of this finding is in the promise that a thick atmosphere and wet environment could have been preserving for a long period and thus the Earth-like climate could withstand the depletion influences long enough. This is the type of encouraging arguments that the community of terraforming enthusiasts are waiting for. Problems are still huge, starting with the enormous amounts of energy required for the journey to Mars and back.

2. Energy requirements for the Earth-Mars- Earth journey

The problem of transportation from Earth to Mars and vice versa are restricted by the high amount of energy that must be spent for one kg of payload to climb into the gravitational field of the Sun up to the Mars level orbit and back to the Earth's orbit, where the difference in the orbital speed must also be zeroed. Astrophysical data from the first paragraph allow computing exactly this amount of energy. With the inertial, reactive propulsion of today, two strategies of transferring from Earth to Mars are possible and they belong to the *high thrust navigation* and to *low thrust navigation*. The energy requirements we consider are those for transfer from a low altitude, circular orbit around the starting planet to the low altitude, circular orbit around the destination planet. The very energy of injection into the corresponding low orbits should be added as a compulsory and given extra quantity, as far as the fragmentation of flight for pausing into a transfer orbit neither adds nor diminishes the energetic consumption of the total flight. It remains to determine the energy consumption for the transfer itself.

The motion into a field of central forces always remains planar and is easily describer in a polar, rotating referential by the equation of Binet (Synge and Griffith, 1949), based on the polar angle θ, named the *true anomaly*. For a gravitational field this equation reads

$$\frac{d^2u}{d\theta^2} + u = \frac{K}{h^2u^2} \tag{1}$$

The solution is a conical curve $r(\theta)$ that may be an ellipse, parabola or hyperbola, depending on the initial position and speed of the particle. The expression of the orbit equation may be found while resuming the Cartesian, planar orthogonal coordinates $\{xOy\}$, where the position vector and the velocity are given by (Mortari, 2001)

$$\mathbf{r} = \frac{p}{1+\varepsilon\cos\theta}\begin{bmatrix}\cos\theta \\ \sin\theta\end{bmatrix}, \quad \mathbf{v} = \sqrt{\frac{K_\oplus}{p}}\begin{bmatrix}-\sin\theta \\ e+\cos\theta\end{bmatrix}. \tag{2}$$

Between the local position on the conical orbit and the corresponding orbital velocity the relation exists, that derives from the conservation of energy or simply by calculating the velocity from the integral of the equation of motion, in this case with the Earth gravitational parameter $K_\oplus$,

$$\mathbf{v} = \sqrt{\frac{K_\oplus}{p}} \cdot \frac{1+e\cos\varphi}{p}\begin{bmatrix}0 & -1 \\ 1 & \dfrac{e}{\sin\varphi}\end{bmatrix} \cdot \mathbf{r}. \tag{3}$$

Multiplying the matrices and considering the case when the orbital eccentricity lies between 0 and 1, that means elliptical orbits, the usual energy equation of the orbital speed appears (Synge and Griffith, 1949)

$$v^2 = K_\oplus\left(\frac{2}{r} - \frac{1}{a}\right). \tag{4}$$

These formulae are only needed to access the energy requirements for an impulsive transfer towards Mars and thus a so-called high thrust interplanetary transfer is resulting. The results supplied this way are bound to some approximations when considering the Earth-Mars actual transfer. They mainly come from the fact that the two planets move along slightly non-coplanar orbits, the angle between the two orbital planes is that given in Table 18.2, namely the inclination of Mars's orbit to the ecliptic is

$$i = 1.84964°.$$

This circumstance induces a very little difference in the transfer energy however and all previous results from the coplanar approximation preserve quite correct.

2.1 Energy for Earth-Mars high thrust transfers

With moderately high thrust propulsion the acceleration for the transfer acts quick and the time of flight gets its actual minimum. The amount of energy is directly given by the equation of motion on the quasi-best, quasi-Hohmann transfer orbit that starts from around the Earth and ends around Mars (figure below).

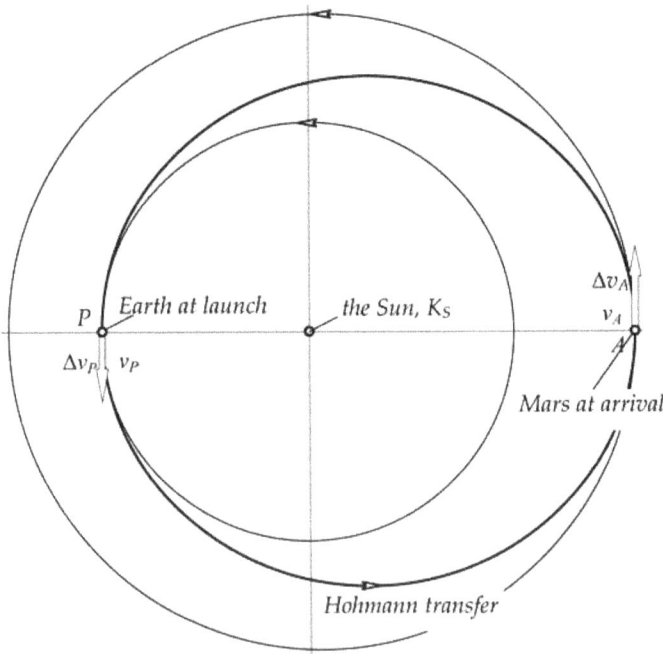

Fig. 3. Minimum energy Earth-Mars transfer

To enter the higher altitude orbit as referred to the Sun, the transfer spacecraft must be provided with extra kinetic energy at the Earth orbit level, exactly equal to the perihelion energy of the transfer Hohmann ellipse, v_P^2. This difference is easily computed by using equation (4).

When we introduce the gravitational parameter of the Sun K_o and the location of the perihelion at Earth orbit level r_P, the extra speed required to the spacecraft is $\Delta v_P = v_P - v_E = 32.746 - 29,7944 = 2.952$ km/s. With the same procedure we observe that at the arrival on Mars orbit level, the spacecraft requires additional energy and speed to remain on the Mars orbit level around the Sun, namely $\Delta v_A = v_M - v_A = 24{,}126 - 21{,}472 = 2.654$ km/s. The total velocity requirement for the spacecraft equals the considerable amount of $\Delta v_\Sigma = 5.606$ km/s.

To this requirement the exit energy from the Earth gravity and the braking energy to avoid fall on Mars must be added. In order to remain with the required extra velocity when the spacecraft leaves the Earth field of gravity, the semi-axis of the departure hyperbola must comply with the value given by

$$v_\infty^2 = K_\oplus \left(\frac{1}{a_{\oplus e}} \right) = 2.952^2 \tag{5}$$

from where the value results,

$$a_{\oplus e} = K_\oplus / 8.714304 = 45759.478 \text{ km} \tag{6}$$

with the corresponding hyperbolic escape velocity at the altitude of a LEO (*Low Earth Orbit*, h=200 km) of

$$v_e^2 = K_\oplus \left(\frac{2}{R_0 + h} + \frac{1}{a_{\oplus e}} \right) = 130.14 \tag{7}$$

from where the value results

$$v_e = 11.408 \text{ km/s}$$

that must be actually supplied to the spacecraft to properly initiate the transfer to Mars. The local parabolic velocity were $v_2 = 11.019$ km/s only.

At the Mars end of the flight, the spacecraft is challenged with a continuous fall towards Mars, as it enters the gravitational field of Mars with a hyperbolic velocity at infinity of 2.654 km/s. Similar computations show that at Mars arrival, namely at an altitude of merely 50 km above the Martian surface, where the aerodynamic breaking begins to manifest, the velocity relative to Mars raises to 5.649 km/s and must be completely slowed down. These transfer velocities are ending in 17.057 km/s.

The high amount of energy at approach and landing on Mars can be diminished by air breaking, at least in part. It only remains then to assure the launching velocity from Earth, with its considerable amount. In terms of energy, this requirement equals 65 MJ for each kilogram of end-mass in hyperbolic orbit, to which the energy for lifting that kilogram from the ground to the altitude of the orbit must be considered. This extra energy of less than 2 MJ looks negligible however, as compared to the kinetic energy requirement from above.

It must be considered that a value of 16 MJ only is required for the return travel to Earth from MLO (Mars low orbit). It is thus hard to understand the reasoning for the so-called *one-way trips to Mars*, or to be open "*no return trips to Mars*", warmly proposed by some experts in Mars colonization from Earth.

3. A thick atmosphere facing erosion

Long term events in the outer atmosphere of the planet are related to the interaction of the solar radiation and particles flow (solar wind) with the very rarefied, heterogeneous fluid envelope of the planet. A model of an electrically charged gas was recently used by Tanaka et al. (2009) to approach the erosion rate of the Martian atmosphere under high UV solar radiation. We add now the double phase fluid model that covers the dust dispersion proved even in 1975 by the Viking spacecraft to flow at high altitudes into the tinny atmosphere. First the equations of conservation and motion for the heterogeneous and rarefied fluid should be introduced.

3.1 Equations governing the exo-atmospheric erosion
While writing the equations of motion of a material particle of the rarefied gas and its condensed, particle content the relative referential related to the moving planet is considered. The equations of motion, referenced to a non-inertial coordinate system, depend on the relative acceleration of the particles and thus on the *transport* and *complementary* acceleration terms, given by Rivals and Coriolis formulae (Rugescu, 2003)

$$\mathbf{a}_t = \mathbf{a}_M + \left(\mathbf{\Omega}_t^2 + \mathbf{\Omega}_t'\right)\cdot \mathbf{x},$$
$$\mathbf{a}_C = 2\mathbf{\Omega}_t \wedge \mathbf{v},$$
(8)

where $\mathbf{\Omega}_t$ is the anti-symmetric matrix of the transport velocity (rotation). Four distinct types of time derivatives exist when the relative fluid motion is described (Rugescu, R. D., 2000). With an apostrophe for the absolute, total (material) time derivative and d/dt for the local total time derivative the vectors and associated matrices of relative motion are

$$\mathbf{r} = \begin{pmatrix} x \\ y \\ z \end{pmatrix}, \quad \mathbf{v} \equiv \begin{pmatrix} u \\ v \\ w \end{pmatrix} = \frac{d\mathbf{r}}{dt}, \quad \mathbf{a} \equiv \begin{pmatrix} a \\ b \\ c \end{pmatrix} = \frac{d^2\mathbf{r}}{dt^2},$$

$$\boldsymbol{\omega}_t = \begin{pmatrix} \omega_x \\ \omega_y \\ \omega_z \end{pmatrix}, \quad \mathbf{\Omega}_t = \begin{bmatrix} 0 & -\omega_z & \omega_y \\ \omega_z & 0 & -\omega_x \\ -\omega_y & \omega_x & 0 \end{bmatrix}.$$
(9)

Only the centripetal component $\mathbf{a}_c = -\mathbf{\Omega}_t^2 \cdot \mathbf{r}$ preserves the direction of the position vector $\mathbf{r}$, the other components presenting ortho-normal properties.

Consequently, the global effect of the relative, non-inertial motion upon the acoustical behavior of the fluid can only be described in a computational 3-Dimensional scheme, although partial effects can be evaluated under simpler assumptions.

The twin viscosity coefficient model observes the reology of the fluid with coefficients of viscosity introduced. The following NS-type, laminar, unsteady hyperbolic matrix equation system is written for both the gas fraction and the condensed part of the fluid mixture, and stands for example as the background of the *Eagle* solver (Tulita et al., 2002), used in many previous aerodynamic investigations at QUB and UPB. A single type of condensed particles is considered, in other words the chemical interactions are seen as minimal and not interfering with the flow process.

To express more precisely the conservation of the gas and particles that fill the same elemental volume dV we observe that a fraction only α of the lateral area of the frontier is occupied by the gas, while the remaider fraction $(1-\alpha)$ is traversed by the condensed particles. Consequently, the area where the pressure acts upon the gas is αA_i for each facet of the cubic volume element (Fig. 4).

In orthogonal and 3D non-inertial coordinates this equations write out for the gas fraction and for the condensate as:

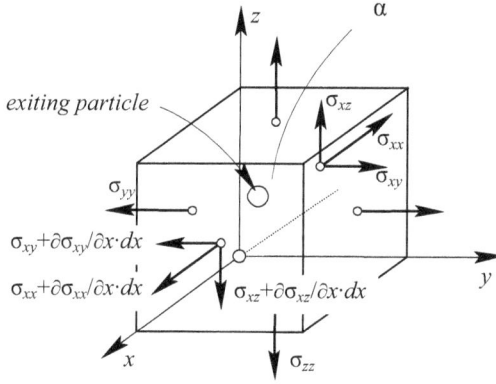

Fig. 4. The heterogeneous fluid

$$\frac{\partial \mathbf{E}}{\partial t} + \frac{\partial \mathbf{F}}{\partial x} + \frac{\partial \mathbf{G}}{\partial y} + \frac{\partial \mathbf{H}}{\partial z} + \mathbf{K} = 0 , \qquad (10)$$

$$\iiint_{A\,TR} \left[\left(\frac{\partial \mathbf{E}}{\partial t} + \frac{\partial \mathbf{F}}{\partial x} + \frac{\partial \mathbf{G}}{\partial y} + \frac{\partial \mathbf{H}}{\partial z} \right) M_p f + K \right] dR dT\, dA = 0 , \qquad (11)$$

where the vectors are used as defined by the following expressions, in connection with the above system (3) and (4):

$$\mathbf{E} = \begin{bmatrix} \rho \\ \rho u \\ \rho v \\ \rho w \\ \rho h - p \\ \rho_c \end{bmatrix} , \qquad (12)$$

$$\mathbf{F} = \begin{bmatrix} \rho u \\ \rho u^2 + \alpha p - 2\mu\alpha \dfrac{\partial u}{\partial x} - \alpha\left(\dfrac{\mu}{3} + \mu' \right) \nabla \cdot \mathbf{V} \\ -\alpha\mu\left(\dfrac{\partial v}{\partial z} + \dfrac{\partial w}{\partial y} \right) \\ -\alpha\mu\left(\dfrac{\partial w}{\partial x} + \dfrac{\partial u}{\partial z} \right) \\ \rho h u - \lambda \dfrac{\partial T}{\partial x} \\ \displaystyle\iiint_{v_p T R} M_p(R) f\, dR\, dT\, dv_p \end{bmatrix} , \qquad (13)$$

$$\mathbf{G} = \begin{bmatrix} \rho v \\ -\mu\alpha\left(\dfrac{\partial u}{\partial y}+\dfrac{\partial v}{\partial x}\right) \\ \rho v^2 + \alpha p - 2\alpha\mu\dfrac{\partial v}{\partial y} - \alpha\left(\dfrac{\mu}{3}+\mu'\right)\nabla\cdot\mathbf{V} \\ -\alpha\mu\left(\dfrac{\partial w}{\partial y}+\dfrac{\partial v}{\partial z}\right) \\ \rho h v - \lambda\dfrac{\partial T}{\partial y} \\ \iiint\limits_{\mathbf{v}_p T R} M_p(R)f\,dR\,dT\,d\mathbf{v}_p \end{bmatrix}, \tag{14}$$

$$\mathbf{H} = \begin{bmatrix} \rho w \\ -\mu\alpha\left(\dfrac{\partial u}{\partial z}+\dfrac{\partial w}{\partial x}\right) \\ -\alpha\mu\left(\dfrac{\partial v}{\partial z}+\dfrac{\partial w}{\partial y}\right) \\ \rho w^2 + \alpha p - 2\alpha\mu\dfrac{\partial w}{\partial z} - \alpha\left(\dfrac{\mu}{3}+\mu'\right)\nabla\cdot\mathbf{V} \\ \rho h w - \lambda\dfrac{\partial T}{\partial z} \\ \iiint\limits_{\mathbf{v}_p T R} M_p(R)f\,dR\,dT\,d\mathbf{v}_p \end{bmatrix}, \tag{15}$$

$$\mathbf{K} = \begin{bmatrix} 0 \\ \gamma_{Mx} + z\omega_y' - y\omega_z' + \omega_x(y\omega_y + z\omega_z) - x(\omega_y^2 + \omega_z^2) + \\ \qquad +2(w\omega_y - v\omega_z) - \rho f_x + S_x \\ \gamma_{My} + x\omega_z' - z\omega_x' + \omega_y(z\omega_z + x\omega_x) - y(\omega_z^2 + \omega_x^2) + \\ \qquad +2(u\omega_z - w\omega_x) - \rho f_y + S_y \\ \gamma_{Mz} + y\omega_x' - x\omega_y' + \omega_z(x\omega_x + y\omega_y) - z(\omega_x^2 + \omega_y^2) + \\ \qquad +2(v\omega_x - u\omega_y) - \rho f_z + S_z \\ -\mathbf{V}\cdot\nabla p - \boldsymbol{\Pi}{:}\nabla\circ\mathbf{V} + S \end{bmatrix} \tag{16}$$

The following notations were used in the expression of the vector $\mathbf{K}$, referring to the distribution of condensed particles:

$$S_x \equiv \iiint\limits_{u_p T R}\left(\frac{9}{2}\frac{\mu_g c}{\rho_p R^2} - 3\frac{d\ln R}{dt} - \frac{d\ln f}{dt}\right)(u-u_p)M_p f(x,r,T,u_p,t)\,dR\,dT\,du_p \ ,$$

$$S_y \equiv \iiint_{v_p\ TR} \left(\frac{9}{2} \frac{\mu_g c}{\rho_p R^2} - 3\frac{d\ln R}{dt} - \frac{d\ln f}{dt} \right)(v - v_p)M_p f(x,r,T,u_p,t)dR\,dT\,dv_p\ ,$$

$$S_z \equiv \iiint_{w_p\ TR} \left(\frac{9}{2} \frac{\mu_g c}{\rho_p R^2} - 3\frac{d\ln R}{dt} - \frac{d\ln f}{dt} \right)(w - w_p)M_p f(x,r,T,u_p,t)dR\,dT\,dw_p\ ,$$

$$S \equiv \iiint_{v_p\ TR} \left\{ \left(e_p + v_p^2/2\right)D_p - \left[A_p(\mathbf{v} - \mathbf{v}_p)^2 + B_p \right]M_p f(x,r,T,u_p,t) \right\} dR\,dT\,d\mathbf{v}_p\ .$$

The equations involving the solid phase (dust particles) depend upon the distribution of the particles along their size, temperature, velocity and their distribution in space as a variable of time. Consequently some extra terms in the four expressions from above are to be written for those functions as follows:

$$A = \begin{bmatrix} u_p \\ v_p \\ w_p \\ \mathbf{v}_p \end{bmatrix},\ E = \begin{bmatrix} u_p \\ v_p \\ w_p \\ M_p f(h_p + v_p^2) \end{bmatrix},\ F = \begin{bmatrix} u_p^2 \\ u_p v_p \\ u_p w_p \\ M_p f(h_p + v_p^2)u_p \end{bmatrix} \qquad (17)$$

$$G = \begin{bmatrix} u_p v_p \\ v_p^2 \\ v_p w_p \\ M_p f(h_p + v_p^2)v_p \end{bmatrix},\ H = \begin{bmatrix} u_p w_p \\ v_p w_p \\ w_p^2 \\ M_p f(h_p + v_p^2)w_p \end{bmatrix}, \qquad (18)$$

$$K = \begin{bmatrix} A_p(u - u_p) - f_x \\ A_p(v - v_p) - f_y \\ A_p(w - w_p) - f_z \\ \left[(h_p + v_p^2)\nabla \cdot \mathbf{v}_p + B_p - 2\mathbf{v}_p \cdot \mathbf{f}\right]M_p f \end{bmatrix}. \qquad (19)$$

Other three simplifying notations were used to define the factors from the vector (18.19) as follows:

$$A_p \equiv \frac{9}{2} \frac{\mu_g c}{\rho_p R^2} - 3\frac{d\ln R}{dt} - \frac{d\ln f}{dt} \qquad (20)$$

$$B_p \equiv \frac{3}{\rho_p R_p}\left[\alpha_c(T_p - T) + \sigma_0\left(\varepsilon_p T_p^4 - \varepsilon T^4\right) \right] \qquad (21)$$

$$D_p \equiv \frac{d\left(M_p f_p\right)}{dt} + M_p f_p \nabla \cdot \mathbf{v}_p \qquad (22)$$

While ionization is acting on the rarefied fluid the terms including the external, field forces include the Lorenz forces that appear in the fluid. Coefficients and properties of the field parameters are taken into account within the model developed by Terada (Terada et al. 2009),

Solar wind parameters	Present	4.5 Ga
Solar wind proton density	3 cm^{-3}	1000 cm^{-3}
Solar wind velocity	450 km s^{-1}	2000 km s^{-1}
B$_{IMF}$ (Parker spiral)	2 nT	60 nT
XUV flux (normalized to martian orbit)	1 × (present-day low solar activity)	100 × (present-day moderate solar activity)

Table 3. Solar flux parameters after Terada

A 3-D computational code is under development for the simulation of the exosphere motion and erosion into an inertial reference frame, bound to the planet surface. To image the interaction of the solar flux with the upper atmosphere of Mars, this interaction as given by the Terada simulations is presented in figure 5.

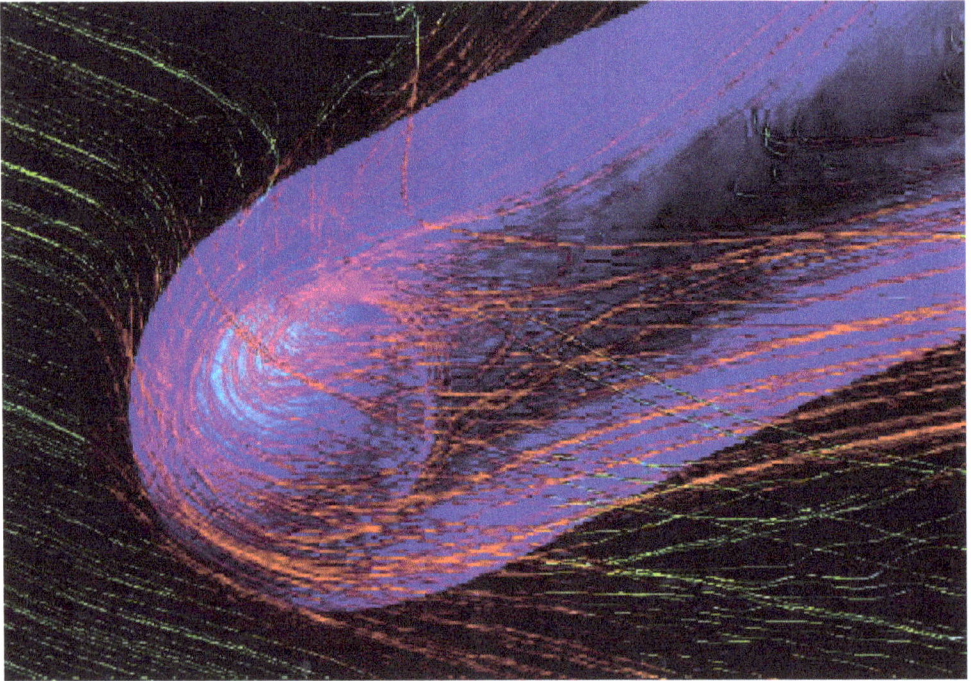

Fig. 5. Ion of O+ erosion simulated by Terada for a Mars atmosphere cca. 4.5 Ga ago. A density of N(O$^+$)=100 cm^{-3} is indicated with a blue transparent isosurface contour.

The authors of the simulation show that if they use the same assumption as Pérez-de-Tejada (1992) or Lammer et al. (2003b) that the cold ions are lost through the entire circular ring area around the planet's terminator, a maximum O+ loss rate of about 1.2×10^{29} s^{-1} is obtained. Integrating this loss rate over a period of Δt=150 million years, a maximum water loss equivalent to a global Martian ocean with a depth of 70 m is obtained, which is quite impressive and shows that the erosion of the atmosphere is extremely severe.

4. Solar-gravitational draught on Mars

The innovative model of the solar-gravitational air accelerator for use on Mars and other celestial bodies with thin atmosphere or without atmosphere is the closed circuit tower in figure 6.

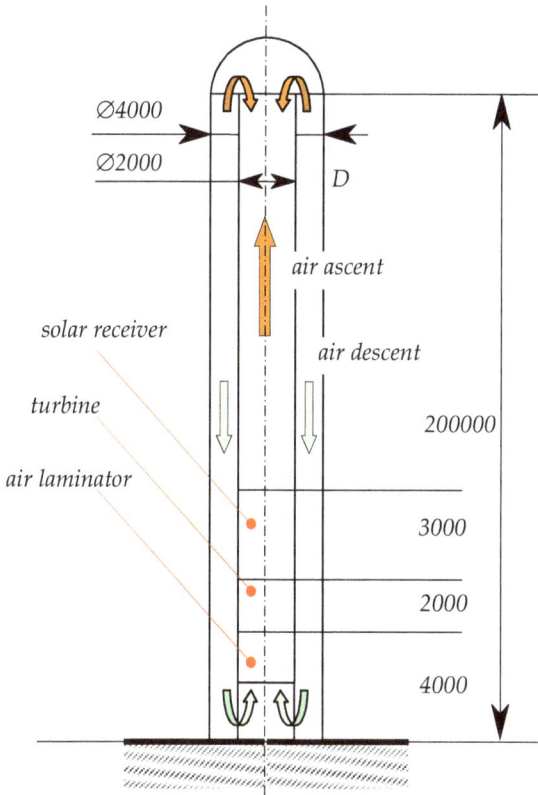

Fig. 6. Closed-circuit thermal-gravity tower for celestial bodies without atmosphere.

The air ascent tunnel is used as a heater with solar radiation collected through the mirror array, while the descent tunnels are used as air coolers to close the gravity draught.

According to the design in Fig. 6, a turbine is introduced in the facility next to the solar receiver, with the role to extract at least a part of the energy recovered from the solar radiation and to transmit it to the electric generator, for converting to electricity. The heat from the flowing air is thus transformed into mechanical energy with the payoff of a supplementary air rarefaction and cooling in the turbine.

The best energy extraction will take place when the air recovers entirely the ambient temperature before the solar heating, although this desire remains for the moment rather hypothetical. To search for the possible amount of energy extraction, the quotient ω is introduced, as further defined. Some differences appear in the theoretical model of the turbine system as compared to the simple gravity draught wind tunnel previously described.

The process of air acceleration at tower inlet is governed by the same energy (Bernoulli) incompressible (constant density ρ_0 through the process) equation as in the previous case,

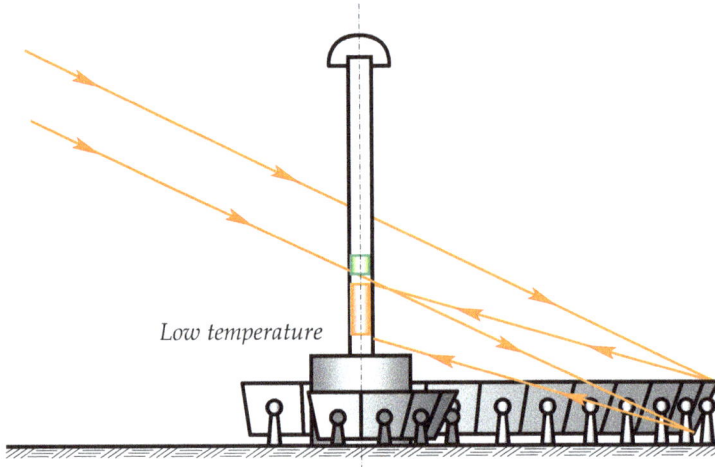

Fig. 7. Solar array concept for the closed-circuit thermal-gravity tower.

$$p_1 = p_0 - \frac{\dot{m}^2}{2\rho_0 A^2} . \tag{23}$$

The air is heated in the solar receiver with the amount of heat q, into a process with dilatation and acceleration of the airflow, accompanied by the usual pressure loss, called sometimes "*dilatation drag*" [8]. Considering a constant area cross-section in the heating solar receiver zone of the tube and adopting the variable γ for the amount of heating rather then the heat quantity itself,

$$\gamma = \frac{\rho_0 - \rho_2}{\rho_0} = 1 - \beta , \tag{24}$$

with a given value for

$$\beta = \frac{T_1}{T_2} < 1 , \tag{25}$$

the continuity condition shows that the variation of the speed is given by

$$c_2 = c_1 / \beta . \tag{26}$$

No global impulse conservation appears in the tower in this case, as long as the turbine is a source of impulse extraction from the airflow. Consequently the impulse equation will be written for the heating zone only, where the loss of pressure due to the air dilatation occurs, in the form of eq. 27,

$$p_2 + \frac{\dot{m}^2}{\rho_2 A^2} = p_1 + \frac{\dot{m}^2}{\rho_0 A^2} - \Delta p_R . \tag{27}$$

A possible pressure loss due to friction into the lamellar solar receiver is considered through Δp_R. The dilatation drag is thus perfectly identified and the total pressure loss Δp_Σ from outside up to the exit from the solar heater is present in the expression

$6 - outer\ air$

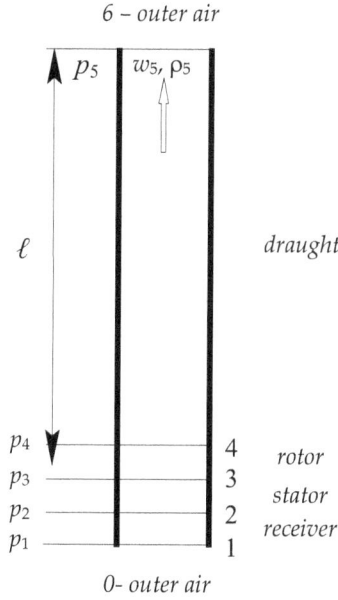

Fig. 8. Main stations in the turbine cold-air draught tower.

$$p_2 = p_0 - \frac{\dot{m}^2}{2\rho_0 A^2} - \frac{\dot{m}^2}{\rho_2 A^2} + \frac{\dot{m}^2}{\rho_0 A^2} - \Delta p_R \equiv p_0 - \Delta p_\Sigma . \tag{28}$$

Observing the definition of the rarefaction factor in (24) and using some arrangements the equation (28) gets the simpler form

$$p_2 = p_0 - \frac{\dot{m}^2}{\rho_0 A^2} \cdot \frac{\gamma+1}{2(1-\gamma)} - \Delta p_R . \tag{29}$$

The thermal transform further into the turbine stator grid is considered as isentropic, where the amount of enthalpy of the warm air is given by

$$q = \frac{p_1}{\rho_1} \cdot \frac{1-\beta}{\beta} + \frac{\dot{m}^2}{\rho_1^2 A^2} \cdot \left[\frac{1-\frac{\Gamma}{2}\beta}{\beta} + \frac{\frac{\Gamma}{2}-1}{\beta^2} \right] - \frac{\Delta p_R}{\rho_1} \cdot \frac{1}{\beta} .$$

If the simplifying assumption is accepted that, under this aspect only, the heating progresses at constant pressure, then a far much simpler expression for the enthalpy fall in the stator appears,

$$\Delta h_{23} = \omega q = \omega c_p T_2 \gamma . \tag{30}$$

To better describe this process a choice between a new rarefaction ratio of densities ρ_3/ρ_2 or the energy quota ω must be engaged and the choice is here made for the later. Into the isentropic stator the known variation of thermal parameters occurs,

$$\frac{T_3}{T_2} = 1 - \omega\gamma \ , \tag{31}$$

$$\frac{p_3}{p_2} = (1 - \omega\gamma)^{\frac{\kappa}{\kappa-1}} \ , \tag{32}$$

$$\frac{\rho_3}{\rho_2} = (1 - \omega\gamma)^{\frac{1}{\kappa-1}} \ . \tag{33}$$

The air pressure at stator exit follows from combining (32) and (29) to render

$$p_3 = \left[p_0 - \frac{\dot{m}^2}{\rho_0 A^2} \cdot \frac{\gamma+1}{2(1-\gamma)} - \Delta p_R \right] (1 - \omega\gamma)^{\frac{\kappa}{\kappa-1}} \ . \tag{34}$$

Considering a Zölly-type turbine the rotor wheel is thermally neutral and no variation in pressure, temperature and density appears. The only variation is in the air kinetic energy, when the absolute velocity of the airflow decreases from c_3 to $c_3 \sin\alpha_1$ and this kinetic energy variation is converted to mechanical work outside. Consequently $\rho_4 = \rho_3$, $p_4 = p_3$, $T_4 = T_3$ and

$$c_4 = \frac{c_1}{(1-\gamma)(1-\omega\gamma)^{\frac{1}{\kappa-1}}} \ . \tag{35}$$

The air ascent in the tube is only accompanied by the gravity up-draught effect due to its reduced density, although the temperature could drop to the ambient value. We call this quite strange phenomenon the *cold-air draught*. It is governed by the simple gravity form of Bernoulli's equation of energy,

$$p_5 = p_3 - g\rho_3\ell \ . \tag{36}$$

The simplification was assumed again that the air density varies insignificantly during the tower ascent. The value for p_3 is here the one in (35). At air exit above the tower a sensible braking of the air occurs in compressible conditions, although the air density suffers insignificant variations during this process.

The Bernoulli equation is used to retrieve the stagnation pressure of the escaping air above the tower, under incompressible conditions

$$p_6{}^* = p_5 - \frac{\Gamma}{2}\rho_5 c_5^2 = p_5 + \frac{\Gamma}{2} \cdot \frac{\dot{m}^2}{\rho_3 A^2} = p_5 + \frac{\Gamma}{2} \cdot \frac{\dot{m}^2}{\rho_0 A^2} \cdot \frac{\rho_0}{\rho_3} \ . \tag{37}$$

Value for p_5 from (36) and for the density ratio from (24) and (33) are now used to write the full expression of the stagnation pressure as

$$p_6{}^* = (p_0 - \Delta p_R)(1 - \omega\gamma)^{\frac{\kappa}{\kappa-1}} - \frac{\dot{m}^2}{\rho_0 A^2} \cdot \frac{\gamma+1}{2(1-\gamma)} \cdot (1 - \omega\gamma)^{\frac{\kappa}{\kappa-1}} +$$
$$+ \frac{\dot{m}^2}{\rho_0 A^2} \cdot \frac{\Gamma}{2} \cdot \frac{1}{(1-\gamma)\cdot(1-\omega\gamma)^{\frac{1}{\kappa-1}}} - g\rho_4\ell \tag{38}$$

It is observed again that up to this point the entire motion into the tower hangs on the value of the mass flow-rate, yet unknown. The mass flow-rate itself will manifest the value that fulfils now the condition of outside pressure equilibrium, or

$$p_6{}^* = p_0 - g\rho_0\ell .$$ (39)

This way the local altitude air pressure of the outside atmosphere equals the stagnation pressure of the escaping airflow from the inner tower. Introducing (38) in (39), after some re-arrangements the dependence of the global mass flow-rate along the tower, when a turbine is inserted after the heater, is given by the final developed formula:

$$R^2(\gamma) \equiv \frac{\dot{m}^2}{2g\ell\rho_0^2 A^2} = \frac{1-\gamma}{(\gamma+1)(1-\omega\gamma)^{\frac{\kappa+1}{\kappa-1}} - \Gamma}(1-\omega\gamma)^{\frac{1}{\kappa-1}}\left\{ 1 - (1-\gamma)(1-\omega\gamma)^{\frac{1}{\kappa-1}} + \right.$$
$$\left. + \frac{p_0}{g\rho_0\ell}\left[(1-\omega\gamma)^{\frac{\kappa}{\kappa-1}} - 1\right] - \frac{\Delta p_R}{g\rho_0\ell}(1-\omega\gamma)^{\frac{\kappa}{\kappa-1}}\right\}$$ (40)

where the notations are again recollected

$\gamma = \dfrac{p_0 - p_2}{p_0}$, the dilatation by heating in the heat exchanger, previously denoted by r;

ω = the part of the received solar energy which could be extracted in the turbine;

Δp_R = pressure loss into the heater and along the entire tube either.

All other variables are already specified in the previous chapters. It is clearly noticed that by zeroing the turbine effect ($\omega = 0$) the formula (40) reduces to the previous form, or by neglecting the friction, which stays as a validity check for the above computations. For different and given values of the efficiency ω the variation of the mass flow-rate through the tube depends parabolically of the rarefaction factor γ.

Notice must be made that the result in (40) is based on the convention (30). The exact expression of the energy q introduced by solar heating yet does not change this result significantly. Regarding the squared mass flow-rate itself in (40), it is obvious that the right hand term of its expression must be positive to allow for real values of R^2. This only happens when the governing terms present the same sign, namely

$$\left\{(\gamma+1)(1-\omega\gamma)^{\frac{\kappa+1}{\kappa-1}} - \Gamma\right\}\cdot\left\{1-(1-\gamma)(1-\omega\gamma)^{\frac{1}{\kappa-1}} + \frac{p_0}{g\rho_0\ell}\left[(1-\omega\gamma)^{\frac{\kappa}{\kappa-1}} - 1\right] - \frac{\Delta p_R}{g\rho_0\ell}(1-\omega\gamma)^{\frac{\kappa}{\kappa-1}}\right\} : 0 .$$ (41)

The larger term here is the ratio $p_0/(g\rho_0\ell)$, which always assumes a negative sign, while not vanishing. The conclusion results that the tower should surpass a minimal height for a real R^2 and this minimal height were quite huge. Very reduced values of the efficiency ω should be permitted for acceptably tall solar towers. This behavior is nevertheless altered by the first factor in (41) which is the denominator of (30) and which may vanish in the usual range of rarefaction values γ. A sort of thermal resonance appears at those points and the turbine tower works properly well.

5. Reasons and costs for terraforming Mars

Thicken Mars' atmosphere, and make it more like Earth's. Earth's atmosphere is about 78% Nitrogen and 21% Oxygen, and is about 140 times thicker than Mars' atmosphere. Since Mars is so much smaller than Earth (about 53% of the Earth's radius), all we'd have to do is bring about 20% of the Earth's atmosphere over to Mars. If we did that, not only would Earth be relatively unaffected, but the Martian atmosphere, although it would be thin (since the force of gravity on Mars is only about 40% of what it is on Earth), would be breathable, and about the equivalent consistency of breathing the air in Santa Fe, NM. So that's nice; breathing is good.

Mars needs to be heated up, by a lot, to support Earth-like life. Mars is cold. Mars is damned cold. At night, in the winter, temperatures on Mars get down to about -160 degrees! (If you ask, "Celcius or Fahrenheit?", the answer is first one, then the other.) But there's an easy fix for this: add greenhouse gases. This has the effect of letting sunlight in, but preventing the heat from escaping. In order to keep Mars at about the same temperature as Earth, all we'd have to do is add enough Carbon Dioxide, Methane, and Water Vapor to Mars' atmosphere. Want to know something neat? If we're going to move 20% of our atmosphere over there, we may want to move 50% of our greenhouse gases with it, solving some of our environmental problems in the process.

These greenhouse gases would keep temperatures stable on Mars and would warm the planet enough to melt the icecaps, covering Mars with oceans. All we'd have to do then is bring some lifeforms over and, very quickly, they'd multiply and cover the Martian planet in life. As we see on Earth, if you give life a suitable environment and the seeds for growth/regrowth, it fills it up very quickly.

So the prospects for life on a planet with an Earth-like atmosphere, temperature ranges, and oceans are excellent. With oceans and an atmosphere, Mars wouldn't be a red planet any longer. It would turn blue like Earth! This would also be good for when the Sun heated up in several hundred million years, since Mars will still be habitable when the oceans on Earth boil. But there's one problem Mars has that Earth doesn't, that could cause Mars to lose its atmosphere very quickly and go back to being the desert wasteland that it is right now: Mars doesn't have a magnetic field to protect it from the Solar Wind. The Earth's magnetic field, sustained in our molten core, protects us from the Solar Wind.

Mars needs to be given a magnetic field to shield it from the Solar Wind. This can be accomplished by either permanently magnetizing Mars, the same way you'd magnetize a block of iron to make a magnet, or by re-heating the core of Mars sufficiently to make the center of the planet molten. In either case, this allows Mars to have its own magnetic field, shielding it from the Solar Wind (the same way Earth gets shielded by our magnetic field) and allowing it to keep its atmosphere, oceans, and any life we've placed there.

But this doesn't tell us how to accomplish these three things. The third one seems to us to be especially difficult, since it would take a tremendous amount of energy to do. Still, if you wanted to terraform Mars, simply these three steps would give you a habitable planet.

The hypothetical process of making another planet more Earth-like has been called terraforming, and terraforming Mars is a frequently mentioned possibility in terraforming discussions. To make Mars habitable to humans and earthly life, three major modifications are necessary. First, the pressure of the atmosphere must be increased, as the pressure on the surface of Mars is only about 1/100th that of the Earth. The atmosphere would also need the addition of oxygen. Second, the atmosphere must be kept warm. A warm atmosphere would melt the large quantities of water ice on Mars, solving the third problem, the absence of water.

Terraforming Mars by building up its atmosphere could be initiated by raising the temperature, which would cause the planet's vast CO_2 ice reserves to sublime and become atmospheric gas. The current average temperature on Mars is −46 °C (-51 °F), with lows of −87 °C (-125 °F), meaning that all water (and much carbon dioxide) is permanently frozen. The easiest way to raise the temperature seems to be by introducing large quantities of CFCs (chlorofluorocarbons, a highly effective greenhouse gas) into the atmosphere, which could be done by sending rockets filled with compressed CFCs on a collision course with Mars. After impact, the CFCs would drift throughout Mars' atmosphere, causing a greenhouse effect, which would raise the temperature, leading CO_2 to sublimate and further continuing the warming and atmospheric buildup. The sublimation of gas would generate massive winds, which would kick up large quantities of dust particles, which would further heat the planet through direct absorption of the Sun's rays. After a few years, the largest dust storms would subside, and the planet could become habitable to certain types of algae and bacteria, which would serve as the forerunners of all other life. In an environment without competitors and abundant in CO_2, they would thrive. This would be the biggest step in terraforming Mars.

6. Conclusion

The problem of creating a sound source of energy on Mars is of main importance and related to the capacity of transportation from Earth to Mars, very limited in the early stages of Mars colonization, and to the capacity of producing the rough materials in situ. Consequently the most important parameter that will govern the choice for one or another means of producing energy will be the specific weight of the powerplant. Besides the nuclear sources, that most probably will face major opposition for a large scale use, the only applicable source that remains valid is the solar one. As far as the solar flux is almost four times fainter on Mars than on Earth, the efficiency of PVC remains very doubtfull, although it stands as a primary candidate. This is why the construction of the gravity assisted air accelerators looks like a potential solution, especially when rough materials will be available on Mars surface itself. The thermal efficiency of the accelerator for producing a high power draught and the propulsion of a cold air turbine remains very high and attractive. The large area of the solar reflector array is still one of the basic drawbacks of the system, that only could be managed by creating very lightweight solar mirrors, but still very stiff to withstand the winds on Mars surface.

7. References

*** (1977), Scientific Results of the Viking Project, *Journal of Geophysical Research*, vol. 82, no. 28, A.G.U., Washington, D.C.

*** *The Astronomical Almanac*, 2010, U.S. Naval Observatory and H.M. Nautical Almanac Office.

Michael Allison and Megan McEwen, 2000. A post-Pathfinder evaluation of aerocentric solar coordinates with improved timing recipes for Mars seasonal/diurnal climate studies. Planetary and Space Science, 48, 215-235.

Asimov, Isaac (1979), *Civilisations extraterrestres*, Ed. L'Etincelle, Montreal, Quebec, Canada.

Andre L. Berger, 1978. Long Term Variations of Daily Insolation and Quaternary Climatic Changes. Journal of the Atmospheric Sciences, volume 35(12), 2362-2367.

Berger A and Loutre MF, 1992. Astronomical solutions for paleoclimate studies over the last 3 million years. Earth and Planetary Science Letters, 111, 369-382.

Bibring, J.P., and 42 others (2006), Global mineralogical and aqueous Mars history derived from OMEGA/Mars Express data, *Science*, vol. 312, pp. 400–404.

P. Bizony (1998), *The Exploration of Mars-Searching for the Cosmic Origins of Life*, Aurum Press, London.

David Brain (2008), ESA observations indicate Mars and Venus are surprisingly similar, *Thaindian News*, March 6th <esa-observations-indicate-mars-and-venus-are-surprisingly-similar_10024474.html>.

Bullock, M. A., and J. M. Moore (2007), Atmospheric conditions on early Mars and the missing layered carbonates, *Geophys. Res. Lett.*, 34, L19201, doi:10.1029/2007GL030688.

Sylvio Ferraz-Mello (1992), Chaos, resonance, and collective dynamical phenomena in the solar system, *Proceedings of the 152nd Symposium of the IAU*, Angra dos Reis, Brazil, July 15-19, 1991, Springer, ISBN 0792317823, 9780792317821, 416 pp.

Flammarion, Nicolas Camille (1862), *La pluralité des mondes habités*, Didier, Paris.

Jones, Barrie W. (2008), Mars before the space age, *International Journal of Astrobiology*, Volume 7, Number 2, 143-155.

McGovern, Patrick J., and Morgan, Julia K. (2009), Mars Volcanic spreading and lateral variations in the structure of Olympus Mons, *Geology*, 37, pp 139-142.

D. Mortari, "*On the Rigid Rotation Concept in n-Dimensional Spaces*" Journal of the Astronautical Sciences, vol. 49, no. 3, July-September 2001.

Oparin, A. I. and Fesenkov, V. G. (1956), *Jizni vo vselennoi* (*Life in Universe*-Russ.), Ed. Academy of Science of USSR.

Phillips, R.J., Zuber, M.T., Solomon, S.C., Golombek , M.P., Jakosky, B.M., Banerdt, W.B., Smith, D.E., Williams, R.M.E., Hynek, B.M., Aharonson , O., and Hauck, S.A. (2001), Ancient geodynamics and global-scale hydrology on Mars: *Science*, vol. 291, pp. 2587–2591.

Rugescu, R. D. (2003), Sound Pressure Behavior in Relative Fluid Mechanics, *Proceedings of the 10th International Congress on Sound and Vibration (ICSV10)*, Stockholm, Sweden, July 7-10, pp. 3169-3176.

Rugescu, R. D. (2000), *On the Principles of Relative Motion of Continua*, Scientific Bulletin of U.P.B., series A (Applied Mathematics and Physics), 62, 2/2000, pp. 97-108;

Schubert, G., Russell, C.T., and Moore, W.B. (2000), Timing of the martian dynamo, *Nature* 408, pp. 666–667.

Sheehan, W. (1996), *The Planet Mars: A History of Observation & Discovery*, University of Arizona Press.

Smith, D. E., Lerch, F. J., Nerem, R. S., Zuber, M. T., Patel, G. B., Fricke, S. K., Lemoine, F. G. (1993), An Improved Gravity Model for Mars: Goddard Mars Model 1, Journal of Geophysical Research (ISSN 0148-0227), vol. 98, no. E11, p. 20, 871-889.

Standish, E. M. (1998), JPL IOM 312.F-98-048, (DE405/LE405 Ephemeris).

Terada, N., Kulikov, Y. N., Lammer, H., Lichtenegger, H. I. M., Tanaka, T., Shinagawa, H., & Zhang, T. (2009), Atmosphere and Water Loss from Early Mars Under Extreme Solar Wind and Extreme Ultraviolet Conditions, *Astrobiology*, Vol. 9, No. 1, 2009, © Mary Ann Liebert, Inc., DOI: 10.1089/ast.2008.0250

C. Tulita, S. Raghunathan, E. Benard, *Control of Steady Transonic Periodic Flow on NACA-0012 Aerofoil by Contour Bumps*, Department of Aeronautical Engineering, The Queen's University of Belfast, Belfast, Northern Ireland, United Kingdom, 2002;

Synge, J. L., and Griffith, B. A., (1949), Principles of Mechanics, second ed., McGraw-Hill Book Company, Inc., New York, Toronto, London.

Wood, B.E., Müller, H.-R., Zank, G.P., Linsky, J.L., and Redfield, S. (2005) New mass-loss measurements from astrospheric Ly-α absorption, *Astrophys. J.*, 628:L143–L146.

Zhang, M.H.G., Luhmann, J.G., Nagy, A.F., Spreiter, J.R., and Stahara, S.S. (1993), Oxygen ionization rates at Mars and Venus: relative contributions of impact ionization and charge exchange. *J. Geophys. Res.*, 98, pp. 3311–3318.

7

Numerical Simulation of Solar Cells and Solar Cell Characterization Methods: The Open-Source on Demand Program AFORS-HET

Rolf Stangl, Caspar Leendertz and Jan Haschke
Helmholtz-Zentrum Berlin für Materialien und Energie,
Institut für Silizium Photovoltaik, Kekule-Str.5, D-12489 Berlin
Germany

1. Introduction

Within this chapter, the principles of numerical solar cell simulation are described, using AFORS-HET (automat **for** simulation of **het**erostructures). AFORS-HET is a one dimensional numerical computer program for modelling multi layer homo- or heterojunction solar cells as well as some common solar cell characterization methods.

Solar cell simulation subdivides into two parts: optical and electrical simulation. By optical simulation the local generation rate $G(\mathbf{x},t)$ within the solar cell is calculated, that is the number of excess carriers (electrons and holes) that are created per second and per unit volume at the time t at the position $\mathbf{x}$ within the solar cell due to light absorption. Depending on the optical model chosen for the simulation, effects like external or internal reflections, coherent superposition of the propagating light or light scattering at internal surfaces can be considered. By electrical simulation the local electron and hole particle densities $n(\mathbf{x},t), p(\mathbf{x},t)$ and the local electric potential $\varphi(\mathbf{x},t)$ within the solar cell are calculated, while the solar cell is operated under a specified condition (for example operated under open-circuit conditions or at a specified external cell voltage). From that, all other internal cell quantities, such like band diagrams, local recombination rates, local cell currents and local phase shifts can be calculated. In order to perform an electrical simulation, (1) the local generation rate $G(\mathbf{x},t)$ has to be specified, that is, an optical simulation has to be done, (2) the local recombination rate $R(\mathbf{x},t)$ has to be explicitly stated in terms of the unknown variables n, p, φ, $R(\mathbf{x},t) = f(n, p, \varphi)$. This is a recombination model has to be chosen. Depending on the recombination model chosen for the simulation, effects like direct band to band recombination (radiative recombination), indirect band to band recombination (Auger recombination) or recombination via defects (Shockley-Read-Hall recombination, dangling-bond recombination) can be considered.

In order to simulate a real measurement, the optical and electrical simulations are repeatedly calculated while changing a boundary condition of the problem, which is specific to the measurement. For example, the simulation of a i-V characteristic of a solar cell is done by calculating the internal electron and hole current (the sum of which is the total current) as a function of the externally applied voltage.

Most solar cells, which are on the market today, can be described as a one dimensional sequence of different semiconductor layers. If they are uniformly illuminated, a one dimensional solar cell modelling is sufficient (the internal electron/hole current can flow only in one direction). This is the case for most wafer based silicon solar cells as well as for most thin film solar cells on glass as long as the integrated series connection shall not be explicitly modelled, see Fig.1 (left).

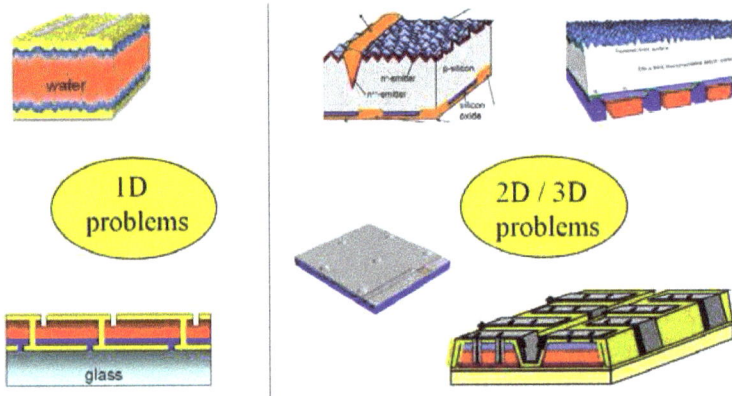

Fig. 1. solar cell structures which can be treated as a one dimensional problem (left), or which have to be treated as a two or even three dimensional problem (right).

However, in order to minimize contact recombination, stripe- or point-like metallic contacts which are embedded within an insulating passivation layer (i.e. silicon nitride, silicon oxide) are sometimes introduced. These contacts can either be placed on both sides of the solar cell or favourably only at the rear side of the solar cell, thereby avoiding shadowing due to the contacts. In these cases, the resulting solar cells have to be modelled as two or even three dimensional problems (the internal electron/hole current can flow in 2 or even 3 directions), see Fig.1 (right). In the current version 2.4 of AFORS-HET only 1D simulations are possible; however, there is a 2D mode under development.

Another possibility to reduce contact recombination is the use of heterojunctions, that is different semiconductors are used to form the solar cell absorber (photon collecting area), the electron extracting area and the hole extracting area of the solar cell. Ideally, the excess carriers of the solar cell absorber (electrons and holes) should be selectively attracted/repelled towards the contacts, see Fig. 2. These selective contacts can be either conventionally realized by doping/counter doping of the solar cell absorber, leading to a formation of an internal electric field by which the selective excess carrier separation is achieved. In this case, homojunctions will form, i.e. there are no band offsets, as the absorber and the electron/hole extracting areas of the solar cell consist of the same semiconductor. In principle, if different semiconductors with appropriately matched work functions are used to form the electron/hole extracting areas, heterojunctions can be formed having the same internal electric field as the homojunction, but with additional band offsets that enhance the repelling character of the contacts, see Fig. 2 (right).

A heterojunction solar cell will thus have a higher open circuit voltage compared to a homojunction solar cell. Less excess carriers of the repelled type are transported into the

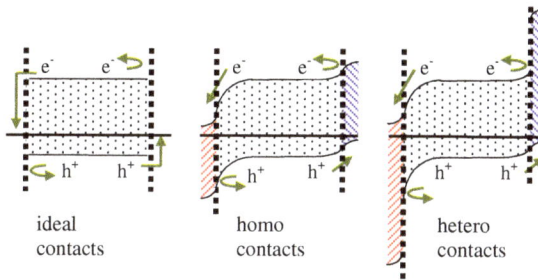

Fig. 2. schematic sketch of selective absorber contacts (band diagrams of a p-type semiconductor used as an absorber material). Ideal contacts (left), homojunction contacts (middle) and ideally aligned heterojunction contacts (right). NOTE: The dimensions of the x axis are schematic and not in scale!

electron/hole collecting regions, and thus the contact recombination at the metallic contacts is reduced. However, an essential pre-requisite is not to create too many interface defects during the formation of the heterojunction at the interface between the absorber and the electron/hole collecting area, which will otherwise act as additional recombination centres.

A realistic computer program for solar cell modelling should therefore be able to handle homojunctions as well as heterojunctions, and it should be able to consider interface defects and the corresponding interface recombination $R^{it}(t)$. Depending on the physical assumption how to describe an electron/hole transport across a heterojunction interface, a distinct interface model has to be chosen. For example, within the current version of AFORS-HET 2.4 a drift-diffusion and a thermionic emission interface model can be chosen, allowing the placement of interface defects but neglecting tunnelling. Tunneling interface models are under development.

To assure a numerical simulation with reliable results, a good model calibration, i.e. a comparison of simulation results to a variety of different characterisation methods, is necessary. The solar cell under different operation conditions should be compared to the simulations. Also different characterisation methods for the solar cell components, i.e. for the individual semiconductor layers and for any sub stacks should be tested against simulation. Only then the adequate physical models as well as the corresponding model input parameters can be satisfactory chosen. Thus a good solar cell simulation program should be able to simulate the common characterisation methods for solar cells and its components.

In this chapter, we describe AFORS-HET (**a**utomat **for s**imulation of **het**erostructures), a one dimensional numerical computer program to simulate solar cells as well as typical solar cell characterisation methods. Thus a variety of different measurements on solar cell components or on the whole solar cell can be compared to the corresponding simulated measurements in order to calibrate the parameters used in the simulations.

All optical and electrical models, which can be used in AFORS-HET, are discussed and their mathematical and physical background is stated. Furthermore, many solar cell characterisation methods, which can be simulated by AFORS-HET, are sketched. The difference in modelling thick film (wafer based) or thin film solar cells on glass will be investigated in order to choose the appropriate model. The basic input parameters of the corresponding models are described. Some selected results in modelling wafer based amorphous/crystalline silicon solar cells illustrate the concepts of numerical solar cell simulation within practical applications.

2. Brief description of AFORS-HET

The current version 2.4 of AFORS-HET solves the one dimensional semiconductor equations (Poisson´s equation and the transport and continuity equation for electrons and holes) with the help of finite differences under different conditions, i.e.: (a) equilibrium mode (b) steady state mode, (c) steady state mode with small additional sinusoidal perturbations, (d) simple transient mode, that is switching external quantities instantaneously on/off, (e) general transient mode, that is allowing for an arbitrary change of external quantities. A multitude of different physical models has been implemented. The generation of electron/hole pairs (optical models of AFORS-HET) can be described either by Lambert-Beer absorption including rough surfaces and using measured reflection and transmission files, or by calculating the plain surface incoherent/coherent multiple internal reflections, using the complex indices of reflection for the individual layers. Different recombination models can be considered within AFORS-HET: radiative recombination, Auger recombination, Shockley-Read-Hall and/or dangling-bond recombination with arbitrarily distributed defect states within the bandgap. Super-bandgap as well as sub-bandgap generation/recombination can be treated. The following interface models for treating heterojunctions are implemented: Interface currents can be modelled to be either driven by drift diffusion or by thermionic emission. A band to trap tunnelling contribution across a hetero-interface can be considered. The following boundary models can be chosen: The metallic contacts can be modelled as flatband or Schottky like metal/semiconductor contacts, or as metal/insulator/semiconductor contacts. Furthermore, insulating boundary contacts can also be chosen.

Thus, all internal cell quantities, such as band diagrams, quasi Fermi energies, local generation/recombination rates, carrier densities, cell currents and phase shifts can be calculated. Furthermore, a variety of solar cell characterisation methods can be simulated, i.e.: current voltage, quantum efficiency, transient or quasi-steady-state photo conductance, transient or quasi-steady-state surface photovoltage, spectral resolved steady-state or transient photo- and electro-luminescence, impedance/admittance, capacitance-voltage, capacitance-temperature and capacitance-frequency spectroscopy and electrical detected magnetic resonance. The program allows for arbitrary parameter variations and multidimensional parameter fitting in order to match simulated measurements to real measurements.

AFORS-HET, version 2.4, is an open source on demand program. If you want to contribute send an e-mail to AFORS-HET@helmholtz-berlin.de, specifying in detail what you would like to implement. It is distributed free of charge and it can be downloaded via internet:

http://www.helmholtz-berlin.de/forschung/enma/si-pv/projekte/asicsi/afors-het/index_en.html

3. Basic input parameter of AFORS-HET and associated physical models

3.1 Optical parameter (super bandgap generation optical models)

The incoming spectral photon flux $\Phi_0(\lambda, t)$, that is the number of incident photons of wavelength λ at the time t, has to be stated. In order to calculate the local super-bandgap generation rate $G(x, t)$ within the semiconductor stack, that is the number electrons and holes that are created per second and per unit volume at the time t at the position x due to super-bandgap light absorption, there are two optical models available: (1) Lambert-Beer absorption and (2) coherent/incoherent internal multiple reflections. For both models, the thicknesses L_i and the dielectric properties of the semiconductor layers have to be specified, i.e. the complex refractive indices, $\tilde{n}_i(\lambda) = n_i(\lambda) - i\,k_i(\lambda)$ with refractive index $n(\lambda)$ and extinction coefficient $k(\lambda)$. If the model Lambert-Beer absorption is chosen, a measured reflectivity $R(\lambda)$ of the semiconductor stack can be specified, and the resulting absorption $A(\lambda, x, t)$ within the semiconductor stack will be calculated, assuming Lambert Beer absorption by using the specified values for $k_i(\lambda)$ only and performing a ray tracing in order to account for textured surfaces and multiple bouncing of the radiation within the stack. If the model coherent/incoherent internal multiple reflections is chosen, the reflectivity $R(\lambda)$, the transmisivity $T(\lambda)$ and the absorption $A(\lambda, x, t)$ of the semiconductor stack is calculated from the specified values $n_i(\lambda)$, $k_i(\lambda)$, assuming plain surfaces within the stack but taking coherent internal multiple reflections into account, if desired. For both models, $G(x, t)$ is calculated from $A(\lambda, x, t)$ by integration over all wavelengths of the incident spectrum. In order to model optical sub-bandgap generation, optical electron/hole capture cross sections $\sigma_{n,opt} \neq 0$, $\sigma_{p,opt} \neq 0$ for the Shockley-Read-Hall defects have to be specified.

3.2 Layer parameter (semiconductor bulk models)

For each semiconductor layer, the thickness L, the electron/hole mobilities μ_n, μ_p, the effective valence/conduction band densities N_V, N_C, the electron/hole thermal velocities v_n, v_p, the electron affinity χ, the relative dielectric constant ε, the doping profile $N_D(x)$, $N_A(x)$ and the bandgap E_g of the semiconductor has to be specified. In order to describe recombination within the semiconductor, up to four different recombination models can be chosen, (1) radiative recombination, (2) Auger recombination, (3) Shockley-Read-Hall recombination, (4) dangling bond recombination. For radiative recombination, the radiative band to band rate constant r^{bb} has to be specified (Sze & Kwok, 2007). For Auger recombination, the electron/hole Auger rate constants r_n^{Aug}, r_p^{Aug} have to be specified (Sze & Kwok, 2007). For Shockley-Read-Hall recombination, the defect density distribution within the bandgap of the semiconductor $N_{trap}(E)$ and two capture cross sections σ_n, σ_p and if needed also two optical capture cross sections σ_n^{opt}, σ_p^{opt} for the electron/hole capture have to be specified (Sze & Kwok, 2007). For dangling bond recombination, the defect distribution within the bandgap of the semiconductor $N_{trap}(E)$, four capture cross sections σ_n^+, σ_p^0, σ_n^0, σ_p^- and the correlation energy U have to be specified (Sah & Shockley, 1958). Optical capture is not yet implemented in case of dangling bond recombination.

3.3 Interface parameter (semiconductor/semiconductor interface models)

The electron/hole current transport across a semiconductor/semiconductor interface can be described by three different interface models, (1) no interface, (2) drift diffusion interface, (3) thermionic emission interface. If no interface is chosen, no additional interface defects can be specified. Otherwise, an interface defect distribution $N_{trap}^{it}(E)$ can be specified. If the drift diffusion interface is chosen, an interface thickness L_{it} and interface capture cross sections σ_n^{it}, σ_p^{it} have to be specified. For both models (1) and (2), transport across the semiconductor/semiconductor interface is treated according to the drift-diffusion approximation like in the bulk of the semiconductor layers (Sze & Kwok, 2007). If the thermionic emission interface is chosen, the interface is regarded to be infinitively thin and four capture cross sections $\sigma_{n,I}^{it}$, $\sigma_{n,II}^{it}$, $\sigma_{p,I}^{it}$, $\sigma_{p,II}^{it}$ and if needed also four optical capture cross sections $\sigma_{n,I}^{it,opt}$, $\sigma_{n,II}^{it,opt}$, $\sigma_{p,I}^{it,opt}$, $\sigma_{p,II}^{it,opt}$ for electron/hole capture from both sides of the interface have to be specified. Transport across the interface is then treated according to the theory of thermionic emission (Sze & Kwok, 2007).

3.4 Boundary parameter (back/front contact to semiconductor boundary models)

The boundaries of the semiconductor stack may either be metallic (usually constituting the contacts of the solar cell) or they may be insulating in order to simulate some specific measurements requiring insulator contacts. Four different boundary models can be chosen: (1) flatband metal/semiconductor contact, (2) Schottky metal/semiconductor contact, (3) insulator contact, (4) metal/insulator/semiconductor contact. If choosing the flatband metal/semiconductor contact, there will be no band banding induced within the semiconductor due to the contact (flatband contact). The electron/hole surface recombination velocities $S_n^{front/back}$, $S_p^{front/back}$ of the metallic contact have to be specified (Sze & Kwok, 2007). If choosing the Schottky metal/semiconductor contact, an additional work function $\phi^{front/back}$ of the metal contact has to be specified. A depletion or accumulation layer within the semiconductor due to the contact will then form according to Schottky theory (Sze & Kwok, 2007). If choosing the insulator/semiconductor or the metal/insulator/semiconductor contact, interface states between the insulator and the semiconductor can be stated, that is an interface defect distribution $N_{trap}^{it}(E)$ and interface capture cross sections σ_n^{it}, σ_p^{it} have to be specified (Kronik & Shapira, 1999). In case of the metal/insulator/semiconductor contact an additional interface capacity $C^{front/back}$ has to be specified (Kronik & Shapira, 1999). Due to the interface defects a band bending within the semiconductor can form.

3.5 Circuit elements

A series resistance R_s, a parallel resistance R_p, a parallel capacitance C_p and in case of an metal/insulator/semiconductor contact also a series capacitance C_s can be specified. If circuit elements are specified, the internal cell voltage V_{int} and the internal cell current I_{int} of the semiconductor stack will differ from the external cell voltage V_{ext} and external cell current I_{ext} of the modeled device.

3.6 External parameters

External parameters are defined to be parameters which are externally applied to the device under consideration and which can also be easily varied in a real experiment. These are the temperature T of the device, a spectral and a monochromatic illumination source leading to the spectral photon flux $\Phi_0(\lambda,t)$ required for the optical simulations, and the external cell voltage $V_{ext}(t)$ or the external cell current $I_{ext}(t)$ which is applied to the device. The remaining quantity, i.e. the external cell current $I_{ext}(t)$ or the external cell voltage $V_{ext}(t)$ respectively, will be calculated.

4. Mathematical description of the DGL system solved by AFORS-HET

In the following, the differential equations and corresponding boundary conditions, which are solved by AFORS-HET under the various conditions, are stated.

An arbitrary stack of semiconductor layers can be modeled. Within each semiconductor layer the Poisson equation and the transport and continuity equations for electrons and holes have to be solved. At each semiconductor/semiconductor interface and at the front and back side boundary of the stack the current transport through these interfaces/boundaries can be described by different physical models. It results a highly non-linear coupled system of three differential equations with respect to time and space derivatives. The electron density $n(x,t)$, the hole density $p(x,t)$, and the electric potential $\varphi(x,t)$ are the independent variables, for which this system of differential equations is solved. It is solved according to the numerical discretisation scheme as outlined by Selberherr (Selberherr, 1984) in order to linearize the problem and using the linear SparLin solver which is available in the internet (Kundert et. al., 1988).

It can be solved for different calculation modes: (1) EQ calculation mode, describing thermodynamic equilibrium at a given temperature, (2) DC calculation mode, describing steady-state conditions under an external applied voltage or current and/or illumination, (3) AC calculation mode, describing small additional sinusoidal modulations of the external applied voltage/illumination, and (4) TR calculation mode, describing transient changes of the system, due to general time dependent changes of the external applied voltage or current and/or illumination.

In case of using the EQ or the DC calculation mode, all time derivatives vanish, resulting in a simplified system of differential equations. The system of differential equations is then solved for the time independent, but position dependent functions, $n^{EQ/DC}(x)$, $p^{EQ/DC}(x)$, $\varphi^{EQ/DC}(x)$.

$$n(x,t) = n^{EQ}(x), \qquad n(x,t) = n^{DC}(x)$$

$$p(x,t) = p^{EQ}(x) \qquad p(x,t) = p^{DC}(x)$$

$$\varphi(x,t) = \varphi^{EQ}(x) \qquad \varphi(x,t) = \varphi^{DC}(x)$$

In case of using the AC calculation mode, it is assumed that all time dependencies can be described by small additional sinusoidal modulations of the steady-state solutions. All time dependent quantities are then modelled with complex numbers (marked by a dash ~), which allows to determine the amplitudes and the phase shifts between them. I.e., for the independent variables of the system of differential equations, one gets:

$$n(x,t) = n^{DC}(x) + \tilde{n}^{AC}(x)\, e^{i\omega t}$$

$$p(x,t) = p^{DC}(x) + \tilde{p}^{AC}(x)\, e^{i\omega t}$$

$$\varphi(x,t) = \varphi^{DC}(x) + \tilde{\varphi}^{AC}(x)\, e^{i\omega t}$$

In case of using the TR calculation mode, the description of the system starts with a steady-state (DC-mode) simulation, specifying an external applied voltage or current and/or illumination. An arbitrary evolution in time of the external applied voltage or current and/or illumination can then be specified by loading an appropriate file. Then, the time evolution of the system, i.e. the functions $n(x,t)$, $p(x,t)$, $\varphi(x,t)$ during and after the externally applied changes are calculated.

4.1 Optical calculation: super bandgap generation models
In order to describe the generation rate $G_n(x,t)$, $G_p(x,t)$ of electrons and holes due to photon absorption within the bulk of the semiconductor layers, a distinction between super-bandgap generation (for photons with an energy $E_{photon} = hc/\lambda \geq E_g$) and sub-bandgap generation (for photons with an energy $E_{photon} = hc/\lambda \leq E_g$) is made ($\lambda$: photon wavelength h : Planck's constant, c : velocity of light, E_g : bandgap of the semiconductor layer in which the photon absorption takes place). Only the super-bandgap generation rate is calculated by optical modelling as it is independent of the local particle densities $n(x,t)$, $p(x,t)$. Sub-bandgap generation depends on the local particle densities and must therefore be calculated within the electrical modeling part.

The optical super-bandgap generation rate is equal for electrons and holes $G(x,t) = G_n(x,t) = G_p(x,t)$. It can either be imported by loading an appropriate file (using external programs for its calculation) or it can be calculated within AFORS-HET.

So far, two optical models are implemented in AFORS-HET, i.e. the optical model Lambert-Beer absorption and the optical model coherent/incoherent internal multiple reflections. The first one takes textured surfaces and multiple internal boundary reflections into account (due to simple geometrical optics) but neglects coherence effects. It is especially suited to treat wafer based crystalline silicon solar cells. The second takes coherence effects into account, but this is done only for plain surfaces. If coherence effects in thin film solar cells are observable it may be used.

4.1.1 Optical model: Lambert-Beer absorption
Using this model, the absorption within the semiconductor stack will be calculated assuming simple Lambert-Beer absorption, allowing for multiple for and backward traveling of the incoming light, however disregarding coherent interference. A (measured) reflectance and absorptance file of the illuminated contact $R(\lambda)$, $A(\lambda)$ can be loaded or constant values can be used. The incoming spectral photon flux $\Phi_0(\lambda,t)$ is weighted with the contact reflection and absorption, i.e. the photon flux impinging on the first semiconductor layer is given by $\Phi_0(\lambda,t)R(\lambda)A(\lambda)$. To simulate the extended path length caused by a textured surface, the angle of incidence δ of the incoming light can be adjusted. On a textured Si wafer with <111> pyramids, this angle is $\delta=54.74°$, whereas $\delta=0°$ equals normal incidence. The angle γ in which the light travels through the layer stack depends on the wavelength of the incoming light and is calculated according to Snellius' law:

$$\gamma(\lambda) = \delta - \arcsin\left\{\sin(\delta) \cdot \frac{1}{n(\lambda)}\right\},$$

whereas $n(\lambda)$ is the wavelength dependent refraction index of the first semiconductor layer at the illuminated side. Note, that within this model, the change in $\gamma(\lambda)$ is neglected, when the light passes a semiconductor/semiconductor layer interface with two different refraction indices. Thus it is assumed that all photons with a specified wavelength cross the layer stack under a distinct angle γ.

Photon absorption is then calculated from the spectral absorption coefficient $\alpha_x(\lambda) = 4\pi\,k(\lambda)/\lambda$ of the semiconductor layer corresponding to the position x within the stack, which is calculated from the provided extinction coefficient $k(\lambda)$ of the layer. The super bandgap electron/hole generation rate for one single run trough the layer stack (no multiple passes) is then given by:

$$G(x,t) = \int_{\lambda_{min}}^{\lambda_{max}} d\lambda \; \Phi_0(\lambda,t) R(\lambda) A(\lambda) \, \alpha_x(\lambda) \, e^{\frac{-\alpha_x(\lambda)\,x}{\cos(\gamma)}} \; .$$

The minimum and maximum wavelengths λ_{min}, λ_{max} for the integration are generally provided by the loaded spectral range of the incoming spectral photon flux, $\Phi_0(\lambda,t)$. However, if necessary, λ_{max} is modified in order to ensure that only super-bandgap generation is considered: $\lambda_{max} \leq hc / E_g$.

To simulate the influence of light trapping mechanisms, internal reflections at both contacts can be additionally specified. They can either be set as a constant value or wavelength dependant (a measured or calculated file can be loaded). The light then passes through the layer stack several times as defined by the user, thereby enhancing the absorptivity of the layer stack (the local generation rate). The residual flux after the defined number of passes is added to the transmitted flux at the contact, at which the calculation ended (illuminated or not-illuminated contact), disregarding the internal reflection definitions at this contact.

This model was designed to estimate the influence of light trapping of crystalline silicon solar cells and to adapt the simulation to real measurements. However, it neglects the internal multiple reflections and refractions within the layer stack.

4.1.2 Optical model: coherent/incoherent internal multiple reflections

Using this model, the absorption within the semiconductor stack will be calculated by modelling coherent or incoherent internal multiple reflections within the semiconductor stack. Additional non-conducting optical layers in front of the front contact/behind the back contact of the solar cell can be assumed, for example in order to model the effect of anti-reflection coatings. Normal incidence of the incoming illumination is assumed.

The reflectance, transmittance and absorptance of all layers (optical layers and the semiconductor layers) is calculated, using the concepts of complex Fresnel amplitudes. Each layer can be specified to be optically coherent or optically incoherent for a particular light beam (incident illumination). A layer is considered to be coherent if its thickness is smaller than the coherence length of the light beam that is incident on the system.

In order to be able to consider coherent effects, the specified incoming illumination $\Phi_0(\lambda,t)$ is modeled by an incoming electromagnetic wave, with a complex electric field component $\tilde{E}_0^+(\lambda,t)$ (front side illumination, electromagnetic wave traveling in positive direction towards the back contact, with $\Phi_0(\lambda,t) = Const \left| \tilde{E}_0^+(\lambda,t) \right|^2$), or $\tilde{E}_{N+1}^-(\lambda,t)$ respectively (back side illumination, electromagnetic wave traveling in negative direction towards the front contact, with $\Phi_0(\lambda,t) = Const \left| \tilde{E}_{N+1}^-(\lambda,t) \right|^2$). The complex electric field components of the travelling wave are raytraced according to the Fresnel formulas, and thus the resulting electromagnetic wave $\tilde{E}(x,\lambda,t)$ at any position x within the layer stack is calculated. An incoherent layer is modeled by a coherent calculation of several electromagnetic waves within that layer (specified by the integer $N_{incoherentIterations}$), assuming some phase shift between them, and averaging over the resulting electric field components.

4.2 Electrical calculation - bulk layers: semiconductor bulk models

Within the bulk of each semiconductor layer, Poisson's equation and the transport equations for electrons and holes are to be solved in one dimension. So far, there are two semiconductor bulk models available, i.e. the bulk model "standard semiconductor" and the bulk model "crystalline silicon". If using the standard semiconductor model, all bulk layer input parameters as specified in Chapter can be individually adjusted. If using the crystalline silicon bulk model, most input parameters for crystalline silicon are calculated from few remaining input parameters, i.e. from the doping and defect densities $N_D(x)$, $N_A(x)$, N_{trap} of crystalline silicon. Thus effects like band gap narrowing or the doping dependence of the mobility or of the Auger recombination of crystalline silicon are explicitly modeled.

Within each layer, a functional dependence in space can be specified for the doping densities $N_D(x)$, $N_A(x)$. These input parameters can be chosen to be (1) constant, (2) linear, (3) exponential, (4) Gaussian like, (5) error function like decreasing or increasing as a function of the space coordinate x.

4.2.1 Bulk model: standard semiconductor

The doping densities $N_D(x)$, $N_A(x)$ of fixed donator/acceptor states at apposition x within the cell are assumed to be always completely ionized. Contrary, defects $N_{trap}(E)$ located at a specific energy E within the bandgap of the semiconductor can be locally charged/uncharged within the system. Defects can be chosen to be either (1) acceptor-like Shockley-Read-Hall defects, (2) donor-like Shockley-Read-Hall defects or (3) dangling bond defects. Depending on the defect-type chosen, these defects can either be empty, singly occupied with electrons or even doubly occupied with electrons (in case of the dangling bond defect). Acceptor-like Shockley-Read-Hall defects are negatively charged, if occupied and neutral, if empty. Donor-like Shockley-Read-Hall defects are positively charged, if empty, and neutral, if occupied. Dangling bond defects are positively charged, if empty, neutral, if singly occupied and negatively charged, if doubly occupied.

Poisson´s equation, which is to be solved within each layer, reads:

$$\frac{\varepsilon_0 \varepsilon_r}{q} \frac{\partial^2 \varphi(x,t)}{\partial x^2} = p(x,t) - n(x,t) + N_D(x) - N_A(x) + \sum_{trap} \rho_{trap}(x,t)$$

q being the electron charge and ε_0, ε_r being the absolute/relative dielectric constant. The defect density of charged defects $\rho_{trap}(x,t)$ will depend on the defect-type of the defect under consideration and on the local particle densities $n(x,t)$, $p(x,t)$ within in the system. It is described by a trap density distribution function $N_{trap}(E)$ of the defect, specifying the amount of traps at an energy position E within the bandgap and by some corresponding defect occupation functions $f_{0,trap}^{SRH}(E,x,t)$, $f_{1,trap}^{SRH}(E,x,t)$, $f_{+,trap}^{DB}(E,x,t)$, $f_{0,trap}^{DB}(E,x,t)$, $f_{-,trap}^{DB}(E,x,t)$, specifying the probability that traps with an energy position E within the bandgap are empty or singly or doubly occupied with electrons. Thus $\rho_{trap}(x,t)$ equates to

$\rho_{trap}(x,t) = -\int dE \, f_{1,trap}^{SRH}(E,x,t) \, N_{trap}(E)$ in case of acceptor-like Shockley-Read-Hall defects,

$\rho_{trap}(x,t) = +\int dE \, f_{0,trap}^{SRH}(E,x,t) \, N_{trap}(E)$ in case of donator-like Shockley-Read-Hall defects,

$\rho_{trap}(x,t) = +\int dE \left(f_{+,trap}^{DB}(E,x,t) - f_{-,trap}^{DB}(E,x,t) \right) N_{trap}(E)$ in case of dangling bond defects.

The explicit formulas for the defect occupation functions $f_{0,trap}^{SRH}(E,x,t)$, $f_{1,trap}^{SRH}(E,x,t)$, $f_{+,trap}^{DB}(E,x,t)$, $f_{0,trap}^{DB}(E,x,t)$, $f_{-,trap}^{DB}(E,x,t)$ are described later within this text.

The one dimensional equations of continuity and transport for electrons and holes, which have to be solved within each layer, read:

$$-\frac{1}{q}\frac{\partial j_n(x,t)}{\partial x} = G_n(x,t) - R_n(x,t) - \frac{\partial}{\partial t}n(x,t)$$

$$+\frac{1}{q}\frac{\partial j_p(x,t)}{\partial x} = G_p(x,t) - R_p(x,t) - \frac{\partial}{\partial t}p(x,t)$$

The electron/hole super-bandgap generation rates $G_n(x,t)$, $G_p(x,t)$ have to be determined by optical modeling, the corresponding recombination rates $R_n(x,t)$, $R_p(x,t)$ are described later in this text. The electron/hole currents $j_n(x,t)$, $j_p(x,t)$ are driven by the gradient of the corresponding quasi Fermi energy $E_{Fn}(x,t)$, $E_{Fp}(x,t)$. Using a Maxwell Boltzmann approximation for the Fermi-Dirac distribution function, the position dependent Fermi energies and the corresponding local electron/hole currents are explicitly:

$$E_{Fn}(x,t) = E_C(x) + kT\ln\frac{n(x,t)}{N_C(x)} = -q\chi(x) + q\varphi(x,t) + kT\ln\frac{n(x,t)}{N_C(x)}$$

$$E_{Fp}(x,t) = E_V(x) - kT\ln\frac{p(x,t)}{N_V(x)} = -q\chi(x) + q\varphi(x,t) - E_g(x) - kT\ln\frac{p(x,t)}{N_V(x)}$$

$$j_n(x,t) = q\,\mu_n\,n(x,t)\,\frac{\partial E_{Fn}(x,t)}{\partial x}$$

$$j_p(x,t) = q\,\mu_p\,p(x,t)\,\frac{\partial E_{Fp}(x,t)}{\partial x}$$

with the corresponding electron/hole mobilities μ_n, μ_p, the electron affinity χ, the bandgap E_g, the conduction/valence band energy E_C, E_V and the effective conduction/valence band density of states N_C, N_V of the semiconductor.

Recombination

Recombination from the conduction band into the valence band may occur directly, i.e. via radiative band to band recombination, $R_{n,p}^{BB}(x,t)$, or via Auger recombination, $R_{n,p}^{A}(x,t)$. It may also occur via defect states located within the bandgap of the semiconductor, i.e. via Shockley-Read-Hall recombination $R_{n,p}^{SRH}(x,t)$ or via dangling bond recombination, $R_{n,p}^{DB}(x,t)$:

$$R_{n,p}(x,t) = R_{n,p}^{BB}(x,t) + R_{n,p}^{A}(x,t) + R_{n,p}^{SRH}(x,t) + R_{n,p}^{DB}(x,t)$$

Optical sub-bandgap generation

Optical sub-bandgap generation (for $hc/\lambda < E_g$) is calculated using Shockley-Read-Hall recombination statistics. A negative electron/hole SHR recombination rate $R_n^{SRH}(x,t)$,

$R_p^{SRH}(x,t)$ means sub-bandgap generation of an electron/hole from a defect state (trap) into the conduction/valence band. Sub-bandgap generation can either be voltage driven and/or be driven by an optical excitation.

The SRH optical emission coefficients $e_{n,optical}^{trap}(E,x,t)$, $e_{p,optical}^{trap}(E,x,t)$ can be calculated from the optical electron/hole capture cross sections $\sigma_{n,optical}^{trap}$, $\sigma_{p,optical}^{trap}$:

$$e_{n,optical}^{trap}(E,x,t) = \int_{\lambda_{min}}^{\lambda_{max}} d\lambda \; \sigma_{n,optical}^{trap} \; N_C \; \Phi(\lambda,x,t) \; \vartheta(E_C - E - hc/\lambda)$$

$$e_{p,optical}^{trap}(E,x,t) = \int_{\lambda_{min}}^{\lambda_{max}} d\lambda \; \sigma_{p,optical}^{trap} \; N_V \; \Phi(\lambda,x,t) \; \vartheta(E - E_V - hc/\lambda)$$

with $\Phi(\lambda,x,t)$: spectral photon flux inside the semiconductor layers, of wavelength λ at the position x and at time t, N_C, N_V: effective conduction/valence band density, E_C, E_V: energy position of the conduction/valence band, and $\vartheta(E)$: step function, $\vartheta(E)=1$ *for* $E \leq 0$, $\vartheta(E)=0$ *for* $E > 0$.

Again, the minimum and maximum wavelengths λ_{min}, λ_{max} for the integration are generally provided by the loaded spectral range of the incoming spectral photon flux, $\Phi_0(\lambda,t)$. However, if necessary, λ_{min} is modified in order to ensure that only sub-bandgap generation is considered: $\lambda_{min} \geq hc/E_g$.

Radiative recombination

The radiative band to band rate constant r^{BB} has to be specified in order to equate the radiative band to band recombination rates $R_{n,p}^{BB}(x,t)$. The resulting electron and hole recombination rates are always equal:

$$R_{n,p}^{BB}(x,t) = r^{BB} \left\{ n(x,t)\,p(x,t) - N_C\,N_V\,e^{-E_g/kT} \right\}$$

In case of using the DC or AC calculation mode and neglecting second order terms in case of the AC calculation mode, this simplifies to

$$R_{n,p}^{BB}(x) = r^{BB} \left\{ n^{DC}(x)\,p^{DC}(x) - N_C\,N_V\,e^{-E_g/kT} \right\}$$

$$R_{n,p}^{BB}(x,t) = R_{n,p}^{BB}(x) + \tilde{R}_{n,p}^{BB}(x)\,e^{i\omega t}$$

$$\tilde{R}_{n,p}^{BB}(x) = r^{BB}\,n^{DC}(x)\,\tilde{p}^{AC}(x) + r^{BB}\,p^{DC}(x)\,\tilde{n}^{AC}(x)$$

Auger recombination

The electron/hole Auger rate constants r_n^A, r_p^A have to be specified in order to calculate the Auger recombination rates $R_{n,p}^A(x,t)$. Again, the resulting electron and hole recombination rates are always equal:

$$R_{n,p}^A(x,t) = \left[r_n^A\,n(x,t) + r_p^A\,p(x,t)\right] \left\{ n(x,t)\,p(x,t) - N_C\,N_V\,e^{-E_g/kT} \right\}$$

In case of using the DC or AC calculation mode, neglecting second order terms within the AC calculation mode, this simplifies to

$$R_{n,p}^A(x) = \left[r_n^A \, n^{DC}(x) + r_p^A \, p^{DC}(x) \right] \left\{ n^{DC}(x) \, p^{DC}(x) - N_C \, N_V \, e^{-E_g/kT} \right\}$$

$$R_{n,p}^A(x,t) = R_{n,p}^A(x) + \tilde{R}_{n,p}^A(x) \, e^{i\omega t}$$

$$\tilde{R}_{n,p}^A(x) = \left[r_n^A \, n^{DC}(x)^2 + 2 \, r_p^A \, n^{DC}(x) \, p^{DC}(x) \right] \tilde{p}^{AC}(x) + \left[r_p^A \, p^{DC}(x)^2 + 2 \, r_n^A \, n^{DC}(x) \, p^{DC}(x) \right] \tilde{n}^{AC}(x)$$

Shockley Read Hall recombination

Shockley-Read-Hall recombination (Shockley & Read, 1952) requires specifying the character (acceptor-like or donor-like), the capture cross sections σ_n^{trap}, σ_p^{trap}, $\sigma_{n,optic}^{trap}$, $\sigma_{p,optic}^{trap}$ and the energetic distribution $N_{trap}(E)$ of the defect density within the bandgap of the semiconductor, of each defect. An arbitrary number of defects with either one of the following energetic distributions $N_{trap}(E)$ can be chosen:

1. point like distributed at a single energy E_{trap} within the bandgap:

$$N_{trap}(E) = N_{trap}^{point} \, \delta(E - E_{trap})$$

with N_{trap}^{point} : defect density of the point like defect, $\delta(E)$: delta function

2. constantly distributed within a specific region within the bandgap:

$$N_{trap}(E) = \left(E_{trap}^{end} - E_{trap}^{start} \right) N_{trap}^{const} \, \vartheta(E - E_{trap}^{end}) \, \vartheta(E_{trap}^{start} - E)$$

with E_{trap}^{start}, E_{trap}^{end} : start and end energy of the energy interval within the bandgap, where a constant defect density is assumed, N_{trap}^{const} : constant defect density per energy, $\vartheta(E)$: step function

3. exponentially decaying from the conduction/valence band into the bandgap:

$$N_{trap}(E) = N_{trap}^{C,tail} \, e^{-(E_C - E)/E_{trap}^{C,tail}} \quad , \quad N_{trap}(E) = N_{trap}^{V,tail} \, e^{-(E - E_V)/E_{trap}^{V,tail}}$$

i.e. conduction/valence band tail states, with $N_{trap}^{C,tail}$, $N_{trap}^{V,tail}$: tail state density per energy at the conduction/valence band, $E_{trap}^{C,tail}$, $E_{trap}^{V,tail}$: characteristic decay energy (Urbach energy) of the conduction/valence band tail state,

4. Gaussian distributed within the bandgap:

$$N_{trap}(E) = \frac{N_{trap}^{db}}{\sigma_{trap}^{db} \sqrt{2\pi}} e^{-\frac{\left(E - E_{trap}^{db}\right)^2}{2\sigma_{trap}^{db\,2}}}$$

i.e. dangling bond states, with N_{trap}^{db} : total dangling bond state density, E_{trap}^{db} : specific energy of the Gaussian dangling bond peak, σ_{trap}^{db} : standard deviation of the Gaussian dangling bond distribution.

For each defect, electron/hole capture coefficients $c_{n,p}^{trap}$ are equated

$$c_{n,p}^{trap} = v_{n,p}\, \sigma_{n,p}$$

with $v_{n,p}$: electron/hole thermal velocity, $\sigma_{n,p}$: electron/hole capture cross section of the defect. The corresponding electron/hole emission coefficients $e_{n,p}^{trap}(E,x,t)$ are then given by:

$$e_n^{trap}(E,x,t) = c_n^{trap}\, N_C\, e^{-(E_C-E)/kT} + e_{n,optic}^{trap}(E,x,t)$$

$$e_p^{trap}(E,x,t) = c_p^{trap}\, N_V\, e^{-(E-E_V)/kT} + e_{p,optic}^{trap}(E,x,t)$$

In case of using the DC or AC calculation mode, this simplifies to

$$e_n^{trap}(E,x)=c_n^{trap}\, N_C\, e^{-(E_C-E)/kT} + \int d\lambda\; \sigma_{n,optic}^{trap}\, N_C\, \Phi(\lambda,x)\; \vartheta(E_C-E-hc/\lambda) \quad \text{(DC mode)}$$

$$e_p^{trap}(E,x)=c_p^{trap}\, N_V\, e^{-(E-E_V)/kT} + \int d\lambda\; \sigma_{p,optic}^{trap}\, N_V\, \Phi(\lambda,x)\; \vartheta(E-E_V-hc/\lambda)$$

$$e_{n,p}^{trap}(E,x,t) = e_{n,p}^{trap}(E,x) + \tilde{e}_{n,p}^{trap}(E,x)\, e^{i\omega t} \quad \text{(AC mode)}$$

$$\tilde{e}_n^{trap}(E,x)= \int d\lambda\; \sigma_{n,optic}^{trap}\, N_C\, \tilde{\Phi}(\lambda,x)\; \vartheta(E_C-E-hc/\lambda)$$

$$\tilde{e}_p^{trap}(E,x)= \int d\lambda\; \sigma_{p,optic}^{trap}\, N_V\, \tilde{\Phi}(\lambda,x)\; \vartheta(E-E_V-hc/\lambda)$$

Finally, the Shockley-Read-Hall recombination rate due to the defects is

$$R_n^{SRH}(x,t)= \sum_{trap} \int dE\; \left\{ c_n^{trap}\, n(x,t)\, N_{trap}(E)\, f_{0,trap}^{SRH}(E,x,t) - e_n^{trap}(E,x,t)\, N_{trap}(E)\, f_{1,trap}^{SRH}(E,x,t) \right\}$$

$$R_p^{SRH}(x,t)= \sum_{trap} \int dE\; \left\{ c_p^{trap}\, p(x,t)\, N_{trap}(E)\, f_{1,trap}^{SRH}(E,x,t) - e_p^{trap}(E,x,t)\, N_{trap}(E)\, f_{0,trap}^{SRH}(E,x,t) \right\}$$

In case of using the DC or AC calculation mode, neglecting second order terms and assuming zero optical emission coefficients within the AC calculation mode (actual stage of the AFORS-HET development at the moment) this simplifies to

$$R_n^{SRH}(x)= \sum_{trap} \int dE\; \left\{ c_n^{trap}\, n^{DC}(x)\, N_{trap}(E)\, f_{0,trap}^{SRH}(E,x) - e_n^{trap}(E,x)\, N_{trap}(E)\, f_{1,trap}^{SRH}(E,x) \right\}$$

$$R_p^{SRH}(x)= \sum_{trap} \int dE\; \left\{ c_p^{trap}\, p^{DC}(x)\, N_{trap}(E)\, f_{1,trap}^{SRH}(E,x) - e_p^{trap}(E,x)\, N_{trap}(E)\, f_{0,trap}^{SRH}(E,x) \right\}$$

$$R_{n,p}^{SRH}(x,t)= R_{n,p}^{SRH}(x)+ \tilde{R}_{n,p}^{SRH}(x)\, e^{i\omega t}$$

$$\tilde{R}_n^{SRH}(x)= \sum_{trap} \int dE\; \left\{ c_n^{trap}\, N_{trap}(E)\, f_{0,trap}^{SRH}(E,x)\, \tilde{n}^{AC}(x) - \left(c_n^{trap} + e_n^{trap}(E,x) \right)\, N_{trap}(E)\, \tilde{f}_{1,trap}^{SRH}(E,x) \right\}$$

$$\tilde{R}_p^{SRH}(x)= \sum_{trap} \int dE\; \left\{ c_p^{trap}\, N_{trap}(E)\, f_{1,trap}^{SRH}(E,x)\, \tilde{p}^{AC}(x) + \left(c_p^{trap} + e_p^{trap}(E,x) \right)\, N_{trap}(E)\, \tilde{f}_{1,trap}^{SRH}(E,x) \right\}$$

A positive electron/hole SHR recombination rate means recombination of an electron/hole from the conduction/valence band into the defect state (trap), a negative electron/hole SHR recombination rate means sub-bandgap generation of an electron/hole from the defect state (trap) into the conduction/valence band.

Dangling bond recombination

To calculate charge state and recombination of dangling bond defects in amorphous silicon the most exact description developed by Sah and Shockley (Sah & Shockley, 1958) is used. Three different occupation functions $f_{+,trap}^{DB}(E,x,t)$, $f_{0,trap}^{DB}(E,x,t)$, $f_{-,trap}^{DB}(E,x,t)$ for the positively, neutral and negatively charge states have to be derived, corresponding to the empty, single or double occupied electronic state. Four capture/emission processes with the capture cross sections σ_n^+, σ_p^0, σ_n^0, σ_p^- have to be defined as can be seen in Fig. 3. The two transition energies $E_{0/-}$, $E_{+/0}$ are separated by the correlation energy U, which accounts for the fact that the capture-emission process is influenced by the charge state of the dangling or by rearrangement of the lattice in the surrounding.

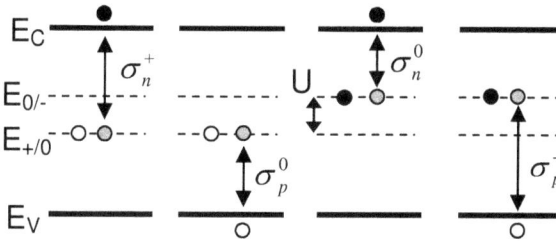

Fig. 3. dangling bond recombination according to Sah and Shockley

For each defect, electron/hole capture coefficients $c_{n/p,+/0/-}^{trap}$ are calculated with the specified electron/hole thermal velocity $v_{n/p}$:

$$c_{n/p,+/0/-}^{trap} = v_{n/p}\, \sigma_{n/p}^{+/0/-}$$

The emission coefficients for the DC calculation mode, neglecting optical emission are given by:

$$e_{n,0}^{trap}(E,x) = \frac{1}{2} c_{n,+}^{trap}\, N_C\, e^{-(E_c-E)/kT}$$

$$e_{p,+}^{trap}(E,x) = 2 c_{p,0}^{trap}\, N_V\, e^{-(E-E_v)/kT}$$

$$e_{n,-}^{trap}(E,x) = 2 c_{n,0}^{trap}\, N_C\, e^{-(E_c-(E+U))/kT}$$

$$e_{p,0}^{trap}(E,x) = \frac{1}{2} c_{p,-}^{trap}\, N_V\, e^{-(E+U-E_v)/kT}$$

Finally the dangling bond recombination coefficients are given by:

$$R_n^{DB}(x) = \sum_{trap} \int dE \left\{ c_{n,+}^{trap}\, n^{DC}(x) N_{trap}(E)\, f_{+,trap}^{DB}(E,x) - e_{n,0}^{trap}(E,x) N_{trap}(E) f_{0,trap}^{DB}(E,x) + \right.$$
$$\left. c_{n,0}^{trap}\, n^{DC}(x) N_{trap}(E)\, f_{0,trap}^{DB}(E,x) - e_{n,-}^{trap}(E,x) N_{trap}(E) f_{-,trap}^{DB}(E,x) \right\}$$

$$R_p^{DB}(x) = \sum_{trap} \int dE \left\{ c_{p,-}^{trap} \, p^{DC}(x) N_{trap}(E) \, f_{-,trap}^{DB}(E,x) - e_{p,0}^{trap}(E,x) N_{trap}(E) f_{0,trap}^{DB}(E,x) + \right.$$

$$\left. c_{p,0}^{trap} \, p^{DC}(x) N_{trap}(E) \, f_{0,trap}^{DB}(E,x) - e_{p,+}^{trap}(E,x) N_{trap}(E) f_{+,trap}^{DB}(E,x) \right\}$$

Dangling bond recombination is still under development. Especially time dependent recombination and optical defect to band emissions are not implemented at the current state of AFORS-HET development.

Defect occupation functions

The defect occupation functions $f_{0,trap}^{SRH}(E,x,t)$, $f_{1,trap}^{SRH}(E,x,t)$, $f_{+,trap}^{DB}(E,x,t)$, $f_{0,trap}^{DB}(E,x,t)$, $f_{-,trap}^{DB}(E,x,t)$ specify the probability for a specific defect (either Shockley-Read-Hall or dangling bond) that traps with an energy position E within the bandgap of the semiconductor are empty or singly or even doubly occupied with electrons.

In case of using the DC or AC calculation mode, they can be explicitly expressed in terms of the local particle densities $n^{DC}(x)$, $p^{DC}(x)$, $\tilde{n}^{AC}(x)$, $\tilde{p}^{AC}(x)$. In case of using the TR calculation mode, the defect occupation functions are generally determined by additional differential equations. Transient DB defect occupation functions have not been implemented in AFORS-HET yet (actual stage of AFORS-HET development).

Shockley Read Hall defect occupation functions

A Shockley-Read-Hall defect can be either empty or occupied by an electron, thus

$$f_{0,trap}^{SRH}(E,x,t) + f_{1,trap}^{SRH}(E,x,t) = 1$$

The Shockley-Read-Hall defect occupation function $f_{1,trap}^{SRH}(E,x,t)$ for electrons will be explicitly stated in case of using the EQ, DC, AC or the TR calculation mode. The Shockley-Read-Hall defect occupation function $f_{0,trap}^{SRH}(E,x,t)$ can then directly be equated.

Generally, a local change of the trapped charge stored in SRH defects must be determined by the difference between the local electron and hole SRH recombination rates:

$$\frac{d}{dt}\rho_{\text{trap}}(x,t) = R_p^{SHR}(x,t) - R_n^{SHR}(x,t)$$

This defines for each defect an additional differential equation for its SHR defect occupation function $f_{1,trap}^{SRH}(E,x,t)$ with respect to its time derivative:

$$\frac{d}{dt} f_{1,trap}^{SRH}(E,x,t) = \left(c_n^{trap} \, n(x,t) + e_p^{trap}(E,x,t) \right)\left(1 - f_{1,trap}^{SRH}(E,x,t)\right) - \left(c_p^{trap} \, p(x,t) + e_n^{trap}(E,x,t) \right) f_{1,trap}^{SRH}(E,x,t) \quad (\#)$$

In case of using the EQ or the DC calculation mode, the time derivative vanishes, and an explicit expression for the SHR defect occupation function, $f_{1,trap}^{SRH,DC}(E,x)$, which is no longer time dependant, can be derived:

$$f_{1,trap}^{SRH,DC}(E,x) = \frac{c_n^{trap} \, n^{DC}(x) + e_p^{trap}(E,x)}{c_n^{trap} \, n^{DC}(x) + e_n^{trap}(E,x) + c_p^{trap} \, p^{DC}(x) + e_p^{trap}(E,x)}$$

Of course, in case of using the EQ calculation mode, the SHR defect occupation function could be also equivalently be described by the Fermi-Dirac distribution function, $f_{1,\,trap}^{SRH,DC}(E,x) = f_{1,\,trap}^{SRH,EQ}(E,x)$, which implicitly determines the position independent Fermi energy E_F.

$$f_{1,\,trap}^{SRH,EQ}(E,x) = \frac{1}{1+e^{\frac{E-E_F}{kT}}}$$

In case of using the AC calculation mode, the differential equation (#) can be explicitly solved, assuming time independent optical emission coefficients within the AC calculation mode (actual stage of the AFORS-HET development at the moment) and assuming the time dependencies $n(x,t) = n^{DC}(x) + \tilde{n}^{AC}(x)\,e^{i\omega t}$, $p(x,t) = p^{DC}(x) + \tilde{p}^{AC}(x)\,e^{i\omega t}$. Neglecting second order terms, one gets for the SHR defect occupation function in the AC calculation mode, $f_{1,\,trap}^{SRH,AC}(E,x,t)$:

$$f_{1,\,trap}^{SRH,AC}(E,x,t) = f_{1,\,trap}^{SRH,DC}(E,x) + \tilde{f}_{1,\,trap}^{SRH,AC}(E,x)\,e^{i\omega t}$$

$$\tilde{f}_{1,\,trap}^{SRH,AC}(E,x) = \frac{c_n^{trap}\,f_{0,\,trap}^{SRH,DC}(E,x)\,\tilde{n}^{AC}(x) - c_p^{trap}\,f_{1,\,trap}^{SRH,DC}(E,x)\,\tilde{p}^{AC}(x)}{c_n^{trap}\,n^{DC}(x) + e_n^{trap}(E,x) + c_p^{trap}\,p^{DC}(x) + e_p^{trap}(E,x) + i\omega}$$

In case of using the TR calculation mode, the transient SRH defect occupation function $f_{1,\,trap}^{SRH,TR}(E,x,t_{i+1})$ at the time step t_{i+1} for an evolution of the system from the time point t_i towards the time point t_{i+1} can be stated by solving the differential equation (#) using a full implicit time discretisation scheme with respect to the particle densities and the emission rates:

$$\frac{d}{dt}f_{1,\,trap}^{SRH,TR}(E,x,t)$$
$$= \left(c_n^{trap}\,n(x,t_{i+1}) + e_p^{trap}(E,x,t_{i+1})\right)\left(1 - f_{1,\,trap}^{SRH,TR}(E,x,t)\right) - \left(c_p^{trap}\,p(x,t_{i+1}) + e_n^{trap}(E,x,t_{i+1})\right)f_{1,\,trap}^{SRH,TR}(E,x,t)$$

An analytical solution of this differential equation leads to:

$$f_{1,\,trap}^{SRH,TR}(E,x,t_{i+1}) = f_{1,\,trap}^{SRH,DCtr}(E,x,t_{i+1}) - \frac{f_{1,\,trap}^{SRH,DCtr}(E,x,t_{i+1}) - f_{1,\,trap}^{SRH,TR}(E,x,t_i)}{e^{dt\left(c_n^{trap}\,n(x,t_{i+1}) + e_n^{trap}(E,x,t_{i+1}) + c_p^{trap}\,p(x,t_{i+1}) + e_p(E,x,t_{i+1})\right)}}$$

with

$$f_{SRH}^{DCtr}(E,x,t_{i+1}) = \frac{c_n^{trap}\,n(x,t_{i+1}) + e_n^{trap}(E,x,t_{i+1})}{c_n^{trap}\,n(x,t_{i+1}) + e_n^{trap}(E,x,t_{i+1}) + c_p^{trap}\,p(x,t_{i+1}) + e_p^{trap}(E,x,t_{i+1})}$$

In the steady-state limit, i.e. for Limes $dt \to \infty$, $dt = t_{i+1} - t_i$, this formula converts to the well known steady state SRH defect occupation function $f_{1,\,trap}^{SRH,DC}(E,x) = f_{1,\,trap}^{SRH,DCtr}(E,x,t_{i+1})$.

Dangling bond defect occupation functions

A dangling bond defect can be either empty or singly or doubly occupied by an electron, hereby being in its positive, neutral or negative charged state, thus

$$f_{+,trap}^{DB}(E,x,t) + f_{0,trap}^{DB}(E,x,t) + f_{-,trap}^{DB}(E,x,t) = 1$$

For the EQ and DC calculation mode the occupation functions are given by:

$$f_{+}^{DB}(E,x) = \frac{P^0 P^-}{N^+ P^- + P^0 P^- + N^+ N^0}$$

$$f_{0}^{DB}(E,x) = \frac{P^- N^+}{N^+ P^- + P^0 P^- + N^+ N^0}$$

$$f_{-}^{DB}(E,x) = \frac{N^0 N^+}{N^+ P^- + P^0 P^- + N^+ N^0}$$

where:

$$N^+ = c_n^+ n^{DC} + e_p^+$$

$$P^0 = c_p^0 p^{DC} + e_n^0$$

$$N^0 = c_n^0 n^{DC} + e_p^0$$

$$P^- = c_p^- p^{DC} + e_n^-$$

Dangling bond defects are still under development and especially time dependent dangling bond occupation functions (to be used in the AC or TR calculation mode) are not implemented at the current state of AFORS-HET development.

4.2.2 Bulk model: crystalline silicon

If the bulk model "crystalline silicon" is used for a semiconductor layer, most layer input parameter as described in chapter are calculated from the doping densities and the defect densities of crystalline silicon.

Thus a doping and temperature dependent material parameterization for crystalline silicon can be undertaken, i.e. it is possible to specify (1) the temperature dependence of the intrinsic carrier concentration of crystalline silicon (intrinsic carrier density model), (2) the doping dependence of the electron/hole mobilities in crystalline silicon (mobility model), (3) the doping dependence of the Auger recombination in crystalline silicon or even its dependence on local excess carriers (Auger recombination models), (4) the doping and the temperature dependence of the Shockley-Read-Hall recombination in crystalline silicon (SRH lifetime model) and (5) doping dependence of the bandgap in crystalline silicon (bandgap narrowing model). All these models are similar to the numerical computer simulation program for crystalline silicon solar cells PC1D (Basore & Clugston, 1997).

Only one single neutral SRH defect at a certain energy position E within the bandgap is assumed. The doping densities N_D, N_A and the amount of traps N_{trap} are specified by the user, all other layer input parameters are calculated according to the above mentioned models.

4.3 Electrical calculation - interfaces: semiconductor/semiconductor interface models

Each interface between two adjacent semiconductor layers can be described by three different interface models: (1) interface model: "no interface", (2) interface model: "drift-diffusion interface" and (3) interface model: "thermionic emission interface". If "no interface" is chosen, the transport across the interface is treated in complete analogy to the "drift diffusion" interface model, however, no interface defects can be specified. The "drift diffusion" interface model models the transport across the heterojunction interface in the same way as in the bulk layers, thereby assuming a certain interface thickness. The "thermionic emission" interface model treats a real interface which interacts with both adjacent semiconductor layers.

4.3.1 Interface model: no interface

Per default, the electron and hole currents across the semiconductor/semiconductor interface are assumed to be driven by drift diffusion, with no interface defects present at the interface.

The drift diffusion model assumes an interface layer of a certain thickness (which is given by the specified grid point to boundary distance within the numerical settings of AFORS-HET). Within this interface layer, the material properties change linearly from semiconductor I to semiconductor II. The elelctron/hole currents across the heterojunction interface can then be treated like in the bulk of a semiconductor layer (drift diffusion driven).

Denoting x_{it}^{I}, x_{it}^{II} the positions directly adjacent to the semiconductor I/II located left/right to the interface, $\mu_{n,p}^{I}$, $\mu_{n,p}^{II}$ the mobilities of the two adjacent semiconductors, and

$$\mu_{n,p}^{it} = \frac{\mu_{n,p}^{I} + \mu_{n,p}^{II}}{2}, \quad n^{it} = \frac{n(x_{it}^{I}) + n(x_{it}^{II})}{2}, \quad p^{it} = \frac{p(x_{it}^{I}) + p(x_{it}^{II})}{2},$$ the corresponding electron/hole

currents across the interface are (like in the bulk):

$$j_n^{it}(t) = q \, \mu_n^{it} \, n^{it}(t) \, \frac{\partial E_{Fn}(x,t)}{\partial x}$$

$$j_p^{it}(t) = q \, \mu_p^{it} \, p^{it}(t) \, \frac{\partial E_{Fp}(x,t)}{\partial x}$$

4.3.2 Interface model: drift-diffusion interface

The electron and hole currents across the semiconductor/semiconductor interface are assumed to be driven by drift diffusion, with additional interface defects present at the interface. Half of the interface states can be occupied by electrons or holes from semiconductor I, the other half from semiconductor II.

The interface defects (given in cm^{-2}) are distributed homogeneously within the interface layer (per cm^{-3}). I.e. the specified interface defect density $N_{it}(E)$ is converted into a homogeneous layer defect density of the interface layer, $N_t(E)$, selectively according to one

of the two following formulas, either $N_t(E) = \dfrac{N_{it}(E)}{d_{it}}$ or $N_t(E) = (N_{it}(E))^{\frac{3}{2}}$. Thus half of the

defects specified are recombination active within semiconductor I, the other half within semiconductor II. The calculation of the electron/hole currents across the heterojunction interface is then performed in the same way as when using the interface model "no interface" described above.

4.3.3 Interface model: thermionic emission interface

The electron and hole currents across the interface are assumed to be driven by thermionic emission (Anderson, 1962; Sze & Kwok, 2007). Additional interface defects can be present at the interface. These states can be filled with electrons or holes from both sides of the interface.

Lets denote x_{it}^I, x_{it}^{II} the positions directly adjacent to the semiconductor I/II located left/right to the interface, $\chi^{I,II}$, $E_g^{I,II}$, $v_{n,p}^{I,II}$ the electron affinity, the bandgap and the thermal velocities of the two semiconductors. According to Anderson theory (Anderson, 1962) the conduction/valence band offsets ΔE_C, ΔE_V, which determine the energetic barrier of the heterojunction interface to be overcome by thermionic emission, are:

$$\Delta E_C = \chi^{II} - \chi^I$$

$$\Delta E_V = E_g^{II} - E_g^I + \chi^{II} - \chi^I$$

The sign convention is such, that ΔE_C, ΔE_V is negative if E_C, E_V drops from the left side to the right side of the interface. The thermionic emission currents across the heterojunction interface $j_n^{TE,I \to II}(t)$, $j_p^{TE,I \to II}(t)$, $j_n^{TE,II \to I}(t)$, $j_p^{TE,II \to I}(t)$ are then explicitly written using the $\vartheta(E)$ step function, $\vartheta(E)=1$ for $E \leq 0$, $\vartheta(E)=0$ for $E > 0$:

$$j_n^{TE,I \to II}(t) = v_n^I \, n\!\left(x_{it}^I,t\right) e^{-\frac{|\Delta E_C|}{kT}\vartheta(-\Delta E_C)} \, , \qquad\qquad j_n^{TE,II \to I}(t) = v_n^{II} \, n\!\left(x_{it}^{II},t\right) e^{-\frac{|\Delta E_C|}{kT}\vartheta(\Delta E_C)}$$

$$j_p^{TE,I \to II}(t) = v_p^I \, p\!\left(x_{it}^I,t\right) e^{-\frac{|\Delta E_V|}{kT}\vartheta(\Delta E_V)} \, , \qquad\qquad j_p^{TE,II \to I}(t) = v_p^{II} \, p\!\left(x_{it}^{II},t\right) e^{-\frac{|\Delta E_V|}{kT}\vartheta(-\Delta E_V)}$$

The netto electron/hole current across the interface due to thermionic emission j_n^{it}, j_p^{it}, is:

$$j_n^{it} = j_n^{TE,I \to II} - j_n^{TE,II \to I} \, , \qquad\qquad j_p^{it} = j_p^{TE,I \to II} - j_p^{TE,II \to I}$$

Additional to the thermionic emission process across the heterojunction interface, there is recombination due to the interface defects. The interface electron/hole recombination rates from both sides of the interface $R_{n,I}^{it}(t)$, $R_{n,II}^{it}(t)$, $R_{p,I}^{it}(t)$, $R_{p,II}^{it}(t)$, are described with SRH recombination:

$$R_{n,I}^{it}(t) = \sum_{trap} \int dE \left\{ c_{n,I}^{it} \, n\!\left(x_{it}^I,t\right) N_{trap}^{it}(E)\!\left(1 - f_{trap}^{it}(E,t)\right) - e_{n,I}^{it}(E,t) N_{trap}^{it}(E) f_{trap}^{it}(E,t) \right\}$$

$$R_{n,II}^{it}(t) = \sum_{trap} \int dE \left\{ c_{n,II}^{it} \, n\!\left(x_{it}^{II},t\right) N_{trap}^{it}(E)\!\left(1 - f_{trap}^{it}(E,t)\right) - e_{n,II}^{it}(E,t) N_{trap}^{it}(E) f_{trap}^{it}(E,t) \right\}$$

$$R_{p,I}^{it}(t) = \sum_{trap} \int dE \left\{ c_{p,I}^{it} \, p\!\left(x_{it}^I,t\right) N_{trap}^{it}(E) f_{trap}^{it}(E,t) - e_{p,I}^{it}(E,t) N_{trap}^{it}(E)\!\left(1 - f_{trap}^{it}(E,x,t)\right) \right\}$$

$$R_{p,II}^{it}(t) = \sum_{trap} \int dE \left\{ c_{p,II}^{it} \, p\left(x_{it}^{II}, t\right) N_{trap}^{it}(E) f_{trap}^{it}(E,t) - e_{p,II}^{it}(E,t) N_{trap}^{it}(E) \left(1 - f_{trap}^{it}(E,x,t)\right) \right\}$$

Thus, recombination at the interface is treated equivalently to bulk SRH recombination, with two exceptions: The interface defect density $N_{trap}^{it}(E)$ is now given in defects per $cm^{-2} \, eV^{-1}$ instead of $cm^{-3} \, eV^{-1}$, consequently, interface recombination is now a recombination current, given in $cm^{-2} \, s^{-1}$, instead of $cm^{-3} \, s^{-1}$. Furthermore, the interface defect distribution function $f_{trap}^{it}(E,t)$ changes compared to the bulk defect distribution function $f_{trap}(E,x,t)$, as the interface states can interact with both adjacent semiconductors. For the EQ or DC calculation mode, one gets explicitly :

$$f_{trap}^{it,DC}(E) = f_{trap}^{it,EQ}(E) =$$

$$\frac{c_{n,I}^{it} \, n\left(x_{it}^{I}\right) + c_{n,II}^{it} \, n\left(x_{it}^{II}\right) + e_{p,I}^{it}(E) + e_{p,II}^{it}(E)}{c_{n,I}^{it} \, n\left(x_{it}^{I}\right) + c_{n,II}^{it} \, n\left(x_{it}^{II}\right) + c_{p,I}^{it} \, p\left(x_{it}^{I}\right) + c_{p,II}^{it} \, p\left(x_{it}^{II}\right) + e_{n,I}^{it}(E) + e_{n,II}^{it}(E) + e_{p,I}^{it}(E) + e_{p,II}^{it}(E)}$$

Using the AC calculation mode, one gets:

$$f_{trap}^{it,AC}(E,t) = f_{trap}^{it,DC}(E) + \tilde{f}_{trap}^{it,AC}(E) \, e^{i\omega t}$$

$$\tilde{f}_{trap}^{it,AC}(E) =$$

$$\frac{\left\{1 - f_{trap}^{it,DC}(E)\right\} \left\{c_{n,I}^{it} \, \tilde{n}^{AC}\left(x_{it}^{I}\right) + c_{n,II}^{it} \, \tilde{n}^{AC}\left(x_{it}^{II}\right)\right\} - f_{trap}^{it,DC}(E) \left\{c_{p,I}^{it} \, \tilde{p}^{AC}\left(x_{it}^{I}\right) + c_{p,II}^{it} \, \tilde{p}^{AC}\left(x_{it}^{II}\right)\right\}}{c_{n,I}^{it} \, n\left(x_{it}^{I}\right) + c_{n,II}^{it} \, n\left(x_{it}^{II}\right) + c_{p,I}^{it} \, p\left(x_{it}^{I}\right) + c_{p,II}^{it} \, p\left(x_{it}^{II}\right) + e_{n,I}^{it}(E) + e_{n,II}^{it}(E) + e_{p,I}^{it}(E) + e_{p,II}^{it}(E) + i\omega}$$

Within the actual stage of AFORS-HET development, interface states described by thermionic emission are only implemented within the EQ, DC and AC calculation mode, i.e. the transient defect distribution function of such states has not been implemented yet.

The heterojunction interface itself is treated as a boundary condition for the differential equations describing the semiconductor layers. Thus, six boundary conditions for the potential and the electron/hole currents at each side of the interface have to be stated, i.e.:

The potential is assumed to be equal on both sides of the interface (thereby neglecting interface dipoles):

1. $\varphi\left(x_{it}^{I}\right) = \varphi\left(x_{it}^{II}\right)$

The total charge stored in the interface states is equal to the difference in the dielectric displacements (a consequence of the Gauss law applied to the Poisson equation)

2. $\varepsilon_0 \varepsilon_r^{I} \left. \dfrac{\partial \varphi(x)}{\partial x}\right|_{x_{it}^{I}} - \varepsilon_0 \varepsilon_r^{II} \left. \dfrac{\partial \varphi(x)}{\partial x}\right|_{x_{it}^{II}} = q \sum_{defects} \rho_{it}$

The total current across the heterojunction interface $j_{ges}^{it}(t) = j_n^{it}(t) + j_p^{it}(t)$ under steady-state conditions is equal to the constant (that is position independent) total current left (or right) to the interface

3. $j_n^{it}(t) + j_p^{it}(t) = j_n\left(x_{it}^{I}, t\right) + j_p\left(x_{it}^{I}, t\right)$

The total electron/hole recombination rate from both sides of the interface is equal for electrons and holes (valid only for EQ, DC and AC calculation mode)

4. $R_{n,I}^{it}(t) + R_{n,II}^{it}(t) = R_{p,I}^{it}(t) + R_{p,II}^{it}(t)$

The electron/hole current left to the interface is equal to the netto electron/hole current across the heterojunction interface plus the interface recombination current

5. $j_n(x_{it}^I,t) = j_n^{it}(t) + R_{n,I}^{it}(t)$

6. $j_p(x_{it}^I,t) = j_p^{it}(t) - R_{p,I}^{it}(t)$

4.4 Electrical calculation - boundaries: front/back contact to semiconductor models

The electrical front/back contacts of the semiconductor stack are usually assumed to be metallic, in order to be able to withdraw a current. However, they may also be insulating in order to be able to simulate some specific measurement methods like for example quasi steady state photoconductance (QSSPC) or surface photovoltage (SPV). So far, four different boundary models for the interface between the contact and the semiconductor adjacent to the contact can be chosen: (1) "flatband metal/semiconductor contact" (2) "Schottky metal/semiconductor contact", (3) "insulator/semiconductor contact", (4) "metal/insulator/semiconductor contact". The boundaries serve as a boundary condition for the system of differential equations describing the semiconductor stack, thus three boundary conditions for the potential and the electron/hole currents at the front and at the back side of the stack have to be stated.

4.4.1 Boundary model: flatband metal/semiconductor contact

Per default, an idealized flatband metal/semiconductor contact is assumed at the boundaries. That is, only the effective electron/hole surface recombination velocities $S_{n/p}^{front/back}$ have to be specified. The metal work function of the front/back contact, $\phi^{front/back}$ is calculated in a way, that flatband conditions are reached according to Schottky theory (Sze & Kwok, 2007). Normally, flatband conditions are calculated within the thermal equilibrium EQ calculation mode, however, in case of using the DC, AC or TR calculation mode with an external illumination (optical super bandgap generation) enabled, they are recalculated in order to ensure flatband conditions independent from the applied illumination.

The interface between the metallic front/back contact and the semiconductor is treated as a boundary condition for the differential equations describing the semiconductor layers. Thus, for each contact, three boundary conditions involving the potential and electron/hole densities adjacent to the contact have to be stated. Denoting x_{it}^{front}, x_{it}^{back} the position within the semiconductor directly adjacent to the metallic contact, these are:
The electric potential is fixed to zero at one contact (for example the back contact).

1.a $\varphi(x_{it}^{back},t) = 0$

At the other contact (for example the front contact) the external applied cell voltage $V_{ext}(t)$ or the external applied current density $j_{ext}(t)$ through the cell is specified (voltage controlled or current controlled calculation). The external solar cell resistances, i.e. the series resistance R_{ext}^S and the parallel resistance R_{ext}^P, which can optionally be specified, will affect the internal cell voltage $V_{int}(t)$ at the boundary of the semiconductor stack and also the position independent internal current density $j_{int}(t)$ through the semiconductor stack.

In case of a voltage controlled calculation, the internal cell voltage can be expressed by the specified external cell voltage $V_{ext}(t)$ and the position independent internal cell current $j_{int}(t) = Const(t) = j_n(x,t) + j_p(x,t)$:

$$V_{int}(t) = V_{ext}(t) + \frac{j_{int}(t)\,R_{ext}^S}{1 + \frac{R_{ext}^S}{R_{ext}^P}} = V_{ext}(t) + \frac{\left\{ j_n\left(x_{it}^{front},t\right) - j_p\left(x_{it}^{front},t\right) \right\} R_{ext}^S}{1 + \frac{R_{ext}^S}{R_{ext}^P}}$$

Thus the potential at other contact can be specified:

1.b $\quad \varphi\left(x_{it}^{front},t\right) = \phi^{front} - \phi^{back} + V_{int}(t)$

In case of a current controlled calculation, the internal cell current density can be expressed by the specified external cell current density $j_{ext}(t)$ and the internal cell voltage $V_{int}(t) = \varphi\left(x_{it}^{front},t\right) - \varphi\left(x_{it}^{back},t\right) - \phi^{front} + \phi^{back}$:

$$j_{int}(t) = j_{ext}(t) + \frac{V_{int}(t)}{R_{ext}^P} = j_{ext}(t) + \frac{\varphi\left(x_{it}^{front},t\right) - \varphi\left(x_{it}^{back},t\right) - \phi^{front} + \phi^{back}}{R_{ext}^P}$$

Thus the position independent total internal cell current $j_n(x,t) + j_p(x,t) = Const(t)$ can be specified:

1.b $\quad j_n\left(x_{it}^{front},t\right) + j_p\left(x_{it}^{front},t\right) = j_{int}(t)$

Furthermore, the electron and hole particle densities at the interface, or the electron/hole currents into the metal contacts can be specified for both contacts.

In the EQ calculation mode, the majority carrier density at the interface under equilibrium $n^{EQ}\left(x_{it}^{front/back}\right)$ or $p^{EQ}\left(x_{it}^{front/back}\right)$ is given by the majority barrier height $\phi_{Bn}^{Schottky} = q\left\{ \phi^{front/back} - \chi^{front/back} \right\}$, $\phi_{Bp}^{Schottky} = q\left\{ E_g - \phi^{front/back} + \chi^{front/back} \right\}$ of the metal/semiconductor contact (with $\chi^{front/back}$ being the electron affinity of the semiconductor adjacent to the front/back contact):

2.a, 2b $\quad n^{EQ}\left(x_{it}^{front/back}\right) = N_C\, e^{-\frac{\phi_{Bn}^{Schottky}}{kT}}$ $\qquad$ or $\qquad$ $p^{EQ}\left(x_{it}^{front/back}\right) = N_V\, e^{-\frac{\phi_{Bp}^{Schottky}}{kT}}$

The corresponding minority carrier density under equilibrium $p^{EQ}\left(x_{it}^{front/back}\right)$ or $n^{EQ}\left(x_{it}^{front/back}\right)$ is then given by the mass action law:

3.a, 3.b $\quad p^{EQ}\left(x_{it}^{front/back}\right) = \frac{N_C N_V\, e^{-\frac{E_g}{kT}}}{n^{EQ}\left(x_{it}^{front/back}\right)}$ $\qquad$ or $\qquad$ $n^{EQ}\left(x_{it}^{front/back}\right) = \frac{N_C N_V\, e^{-\frac{E_g}{kT}}}{p^{EQ}\left(x_{it}^{front/back}\right)}$

As flatband conditions are chosen, the metal work function is calculated to give a zero build in voltage due to the metal/semiconductor contact:

$$\phi^{front/back} = E_C - E_F\left(x_{it}^{front/back}\right) + \chi^{front/back}$$

In all other calculation modes (DC, AC, TR), the electron/hole currents into the metal contact, $j_n^{it,\,front/back}(t)$, $j_p^{it,\,front/back}(t)$ are specified:

2.a $\quad j_n^{it,\,front}(t) = \quad q\, S_n^{front}\, \left\{ n\!\left(x_{it}^{front},t\right) - n^{EQ}\!\left(x_{it}^{front}\right) \right\}$

2.b $\quad j_n^{it,\,back}(t) = -q\, S_n^{back}\, \left\{ n\!\left(x_{it}^{back},t\right) - n^{EQ}\!\left(x_{it}^{back}\right) \right\}$

3.a $\quad j_p^{it,\,front}(t) = -q\, S_p^{front}\, \left\{ p\!\left(x_{it}^{front},t\right) - p^{EQ}\!\left(x_{it}^{front}\right) \right\}$

3.b $\quad j_p^{it,\,back}(t) = \quad q\, S_p^{back}\, \left\{ p\!\left(x_{it}^{back},t\right) - p^{EQ}\!\left(x_{it}^{back}\right) \right\}$

Furthermore, if using the DC, AC or TR calculation mode with an external illumination (optical super bandgap generation) enabled, an illumination dependent metal work function is calculated, in order to ensure illumination independent flatband conditions: Assuming a zero internal current density (no netto current through the semiconductor stack), the metal work function is now iteratively calculated from the majority quasi Fermi energy $E_{Fn}\!\left(x_{it}^{front},t\right)$ or $E_{Fp}\!\left(x_{it}^{front},t\right)$ instead from the Fermi energy, in order to ensure a zero build in voltage due to the metal/semiconductor contact

$$\phi^{front\,/\,back}(t) = E_C - E_{Fn}\!\left(x_{it}^{front\,/\,back},t\right) + \chi^{front\,/\,back} \qquad \text{or}$$

$$\phi^{front\,/\,back}(t) = E_C - E_{Fp}\!\left(x_{it}^{front\,/\,back},t\right) + \chi^{front\,/\,back}$$

4.4.2 Boundary model: Schottky metal/semiconductor contact

This boundary model can describe metal/semiconductor contacts, which drive the semiconductor into depletion or into accumulation (Sze & Kwok, 2007). Explicit values of the metal work function $\phi^{front\,/\,back}$ can be specified in order to fix the majority barrier height of the metal/semiconductor contact $\phi_{Bn}^{Schottky} = q\left\{ \phi^{front\,/\,back} - \chi^{front\,/\,back} \right\}$, $\phi_{Bp}^{Schottky} = q\left\{ E_g - \phi^{front\,/\,back} + \chi^{front\,/\,back} \right\}$. Otherwise, this boundary model is totally equivalent to the flatband metal/semiconductor boundary model described above.

4.4.3 Boundary model: insulator contact

If the boundary of the semiconductor stack is considered to be insulating, additional interface states can be defined, as according to (Kronik & Shapira, 1999). They are treated equivalent to the bulk, but with densities given in cm^{-2} instead of cm^{-3}. For an insulator/semiconductor contact at the front the three boundary conditions are:

1. $\qquad 0 = -\varepsilon_0 \varepsilon_r \left.\frac{\partial \varphi(x,t)}{\partial x}\right|_{x_{front}} - q\sum_{trap} \rho^{trap}$

2. $\qquad 0 = j_n\!\left(x_{front}\right) - R_{it,n}^{front}$

3. $\qquad 0 = -j_p\!\left(x_{front}\right) - R_{it,p}^{front}$

4.4.4 Boundary model: metal/insulator/semiconductor contact

In case of using a metal/insulator/semiconductor MIS contact, the insulator capacity C has to be additionally specified. At the insulator/semiconductor interface additional interface defects can be defined, which are treated equivalent to the bulk but with densities given in cm^{-2} instead of cm^{-3}, as according to (Kronik & Shapira, 1999). Depending on whether the MIS contact is defined on only one or on both boundaries two different cases have to be discussed. If both boundaries have an MIS contact, the capacities C^{front}, C^{back} of the front and back boundaries can be defined separately. Furthermore, one has to define the voltage fraction f that drops at the front MIS contact compared to the fraction that drops at the back MIS contact. For a given external voltage this defines how the different metal layers are charged. Time dependent boundary conditions (AC or TR calculation mode) for the MIS contact on are not implemented at the current state of AFORS-HET development.

Both semiconductor potentials $\varphi(x_{front})$ and $\varphi(x_{back})$ at the front/back boundary of the semiconductor stack and the metal work functions $\phi^{front/back}$ of the front/back contact enter the boundary condition for the electric potential. Also enters the net charge ρ_{it} of the interface, which has to be calculated by summing over all interface defects. The electron/hole currents into the interface defects $j_n(x_{front})$, $j_p(x_{front})$, are given by the recombination rates $R_{it,n}^{front}$, $R_{it,p}^{front}$ of the interface defects. The three boundary conditions for a MIS contact read:

1. $$0 = C\left\{V_{ext} - \left[\varphi(x_{back}) - \varphi(x_{front})\right] + \left[\phi^{back} - \phi^{front}\right]\right\} - \varepsilon_0\varepsilon_r\left.\frac{\partial\varphi(x)}{\partial x}\right|_{x_{front}} - q\sum_{trap}\rho_{it}^{trap}$$

2. $$0 = j_n(x_{front}) - R_{it,n}^{front}$$

3. $$0 = -j_p(x_{front}) - R_{it,p}^{front}$$

in the case that only one MIS contact at the front boundary is chosen, and

1.a $$0 = f\,C^{front}\left\{V_{ext} - \left[\varphi(x_{back}) - \varphi(x_{front})\right] + \left[\phi^{back} - \phi^{front}\right]\right\} - \varepsilon_0\varepsilon_r\frac{\partial\varphi(x)}{\partial x}\bigg|_{x_{front}} - q\sum_{trap}\rho_{it}^{trap}$$

2.a $$0 = j_n(x_{front}) - R_{it,n}^{front}$$

3.a $$0 = -j_p(x_{front}) - R_{it,p}^{front}$$

1.b $$0 = (1-f)C^{back}\left\{V_{ext} - \left[\varphi(x_{back}) - \varphi(x_{front})\right] + \left[\phi^{back} - \phi^{front}\right]\right\} - \varepsilon_0\varepsilon_r\frac{\partial\varphi(x)}{\partial x}\bigg|_{x_{back}} - q\sum_{trap}\rho_{it}^{trap}$$

2.b $$0 = j_n(x_{back},t) - R_{it,n}^{back}$$

3.b $$0 = -j_p(x_{back},t) - R_{it,p}^{back}$$

in case that two MIS contacts at both boundaries are chosen.

5. Characterization methods simulated by AFORS-HET

In the following it is described how the most common solar cell characterization methods are simulated within AFORS-HET, i.e. current-voltage (IV), quantum efficiency (QE), quasi-steady-state photoconductance (QSSPC), impedance (IMP, ADM, C-V, C-T, C-f), surface photovoltage (ID-SPV, VD-SPV, WD-SPV) and photo-electro-luminescence (PEL).

5.1 Measurement model: current-voltage characteristic (IV)
This measurement varies the external voltage at the boundaries and plots the resulting external current through the semiconductor stack in order to obtain the current-voltage

characteristic of the simulated structure. For each voltage value the total current through the structure (the sum of the electron and hole current at a boundary gridpoint) is calculated. This can be done in the dark or under an illumination. The measurement model can iterate the specific data points maximum-power point (mpp), open-circuit voltage (Voc), short-circuit current (Isc) and thus calculate the fill-factor FF and the efficiency Eff of the solar cell, whereas the illumination power density $P_{illumination}$ in W/cm^2 is calculated from the incident photon spectrum:

$$FF = \frac{V_{mpp} I_{mpp}}{V_{oc} I_{sc}}, \qquad Eff = \frac{V_{mpp} I_{mpp}}{P_{illumination}} = \frac{FF\ V_{oc}\ I_{sc}}{P_{illumination}}$$

5.2 Measurement model: quantum efficiency (QE)

In order to simulate quantum efficiencies, the semiconductor stack is additionally illuminated with a monochromatic irradiation at a certain wavelength λ, and the difference ΔI_{SC}^{irrad} of the resulting short circuit current with and without the additional irradiation is computed. A quantum efficiency $QE(\lambda)$ can then defined as

$$QE(\lambda) = \frac{number.of.electrons.in.the.external.circuit}{number.of.photons} = \frac{\Delta I_{SC}^{irrad}/q}{number.of.photons}$$

Different quantum efficiencies are calculated, depending on the number of photons which are considered: (1) external quantum efficiency (EQE): all photons of the additional irradiation, which are incident on the semiconductor stack, whether they are reflected, absorbed or transmitted, are counted. (2) internal quantum efficiency (IQE): only the absorbed photons of the additional irradiation are counted. Note, that like in a real measurement, photons which are absorbed in the contacts are also counted, despite the fact that they do not contribute to the current. (3) corrected internal quantum efficiency (IQE1): only the photons of the additional irradiation which are absorbed in the semiconductor stack are counted.

5.3 Measurement model: quasi steady state photoconductance (QSSPC)

The excess carrier density dependant lifetimes $\tau_{n\_all}(\Delta n)$, $\tau_{p\_all}(\Delta p)$ for a semiconductor stack of the thickness L under a given external illumination are calculated according to the following equations:

$$\tau_{n\_all}(\Delta n) = \frac{\Delta n}{\Delta G} \qquad\qquad \tau_{p\_all}(\Delta p) = \frac{\Delta p}{\Delta G}$$

$$\Delta n = \left(\int dx \{ n_{illuminated}(x) \} - \int dx \{ n_{dark}(x) \} \right) / L$$

$$\Delta p = \left(\int dx \{ p_{illuminated}(x) \} - \int dx \{ p_{dark}(x) \} \right) / L$$

$$\Delta G = \left(\int dx \{ G_{illuminated}(x) \} - \int dx \{ G_{dark}(x) \} \right) / L$$

The average dark and illuminated carrier densities and the average generation rate are calculated by integrating over the whole structure. Thus the excess carrier densities Δn, Δp and the corresponding change in generation rate ΔG can be calculated. Within the measurement model, the external illumination intensity is varied and the resulting excess carrier dependant lifetimes $\tau_{n\_all}(\Delta n)$ and $\tau_{p\_all}(\Delta p)$ are plotted.

For typical structures that have a c-Si layer with low mobility passivation layers at the front and back additionally c-Si carrier lifetimes τ_{n_c-Si} and τ_{p_c-Si} are calculated by only integrating over the c-Si layer. To model the typical QSSPC measurements of passivated c-Si wafers done with the commercially available setup by Sinton Consulting (Sinton & Cuevas, 1996), an effectively measured carrier lifetime τ_{qss} is calculated by the following equation:

$$\tau_{qss} = \frac{\Delta n_{c-Si}\mu_{n\_c-Si} + \Delta p_{c-Si}\mu_{p\_c-Si}}{\mu_{n\_c-Si} + \mu_{p\_c-Si}} / \Delta G_{c-Si}$$

5.4 Measurement model: impedance, capacitance (IMP, ADM, C-V, C-T)

Both boundaries must be described by a voltage controlled metal-semiconductor contact. Additional to the time independent external DC voltage V_{ext}^{DC} an alternating sinusoidal AC voltage is superimposed, $\tilde{V}_{ext}(x,t) = V_{ext}^{DC}(x) + V_{ext}^{AC}(x) e^{i\omega t}$, with a small amplitude V_{ext}^{AC} and a given frequency f, $\omega = 2\pi f$. The resulting external current through the semiconductor stack in the limes of a sufficiency small amplitude is calculated, $\tilde{I}_{ext}(x,t) = I_{ext}^{DC}(x) + I_{ext}^{AC}(x) e^{i\omega(t+\delta)} = I_{ext}^{DC}(x) + \tilde{I}_{ext}^{AC}(x) e^{i\omega t}$. It is also sinusoidal and of the same frequency f, with an AC-amplitude I_{ext}^{AC} and a phase shift δ, or with a complex amplitude $\tilde{I}_{ext}^{AC}$ respectively.

The impedance is defined to be the complex resistance of the semiconductor stack, i.e. the quotient of ac-voltage to ac-current. The admittance is defined to be the complex conductivity of the semiconductor stack, i.e. the quotient of ac-current to ac-voltage. It can be equivalently represented by a parallel circuit of a conductance G and a capacitance C.

$$IM\tilde{P} = \frac{V_{ext}^{AC}}{\tilde{I}_{ext}^{AC}}, \qquad AD\tilde{M} = \frac{\tilde{I}_{ext}^{AC}}{V_{ext}^{AC}} = G + i\,2\pi\,f\,C$$

Depending on the measurement chosen, the frequency is varied and the amplitude and phase shift of the impedance is plotted (measurement IMP), or the capacitance, conductance and conductance divided by frequency is plotted (measurement ADM). Furthermore, for a fixed frequency f, the capacitance can be plotted as a function of the external DC-voltage (measurement C-V) or as a function of the temperature (measurement C-T).

5.5 Measurement model: surface photovoltage (ID-SPV, VD-SPV, WD-SPV)

In order to simulate a steady-state surface photovoltage (SPV) signal (Kronik & Shapira, 1999), the front side boundary should usually be a metal-insulator-semiconductor contact. The semiconductor stack is additionally illuminated with a monochromatic irradiation at a certain wavelength and intensity. The potential difference Δf with and without monochromatic illumination at the front (first grid point) and at the back (last grid point) of the stack is computed and as output the SPV signal $V_{SPV\_front/back}$ is calculated.

$$V_{front/back}^{SPV} = \Delta\varphi_{front/back} = \left(\varphi_{front/back}^{illuminated} - \varphi_{front/back}^{dark}\right)$$

Note that only one quantity φ_{front} or φ_{back} will change upon illumination, as the potential is fixed to $\varphi = 0$ either at the front side or at the back side. Depending on the measurement, either the intensity of the monochromatic illumination is varied (ID-SPV, intensity dependant surface photovoltage), or the external voltage is varied (VD-SPV, voltage dependant surface photovoltage), or the wavelength of the monochromatic illumination is varied (WD-SPV, wavelength dependant surface photovoltage).

5.6 Measurement model: photo electro luminescence (PEL)

When an external illumination and/or an external voltage are applied the emitted radiation can be calculated according to the generalized Plank equation (Würfel, 1982).

$$I(\lambda) = 2c \int dx \left\{ \frac{\alpha(\lambda,x)}{\lambda^5} \cdot \frac{1}{\exp\left[\left(\frac{hc}{\lambda} - \left(E_{Fn}(x) - E_{Fp}(x)\right)\right)/kT\right] - 1} \right\}$$

By integration over the whole structure the wavelength dependant emitted intensity to the front and back is calculated taking photon re-absorption into account. For a given absorption coefficient α and a given wavelength λ the spectra $I(\lambda)$ of the emitted photons is determined by the splitting of the quasi-Fermi levels of electrons and holes E_{Fn}, E_{Fp}. The external working conditions like external illumination and/or applied voltage that cause the quasi-Fermi level splitting have to be specified. Furthermore the wavelength region for which the emitted intensity is calculated can be selected.

6. Selected examples on AFORS-HET simulations

To illustrate the concepts of numerical solar cell simulation, some selected examples simulating a simple amorphous/crystalline silicon solar cell are shown. The absorber of the solar cell (designed for photon absorption) is constituted by a 300 μm thick p-doped textured silicon wafer, c-Si, whereas the emitter of the solar cell (designed for minority carrier extraction, that is electron extraction) consists of an ultra thin 10 nm layer of n-doped, hydrogenated amorphous Silicon, a-Si:H, see Fig. 4. In order to support the lateral electron transport, a transparent conductive oxide layer, TCO, is used as a front side contact. For the sake of simplicity, majority carrier extraction that is hole extraction, is realized as a simple metallic flatband contact to the p-type absorber. Please note, that this solar cell structure is not a high efficiency structure, as a back surface field region, BSF, for hole extraction in order to avoid contact recombination, has not been used. However, this structure has been chosen, as it clearly reveals the properties of an amorphous/crystalline heterojunction interface.

This interface is crucial for the performance of an amorphous/crystalline heterojunction solar cell: By an adequate wet-chemical pre-treatment of the wafer prior to the deposition of a-Si:H onto the surface of the silicon wafer, one has to ensure that an a-Si:H/c-Si heterocontact with a low a-Si:H/c-Si interface state density, D_{it}, will form. The influence of D_{it} on the solar cell performance as well as on various solar cell characterisation methods will be shown. Thus a sensitivity analysis of different measurement methods in order to measure an unknown D_{it} is performed my means of numerical simulation.

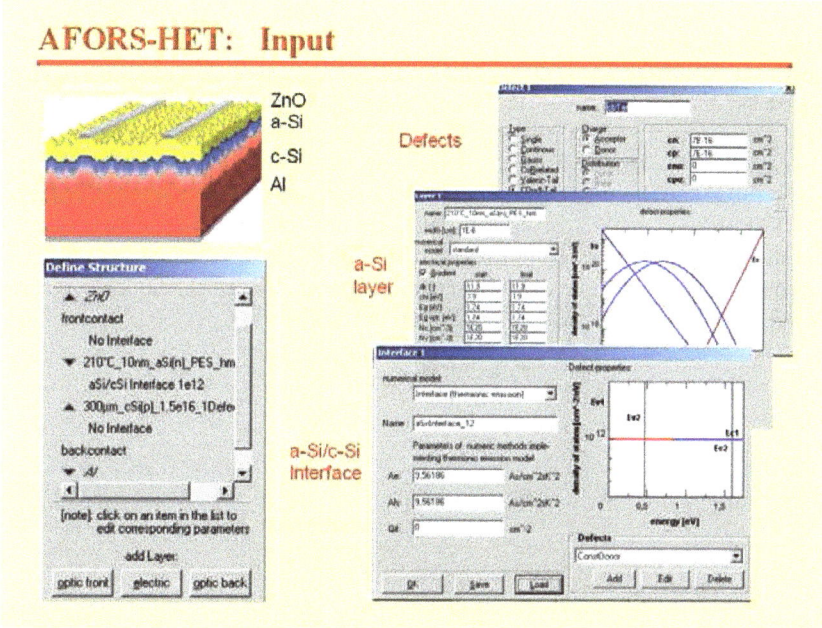

Fig. 4. Screenshots of typical AFORS-HET input: Simulation of TCO/a-Si:H(n)/c-Si(p)/Al heterojunction solar cells. (left) layer sequence, (right) defect distributions $N_{trap}(E)$ of the a-Si:H(n) layer and of the a-Si:H(n)/c-Si(p) interface.

Fig. 4 shows typical screenshots of an AFORS-HET input while modelling the above mentioned TCO/a-Si:H(n)/c-Si(p)/Al heterojunction solar cell. In order to model the c-Si absorber, the bulk model "crystalline silicon" is chosen, specifying the appropriate doping (i.e. $N_A = 1.5 \ 10^{16} \ cm^{-3}$) and the appropriate lifetime of the wafer (i.e. specifying a defect density of a single midgap defect $N_t = 1.\ 10^{10} \ cm^{-3}$, which corresponds to a mean lifetime of $1 \, ms$ as indicated in the input window). In order to model the a-Si:H emitter, the bulk model "standard semiconductor" is chosen, specifying the measured density of state distributions within the bandgap of a-Si:H (Korte & Schmidt, 2008), see Fig. 4. I.e. the measured Urbach tail states and the measured dangling bond states of a-Si:H have to be stated and the doping density N_D has to be adjusted to a value which leads to the measured Fermi level to valence band distance $E_F - E_V = 250 \, meV$ of a-Si:H (Korte & Schmidt, 2008). Furthermore, the electron affinity of a-Si:H has to be adjusted to a value in order to represent the measured valence band offset $\Delta E_V = E_V^{cSi} - E_V^{aSi} = 450 \, meV$ (Korte & Schmidt, 2008). For modeling the a-Si/c-Si interface, the interface model "drift diffusion interface" is chosen, assuming a simple constant distribution of interface defects within the bandgap, exhibiting a donor like character below midgap and an acceptor like character above midgap, see Fig.4. The TCO layer at the front is modelled as an optical layer, thus at the front contact the measured TCO absorption (Schmidt et. al., 2007) as well as the measured solar cell reflection due to the surface texturing (Schmidt et. al., 2007) is specified. Therefore, for the optical calculation the optical model "Lambert-Beer absorption" has to be specified.

6.1 Optical calculation

Fig. 5 shows the resulting spectral absorptions of the incoming AM 1.5 illumination within the different layers of the solar cell: More than half of the low wavelength radiation ($\lambda \leq 350\,nm$) is absorbed within the 80 nm thick TCO layer and is therefore lost for solar energy conversion. Also the defect-rich, ultra-thin a-Si:H emitter is significantly absorbing photons up to $\lambda \leq 600\,nm$. All photons with $\lambda \leq 600\,nm$, which are not absorbed, are reflected. Most photons with $\lambda \geq 800\,nm$, which are not absorbed by the solar cell absorber, are transmitted, some of them are reflected, a few of them are absorbed in the TCO layer due to free carrier absorption. After exceeding the bandgap of the c-Si absorber (for $\lambda \geq 1120\,nm$) there is no more photon absorption in the absorber.

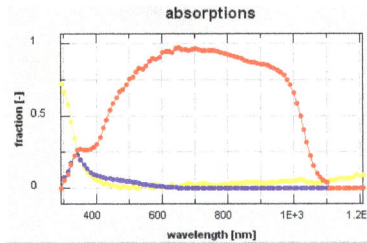

Fig. 5. Screenshot of the spectral absorption within the different solar cell layers (yellow: TCO layer, blue: a-Si:H emitter layer, red: c-Si absorber layer).

6.2 Equilibrium band diagrams

Fig. 6 shows the resulting equilibrium band diagrams (conduction band energy, valence band energy and Fermi energy as a function of the position within the solar cell) assuming different interface state densities D_{it}, after an electrical calculation has been performed.

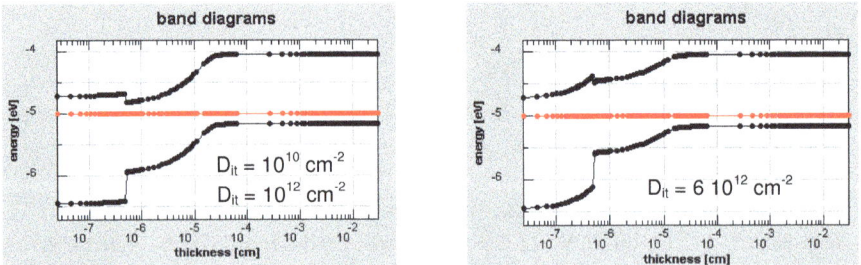

Fig. 6. Screenshots of equilibrium band diagrams (red: Fermi energy, black: valence and conduction band energy) for three different a-Si:H/c-Si interface state densities D_{it}.

Note that the equilibrium band diagram does not change until $D_{it} \geq 2\ 10^{12}\ cm^{-2}$.

6.3 Current-voltage characteristics

However, if one looks at the solar cell performance, i.e. if one calculates the corresponding current-voltage characteristics, D_{it} will reduce the open-circuit voltage of the solar cell for $D_{it} \geq 1\ 10^{10}\ cm^{-2}$, see Fig. 7. Even if interface states in a comparatively low concentration are formed, i.e. $1\ 10^{10}\ cm^{-2} \leq D_{it} \leq 5\ 10^{10}\ cm^{-2}$, this will significantly reduce the solar cell efficiency.

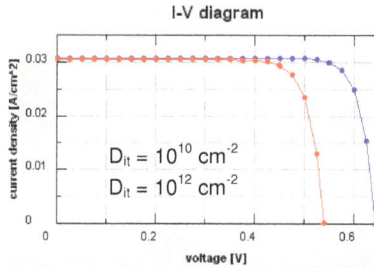

I-V diagram

$D_{it} = 10^{10}\ cm^{-2}$
$D_{it} = 10^{12}\ cm^{-2}$

Fig. 7. Screenshot of a current-voltage simulation under AM 1.5 illumination for two different a-Si:H/c-Si interface state densities D_{it}.

6.4 Quantum efficiency

The influence of D_{it} is not noticeable in a quantum efficiency measurement, as the short-circuit current density is not affected due to a D_{it} variation, and quantum efficiency is a measure for the excess carrier collection efficiency under short circuit conditions. In Fig. 8 internal as well as external quantum efficiency is shown (IQE, EQE), whereas the difference of the two results from the measured reflection losses.

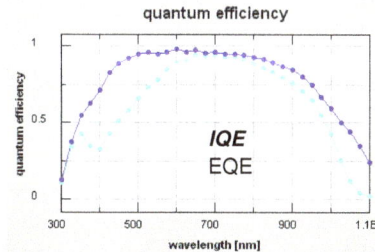

quantum efficiency

IQE
EQE

Fig. 8. Screenshot of a quantum efficiency simulation (there is no difference for different a-Si:H/c-Si interface state densities D_{it}).

6.5 Impedance, capacitance

If one monitors temperature dependent impedance in the dark (i.e. if one calculates the resulting conductance and capacitance as a function of temperature), the onset of the change of the equilibrium band bending due to an increasing D_{it} can be detected. As soon as the increasing D_{it} starts to change the equilibrium band bending, an additional peak in the conductance spectra evolves (Gudovskikh et. al., 2006), see Fig. 9. Thus, dark capacitance-temperature (C-T) measurements are sensitive to interface states only for $D_{it} \geq 2.10^{12}\ cm^{-2}$.

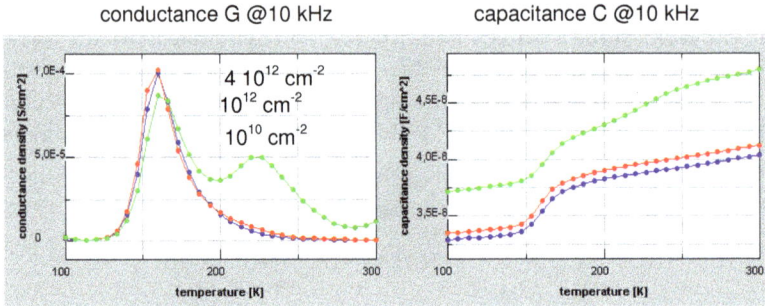

Fig. 9. Screenshot of a capacitance-temperature simulation at an AC frequency of 10 kHz for three different a-Si:H/c-Si interface state densities D_{it} .

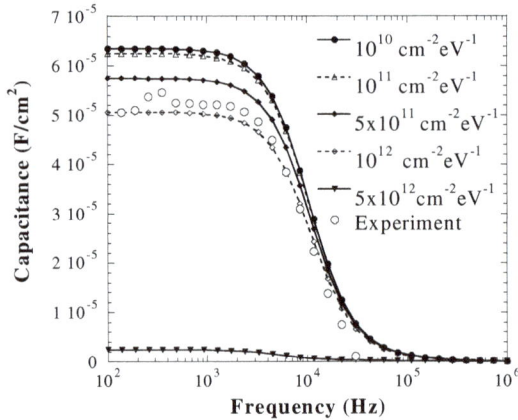

Fig. 10. Comparison of simulated and measured capacitance-frequency measurements under AM1.5 illumination for different a-Si:H/c-Si interface state densities D_{it} . Data from (Gudovskikh et. al., 2006).

In order to enhance the sensitivity towards D_{it} , measurements under illumination have to be performed. Fig. 10 shows an example of an illuminated capacitance-frequency (C-f) measurement, where the corresponding simulations are compared to a real experiment (Gudovskikh et. al., 2006). According to the simulation, the D_{it} of the solar cell under investigation was in the range $D_{it} \approx 8. 10^{11} cm^{-2}$. A sensitivity analysis of this measurement technique indicates a sensitivity towards D_{it} for $D_{it} \geq 1. 10^{11} cm^{-2}$. However, this is still not sufficient in order to characterize well passivated solar cells with a low D_{it} in the range $1\ 10^{10}\ cm^{-2} \leq D_{it} < 1\ 10^{11}\ cm^{-2}$.

6.6 Photoluminescence

Photoluminescence proofs to be quite sensitive to D_{it} . This is because this measurement performs without current extraction. As an example, Fig. 11 shows the simulated steady-state photoluminescence spectra as well as the transient photoluminescence decay (after an integration of the spectra) due to a pulse-like excitation for two different values of $D_{it} = 1.10^{10} cm^{-2}$ and $D_{it} = 1.10^{12} cm^{-2}$. If one integrates the spectra, the simulated measurement signals differ for more than one order of magnitude.

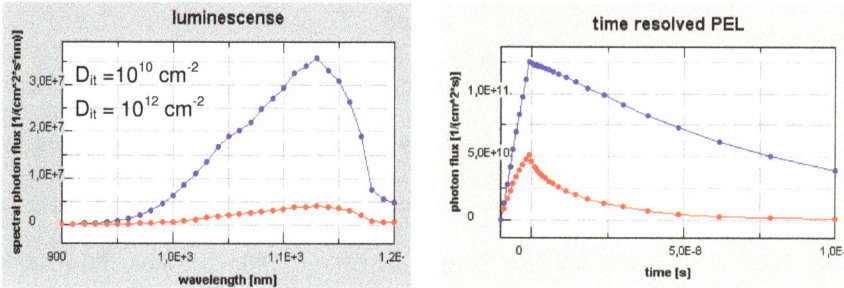

Fig. 11. Screenshots of photoluminescence simulations for two different a-Si:H/c-Si interface state densities D_{it}. (left) steady-state photoluminescence spectra, (right) transient photoluminescence decay after a pulse-like excitation.

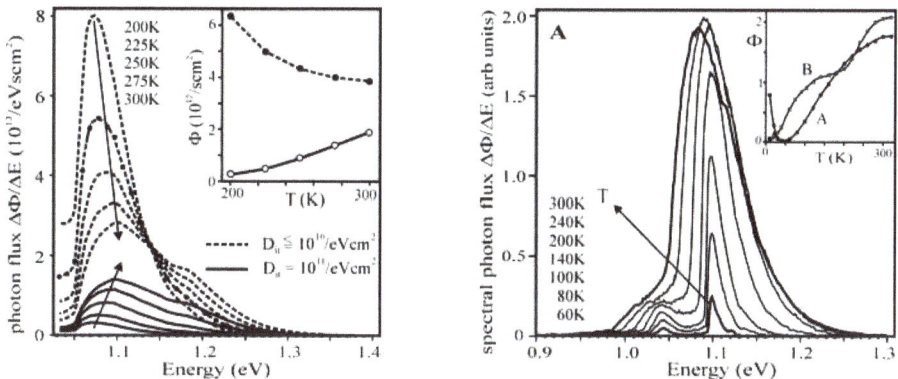

Fig. 12. (left) Simulated temperature dependant photoluminescence measurements for different a-Si:H/c-Si interface state densities D_{it}. (right) Measured temperature dependant photoluminescence. Data from (Fuhs et. al, 2006).

The sensitivity towards D_{it} can even be more enhanced, if one performs temperature dependant photoluminescence measurements, see Fig. 12. Here the character of the measurement even changes if D_{it} is in the range $1\ 10^{10}\ cm^{-2} \leq D_{it} < 1\ 10^{11}\ cm^{-2}$. For $D_{it} \leq 1.10^{10} cm^{-2}$ the spectral emission decreases with increasing temperature, see Fig. 12, thus indicating a non noticeable amount of interface defects, whereas for example for $D_{it} = 1.10^{11} cm^{-2}$ an increasing spectral emission with increasing temperature is observed (Fuhs et. al, 2006).

7. Conclusion

A mathematical description of AFORS-HET, version 2.4, a one dimensional computer program for the simulation of solar cells and solar cell characterization methods has been stated. Some selected examples, simulating amorphous/crystalline silicon heterojunction solar cells and investigating the sensitivity of various measurement methods towards the interface state density D_{it}, were presented.

8. References

Anderson, R. L. (1962). Experiments on Ge-GaAs Heterojunctions, *Solid State Electron.*, 5 (1962), 341-51

Basore, P.; Clugston, D. A. (1997). PC1D, Version 5.9, Copyright 2003, University of New South Wales, latest publication describing the program: PC1D Version 5: 32-bit Solar Cell Simulation on Personal Computers, *26th IEEE Photovoltaic Specialists Conf. (Sept 1997)*

Fuhs, W.; Laades, L.; v.Maydell, K.; Stangl, R.; Gusev, O.B.; Terukov, E.I.; Kazitsyna-Baranovski, S.; Weiser, G. (2006). Band-edge electroluminescence from amorphous/crystalline silicon heterostructure solar cells, *Journal of Non-Crystalline Solids*, 352 (2006) 1884–1887

Gudovskikh, A.S.; Kleider, J.P.; Stangl, R. (2006). New approach to capacitance spectroscopy for interface characterization of a-Si:H/c-Si heterojunctions, *Journal of Non-Crystalline Solids*, 352 (2006) 1213-1216

Korte, L.; Schmidt, M. (2008). Investigation of gap states in phosphorous-doped ultra-thin a-Si:H by near-UV photoelectron spectroscopy, in: *J. Non. Cryst. Sol.* 354 (2008) 2138-2143

Kronik, L.; Shapira, Y. (1999). Surface Photovoltage Phenomena: Theory, Experiment, and Applications, *Surface Science Reports*, 37 (1999), 5-206

Kundert et. al. (1988). A sparse linear equation solver, department of electrical engineering and computer science, Berkeley, CA, USA, 1988, available from: http://www-rab.larc.nasa.gov/nmp/nmpCode.htm

Sah, C.; Shockley, W. (1958). Electron-Hole Recombination Statistics in Semiconductors through Flaws with Many Charge Conditions, *Physical Review*, 109 (1958), 1103

Schmidt, M.; Korte, L.; Laades, A.; Stangl, R.; Schubert, Ch.; Angermann, H.; Conrad, E.; v.Maydell, K. (2007). Physical aspects of a-Si:H/c-Si hetero-junction solar cells, *Thin Solid Films* 515 (2007), p. 7475-7480

Selberherr, S. (1984). Analysis and simulation of semiconductor devices, *Springer Verlag*, 1984

Shockley, W.; Read, W. T. (1952). Statistics of the Recombinations of Holes and Electrons, *Physical Review*, 87 (1952), 835

Sinton, R.A.; Cuevas, A. (1996). Contactless determination of current-voltage characteristics and minority-carrier lifetimes in semiconductors from quasi-steady-state photoconductance data, *Applied Physics Letters*, 69 (1996), 2510-2512

Sze, S. M.; Kwok, K. N. (2007). Physics of Semiconductor Devices, *John Wiley & Sons, Inc., Hoboken, New Jersey*, 2007

Würfel, P. (1982). The chemical potential of radiation, *Journal of Physics C*, 15 (1982), 3967-3985

8

Energy Control System of Solar Powered Wheelchair

Yoshihiko Takahashi, Syogo Matsuo, and Kei Kawakami
Department of Mechanical System Engineering
Department of Vehicle System Engineering
Kanagawa Institute of Technology
Japan

1. Introduction

Independence is a major concern for individuals with severe handicaps. Welfare assistance robotic technology is a popular solution to this concern (e.g. Hashino, 1996; Takahashi, Ogawa, and Machida, 2002 and 2008). Assistance robotic technologies offer potential alternatives to the need for human helpers. People bound to wheelchairs have limited mobility reliant on battery life, which only allows for short distance travel between charges. In addition, recharging batteries is time consuming.

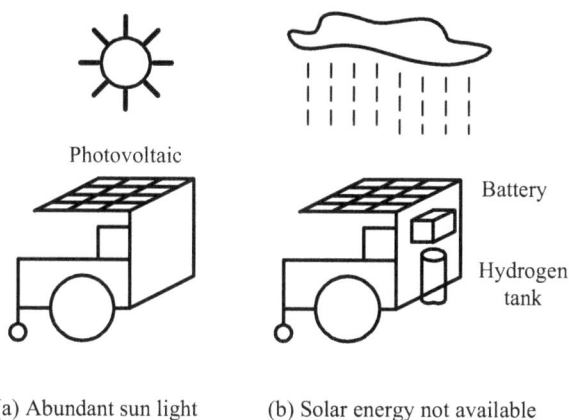

(a) Abundant sun light　　　(b) Solar energy not available

Fig. 1. Running conditions of proposed robotic wheelchair

The aim of this paper is to propose a system which will increase the moving distance of an electrical wheelchair by adding two solar powered energy sources; a small photovoltaic cell and a fuel cell. Fig.1 displays the running conditions of the proposed robotic wheelchair. The control system will ideally give priority to the photovoltaic cell, next to the fuel cell and finally to the battery. When sufficient sun light is available, the photovoltaic cell on the wheelchair roof is used, when it is limited, the fuel cell or the battery is used. The energy control system is designed using a micro computer, and the energy source is quickly changeable. Our objective is that the proposed robotic solar wheelchair will enable users to enjoy increased independence when they are outdoors.

The advantage of using a solar powered energy source is that it produces power without requiring use of fossil fuels. A photovoltaic cell is installed on the roof of the wheelchair, which produces enough power to operate the apparatus when enough sun light is available. The battery is charged using a large photovoltaic cell on the roof of the setup. Hydrogen is produced using a water electrolysis hydrogen generator, and the fuel cell utilizes the produced hydrogen. The large photovoltaic cell also sends electricity to the hydrogen generator.

Photovoltaic cells and fuel cells are representative sustainable technologies (Bialasiewicz, 2008; Okabe et al., 2009; KE Jin et al., 2009). We are able to use two methods to produce hydrogen using these sustainable technologies for our wheelchair. The first method is to generate hydrogen from the electrolysis of water. The next is to use waste biomass which produces biomass ethanol. Hydrogen is produced by steam reforming the ethanol (Takahashi, & Mori, 2006; Essaki et al., 2008; Saxena et al., 2009; Rubin, 2008; Sugano, & Tamiya, 2009). Standard sized fuel cell models are developed with the aim to develop a commercially viable vehicle (Tabo et al., 2004; Kotz, et al., 2001; Rodatz, et al., 2001). Hybrid vehicles using photovoltaic cells and fuel cells are developed in two universities (Konishi, et al., 2008; Obara, 2004). Small fuel cell vehicles were developed (Nishimura, 2008; Takahashi, 2009a and 2009b). A wheelchair with a fuel cell has been developed (Yamamuro, 2003).

This paper will present a robotic wheelchair using solar powered energy sources of the photovoltaic and fuel cell, detail the energy flow concept for charging electricity to the battery and for storing hydrogen to the tank, the mechanical construction, the energy control system, and the experimental results of the running test.

2. Energy flow of proposed robotic wheelchair

A schematic explanation and block diagram of the energy flow used in the proposed robotic wheelchair are shown in Figs. 2 and 3. The energy system used in the robotic wheelchair does not exhaust carbon dioxide as it does not utilize fossil fuels.

The first energy flow line in the schematic diagrams is the line from the photovoltaic cells on the roof of the wheelchair. A cascade connection of two photovoltaic cells (Kyosera, KC-40TJ) of 17.4 V and 43 W in nominal value is utilized as the energy source. The output voltage is reduced to 24 V using a DC-DC converter.

The second energy line is the line from the water electrolysis hydrogen generator. The photovoltaic cell on the setup roof (approximately 10 kW) sends electricity to the water electrolysis hydrogen generator. The generated hydrogen is stored in a metal hydride hydrogen tank of 60 NL. The output pressure of the hydrogen generator is approximately 0.3 MPa. The hydrogen tanks are installed on the wheelchair body after storing hydrogen. A metal hydride tank is used for safety concerns. A fuel cell (Daido Metal, HFC-24100) producing 24 V and 100 W in nominal values is used to generate electricity to the motor.

The third energy flow line is the battery line. The battery is charged with electricity from the photovoltaic cell producing approximately 10 kW on the setup roof, and then installed on the wheelchair body.

The fourth energy flow line is the biomass line. Ethanol is produced from waste biomass. Hydrogen is then generated from the ethanol using a steam reforming hydrogen generator. The generated hydrogen is stored in a 60 NL metal hydride hydrogen tank. The hydrogen

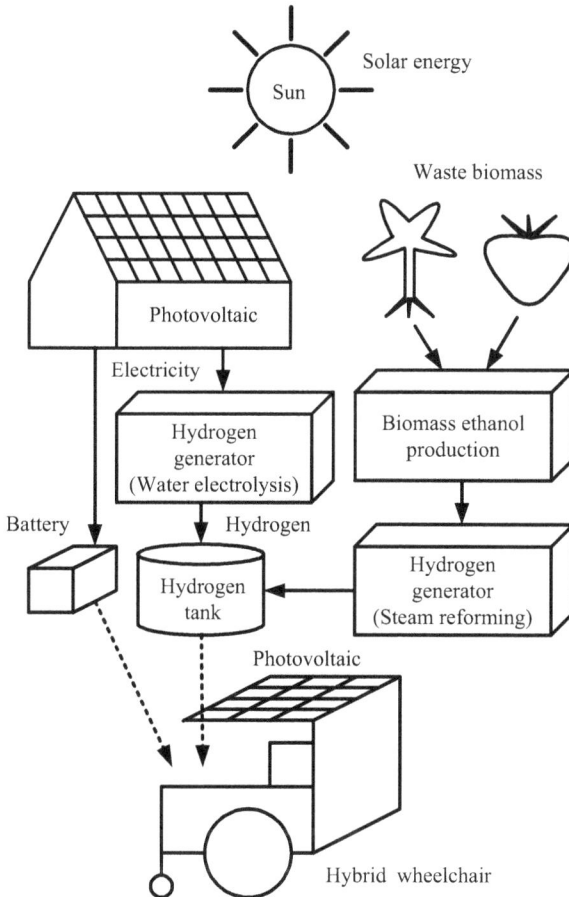

Fig. 2. Schematic explanation of energy flow

tanks are installed on the wheelchair body in the same manner as the second line. Ethanol is safe to handle, and is easy to carry, however, the fourth energy flow line is still a matter under consideration.

The High-Tech Research Center Project for Solar Energy System at the Kanagawa Institute of Technology is conducting research on applications of solar energy. The development of the robotic wheelchair is conducted as a part of the High-Tech Research Center Project. The battery charging and hydrogen storing to the metal hydride are conducted using the facility at the High-Tech Research Center Project.

3. Mechanical construction

Fig. 4 displays the fabricated robotic wheelchair with the photovoltaic and fuel cell. A reinforced YAMAHA JW-1 wheelchair was used as the main body of the experimental set up. In this configuration, the photovoltaic cell, the fuel cell, and the battery are installed on the top, on the back, and under the wheelchair, respectively. The energy control system and

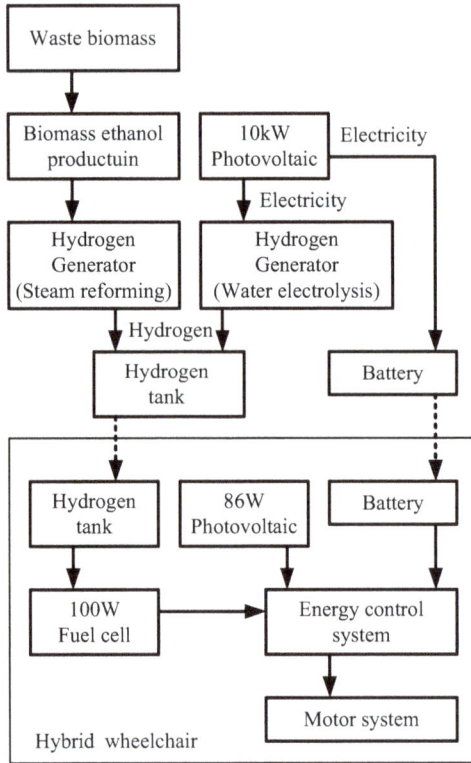

Fig. 3. Block diagram of energy flow

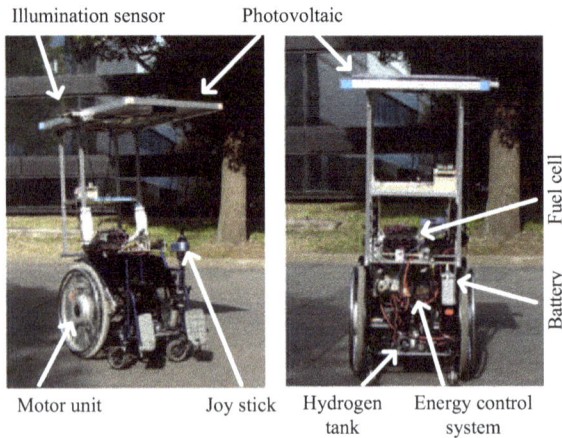

Fig. 4. Fabricated robotic wheelchair with photovoltaic and fuel cell

hydrogen tanks are installed on the back of the wheelchair. Figs.5 (a) and (b) show the photovoltaic and illumination sensor. Fig.6 (a) exhibits the fuel cell (Daido Metal, HFC-24100) and the vibration isolator. Fig.6 (b) shows the metal hydride tanks of 60 NL and 0.3

MPa. The hydrogen pressure is adjusted to 0.08 MPa using a regulator. The main specifications are as follows.

Wheelchair mechanism (Yamaha, JW-1)
    Weight : 13 kg
    Running operation : Joy stick
    Motor : DC24V, 90Wx2
Photovoltaic (Kyosera, KC-40TJ)
    Type : Multi crystal
    Nominal power : 43 W
    Maximum voltage : 17.4 V
    Dimensions : 526x652x54 mm
    Weight : 4.5 kg
Fuel cell (Daido Metal, HFC-24100)
    Nominal power : 100 W
    Nominal voltage : 24 V
    Dimensuions : 160x110x240 mm
    Weight : 3 kg
    Air fans : DC24, 0.94Wx 24

(a) Photovoltaic (b) Illumination sensor

Fig. 5. Photovoltaic and illumination sensor

(a) Fuel cell and vibration isolator (b) Hydrogen tank and regulator

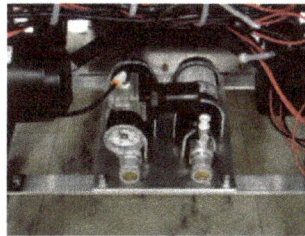

Fig. 6. Fuel cell and hydride tank

4. Energy control system

Fig.7 shows the concept of the energy control system where a micro computer determines the wheelchair condition, and selects the optimum energy source from the three energy sources: the photovoltaic on the wheelchair roof; the fuel cell; or the battery. Solid lines indicate energy flow lines, and dotted lines indicate the control signal flow lines. Fig.8 displays energy control architecture in detail. The switching control system inputs the voltages of the photovoltaic cell, the fuel cell, and the motor drive current, and selects the energy source determined by the wheelchair condition.

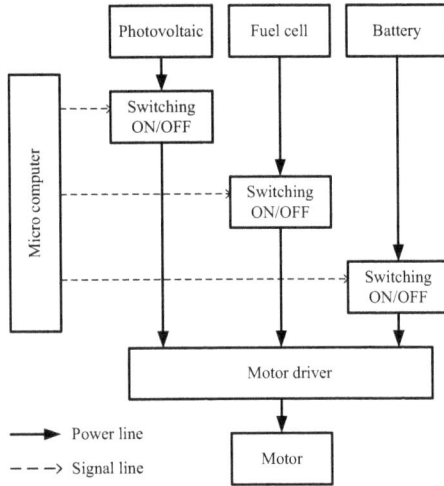

Fig. 7. Concept of energy control system

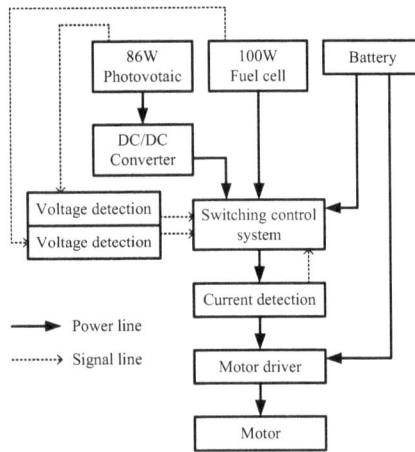

Fig. 8. Detailed energy control architecture

Fig.9 is the software control algorithm of the energy control system. Fig.10 shows the fabricated switching control system of the energy control system where a micro computer controls the entire energy control system, and FETs are used to switch the energy flow. Performance of energy source switching is also tested as this is the first attempt to develop a solar powered wheelchair. The electricity acquired from the photovoltaic cell on the wheelchair roof will be utilized to charge with the battery. Instant power increase using a capacitor will also be required. Improvement of the energy control system must be addressed in future research.

The control system will ideally give priority to the photovoltaic cell then to the fuel cell and then to the battery. Essentially, the switching control is conducted on the motor driving current considering the condition of the photovoltaic and fuel cells. If the motor driving current is below 2.5 A and the photovoltaic voltage is above 30 V, then the photovoltaic is selected. If the motor driving current is below 4.0 A and the fuel cell voltage is above 24 V, then the fuel cell is selected. When the motor driving current is below 20.0 A, then the battery is selected. The following details the software control algorithm of the energy control system.

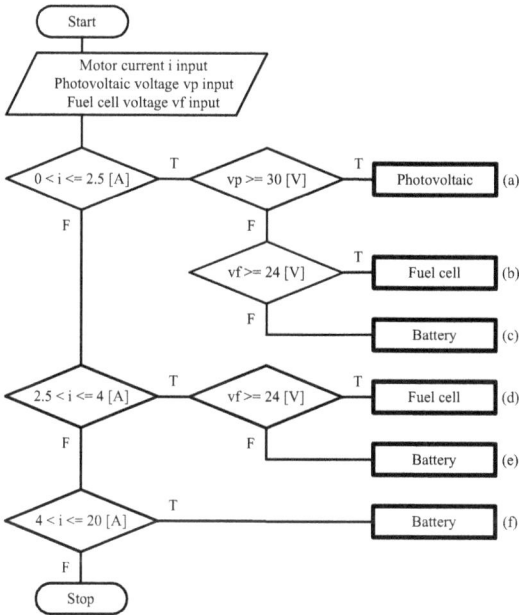

Fig. 9. Software control algorithm of energy control system

Condition (a) :
    When the motor current is over 0.0 A and less than 2.5 A,
    and the photovoltaic voltage is over 30 V,
    then the photovoltaic is selected.
Condition (b) :
    When the motor current is over 0.0 A and less than 2.5 A,
    and the photovoltaic voltage is less than 30 V,
    and the fuel cell voltage is over 24 V,
    then the fuel cell is selected.
Condition (c) :
    When the motor current is over 0.0 A and less than 2.5 A,
    and the photovoltaic voltage is less than 30 V,
    and the fuel cell voltage is less than 24 V,
    then the battery is selected.

Condition (d) :
    When the motor current is over 2.5 A and less than 4.0 A,
    and the fuel cell voltage is over 24 V,
    then the fuel cell is selected.
Condition (e) :
    When the motor current is over 2.5 A and less than 4.0 A,
    and the fuel cell voltage is less than 24 V,
    then the battery is selected.
Condition (f) :
    When the motor current is over 4.0 A and less than 20.0 A,
    then the battery is selected.

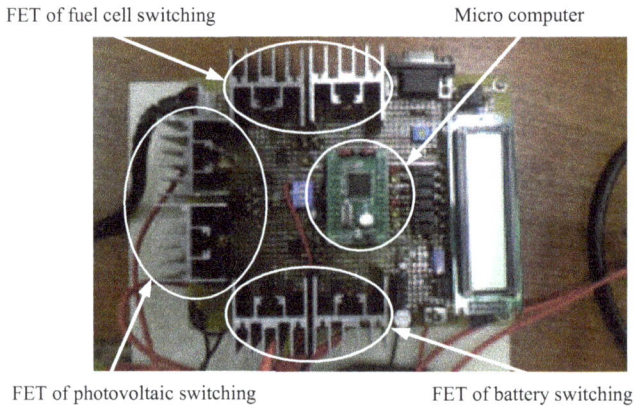

Fig. 10. Fabricated switching control system of energy control system

5. Experimental results

We conducted experiments on energy source selection between the solar panel, the fuel cell, and the battery.

5.1 Experimental conditions

The wheelchair was tested on a flat, straight course. It maintained the same speed maneuvering between two turning points during which joystick operation was required.

Figs.11 to 13 present the experimental results of the running tests using the fabricated wheelchair. Each figure will be detailed later. In each figure, the top graph shows the results of the motor driving current, the photovoltaic (PV) current, and the fuel cell (FC) current. Second, the results of the photovoltaic (PV) voltage, and the fuel cell (FC) voltage. The third figure from the top shows the results of the photovoltaic (PV) power and fuel cell (FC) power. Finally, the last chart shows illumination.

The following four test patterns are conducted;
1. Low Speed mode, Low (approximately 1.4 km/h) (Fig.11)
2. Low Speed mode, High (approximately 2.4 km/h) (Fig.12)
3. High Speed mode, Low (approximately 3.0 km/h) (Fig.13)

4. High Speed mode, High (approximately 3.4 km/h) (Fig.14)

Low Speed and High Speed modes are selected using a switch on the wheelchair frame. "Low" refers to the joy stick declined approximately 10 degrees, "High" refers to the joy stick declined completely. For example, the condition of the "Low Speed mode, Low" denotes that the switch was selected to low speed mode and the joy stick was inclined by approximately 10 degrees. The charts from Figs. 11 through 14 show the results of the conditions: Low Speed mode, Low; Low Speed mode, High; High Speed mode, Low; and High Speed mode, High, respectively.

5.2 Results of "Low Speed mode, Low" condition

Fig.11 (a) shows the experimental results of the current at the "Low Speed mode, Low" condition. The speed was approximately 1.4 km/h. The experimental results show that the average value of the motor driving current was approximately 1.5 A while moving straight on the course. The photovoltaic current was approximately the same value of the motor driving current. The fuel cell current was small. At the turning point, the motor driving current was over 2.5 A, therefore the current of the photovoltaic and fuel cell were lower than the required motor driving current. Energy from the battery was necessary.

Fig.11 (b) shows the experimental results of the voltage. While moving on the straight course, the photovoltaic voltage was reduced, and the fuel cell did not change greatly.

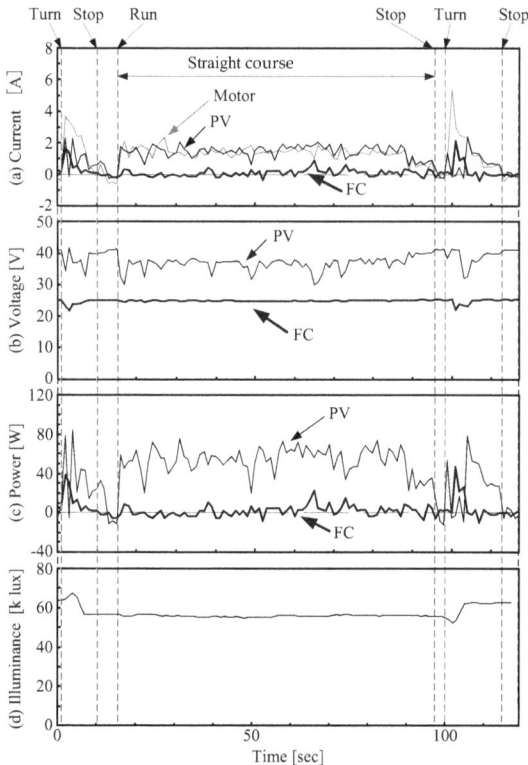

Fig. 11. Experimental results (Low Speed mode, Low)

Fig. 11 (c) shows the experimental results of the power. While moving on the straight course, the photovoltaic power was larger than the fuel cell power. Fig.11 (d) shows the experimental results of the illumination. Illumination was approximately 60 k lux while moving on the straight course. The illumination value changed while turning at the course ends.

5.3 Results of "Low Speed mode, High" condition

Fig.12 (a) shows the experimental results of the current at the "Low Speed mode, High" condition. The speed was approximately 2.4 km/h. The experimental results show that the average value of the motor driving current was approximately 1.8 A while moving on the straight portion of the course. The photovoltaic current was approximately the same value as the motor driving current. The photovoltaic current was occasionally reduced to 0 A. Fuel cell current was negligible. During turns, the motor driving current was over 4.0 A. The photovoltaic and fuel cell current were less than the motor driving current, therefore battery energy was required.

Fig.12 (b) shows the experimental results of the voltage. While moving on the straight part of the course, the photovoltaic voltage was lowered, the change in the fuel cell was negligible. Fig.12 (c) shows the experimental results of the power. The photovoltaic power

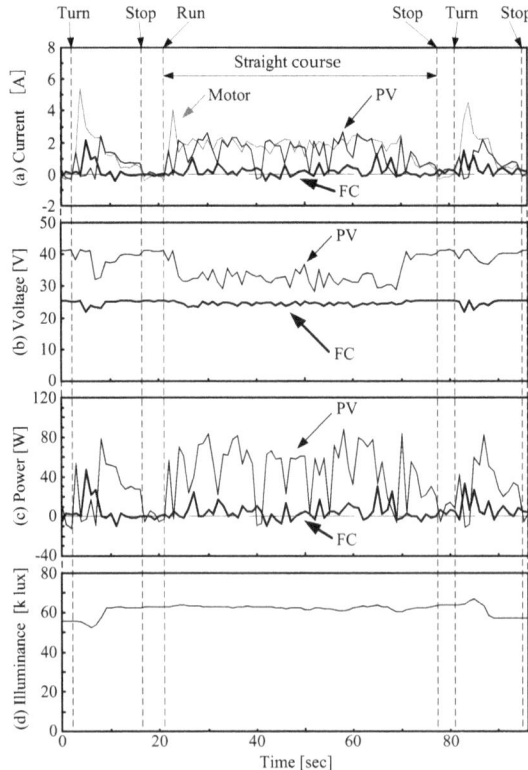

Fig. 12. Experimental results (Low Speed mode, High)

was larger than the fuel cell power. Fig.12 (d) shows the experimental results of the illumination. The illumination was approximately 60 k lux while moving on the straight part of the course. The illumination value changed while turning at the course ends.

5.4 Results of "High Speed mode, Low" condition

Fig.13 (a) shows the experimental results of the current at the "High Speed mode, Low" condition. The speed was approximately 3.0 km/h. The experimental results show that the average value of the motor driving current was approximately 2.0 A while moving straight. The photovoltaic current was approximately the same value as the motor driving current. The photovoltaic current occasionally fell to 0 A. Fuel cell current was minimal. At the turning point, the motor driving current was over 4.0 A, and the current of the photovoltaic and fuel cells were smaller than the motor driving current. Battery energy was required.

Fig.13 (b) shows the experimental results of the voltage. While moving straight, photovoltaic voltage decreased, and the fuel cell did not change prominently. Fig.13 (c) shows the experimental results of the power. While moving straight, photovoltaic power was larger than fuel cell power. Fig.13 (d) shows the experimental results of the illumination. Illumination was approximately 60 k lux while moving straight. It changed while turning at the course ends.

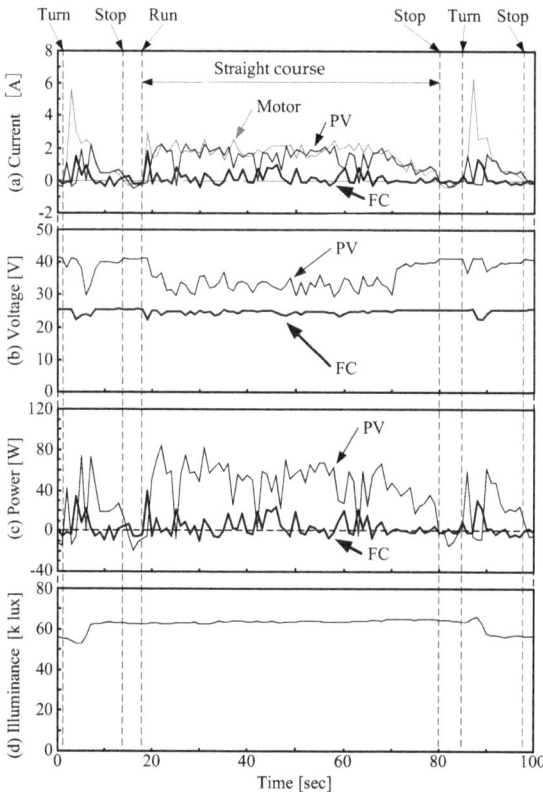

Fig. 13. Experimental results (High Speed mode, Low)

5.5 Results of "High Speed mode, High" condition

Fig.14 (a) shows the experimental results of the current at the "High Speed mode, High" condition. The speed was approximately 3.4 km/h. The experimental results show that the average value of the motor driving current was approximately 2.8 A while moving straight. Photovoltaic current was approximately 0 A. Fuel cell current increased, however the value was smaller than the motor driving current. Combined energy from the fuel cell and the battery was used. At the turning point, the motor driving current was over 4.0 A, and the current of the photovoltaic and fuel cell were smaller than the motor driving current. Energy from the battery was required.

Fig.14 (b) shows the experimental results of the voltage. While moving straight, the photovoltaic voltage was not reduced, however the fuel cell was. Fig.14 (c) shows the experimental results of the power. While moving the straight course, the photovoltaic power was not significant, and the fuel cell power increased. Fig.14 (d) shows the experimental results of the illumination. Illumination was approximately 60 k lux while moving straight. The illumination value changed while turning at the course ends.

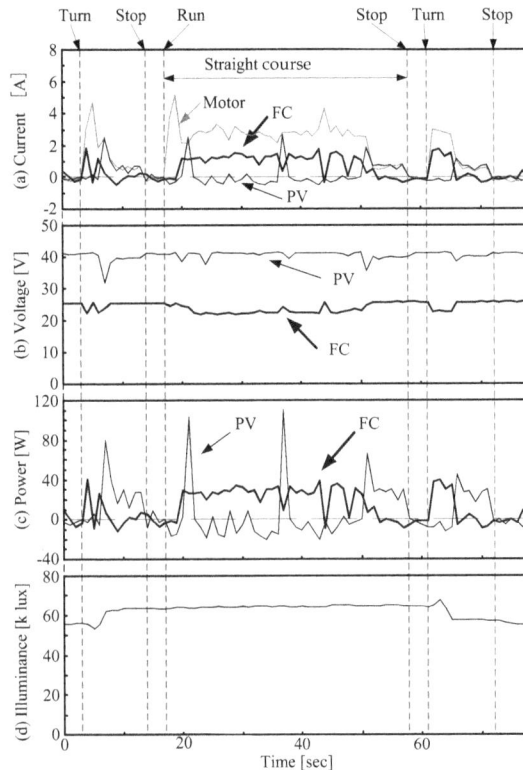

Fig. 14. Experimental results (High Speed mode, High)

5.6 Summary of the experimental results

The proposed robotic solar powered wheelchair uses a small powered photovoltaic of 86 W (at nominal value), and a small powered fuel cell of 100 W (at nominal value). When ample sun light is available, a flat and straight course is used, and the wheelchair travels at a low speed, the robotic wheelchair is able to move primarily powered by the photovoltaic cell. As a result, the solar powered wheelchair is able to travel further distances. When the wheelchair travels at higher speeds, and turns, it requires greater power, therefore it uses the energy from the fuel cell and the battery.

6. Conclusions

A new robotic solar powered wheelchair using three energy sources, a small photovoltaic cell, a small fuel cell, and a battery is proposed in this paper. All three energy sources use solar energy. The photovoltaic cell uses sun light directly. The battery is charged with electricity provided by the large photovoltaic cell installed on the setup roof. Hydrogen for the fuel cell is generated by a water electrolysis hydrogen generator, which is also powered by the same large photovoltaic cell on the building roof. The energy control system selects the optimal energy source to use based on various driving conditions.

It was confirmed from the experimental results that the robotic wheelchair is able to maneuver mainly using the photovoltaic cell when good moving conditions are available (i.e. abundant sun light, a flat and straight course, and low speed). The experimental results demonstrate that the robotic wheelchair is able to increase its moving distance. When moving conditions are not optimal, the robotic solar wheelchair uses energy from the fuel cell and the battery.

Improvements to the energy control system such as charging to the battery from the photovoltaic cell on the wheelchair roof, power increase using a capacitor, and hydrogen generation from waste biomass, must be addressed in future research.

7. Acknowledgments

The authors would like to express their deepest gratitude to the research staff of the High-Tech Research Center Project for Solar Energy Systems at the Kanagawa Institute of Technology for their kind cooperation with the experiments and for their kind advice.

8. References

Hashino, H. (1996); Daily Life Support Robot, *Journal of Robotics Society of Japan,* Vol.14, No.5, pp.614-618

Takahashi, Y., Ogawa, S., and Machida, S., (2002); Mechanical design and control system of robotic wheelchair with inverse pendulum control, Trans. Inst. Meas. Control, vol.24, no.5, pp.355-368.

Takahashi, Y., Ogawa, S., and Machida, S., (2008); Experiments on step climbing and simulations on inverse pendulum control using robotic wheelchair with inverse pendulum control, Trans. Inst. Meas. Control, vol.30, no.1, pp.47-61.

Takahashi, J., And Mori, T., (2006); Hydrogen Production from Reaction of Apple Pomace with Water over Commercial Stream Reforming Ni Catalysis, Journal of Japan Petroleum Institute, vol.49, no.5, pp.262-267.

Essaki, K., Muramatsu, T., and Kato, M., (2008); Hydrogen Production from Ethanol by Equilibrium Shifting Using Lithium Silicate Pellet as CO_2 Absorbent, Journal of Japan Institute of Energy, vol.87, no.1, pp.72-75.

Saxena, R.C., Adhikari, D.K. and Goyal, H.B., (2009); Biomass-Based Energy Fuel Cell through Biochemical Routes, Renew. Sust. Energ. Rev. Vol.13, pp.167-178.

Rubin, E.M., (2008); Genomics of Cellulosic Biofuels, Nature, vol.454, pp.841-845.

Sugano, Y., and Tamiya, E., (2009); A direct Cellulose-Based Fuel Cell System, Journal of Fuel Cell Technology, vol.9, no.1, pp.114-119.

Bialasiewicz, J.T., (2008); Renewable Energy Systems with Photovoltaic Power generations: Operation and Modeling, IEEE Trans. on Industrial Electronics, vol.55, no.7, pp.2752-2758.

Okabe, M., Nakazawa, K., Taruya, K., and Handa, K., (2008); Verification test of solar-powered hydrogen station (SHS) with photovoltaic modules, Honda R&D Technical Review, Vol.20, No.1, pp.67-73.

Ramos-Paja, C.A., Bordons, C., Romero, A., Giral, R., and Martinez-Salamero, L., (2009); Minimum Fuel Cell Consumption Strategy for PEM Fuel Cell, Trans. on Industrial Electronics, vol.56, no.3, pp.685-696.

KE Jin, Xinbo Ruan, MengxiongYang, and Min Xu, (2009); A Hybrid Fuel Cell Power System, Trans on Industrial Electronics, vol.56, no.4, pp.1212-1222.

Tabo, E., Kuzuoka, K., Takada, M., and Yoshida, H., (2004); Fuel cell vehicle technology trends and MMC initiatives, Mitsubishi Motors Technical Review, No.16, pp.51-55.

Kotz, R., Muller, S., Bartschi, M., Schnyder, B., Dietrich, P., Buchi, F.N., Tsukada, A., Scherer, G., Rodatz, P., Garcia, O., Barrade, P., Hermann, V., and Gallay, R., (2001); Supercapacitors for peak-power demand in fuell-cell-driven cars, Electrochemical Society Proceedings, Vol.2001-21, pp.564-575.

Rodatz, P., Garcia, O., Guzzella, L., Buchi, F., Bartschi, M., Tsukada, A., Dietrich, P., Kotz, R., Scherer, and G., Wokaun, A., (2001); Performance and operation characteristics of a hybrid vehicle powered by fuel cells and supercapacitors, Soc. of Automotive Eng. 2003 Congress, SAE Paper 2003-01-0418, pp.1-12.

Konishi, H., Akizuki, M., Ogawa, T., Kojima, H., Yamada, Y., Fujii, H., Matsunaga, N,. Yoshida, Y., Ishida, T., nad Warashina, T., (2008); Development of a Solar and Fuel Cell Powered Hybrid Electrical Vehicle Cocoon 2007, Proc. of 2008 JSME Conf. on Robotics and Mechatronics, 2P1-A18, pp.1-4.

Obara, H., (2004); Progress of Development on the Hybrid Solar Car in Tamagawa University, Journal of Fuel Cell Technology, vol.4, no.2, pp.103-107.

Nishimura, I., (2008); Design and Fabrication of Fuel Cell Vehicle Regarding Manufacturing Education, Proc. of 2008 JSME Conf. on Robotics and Mechatronics, 2P1-A13, pp.1-4.

Takahashi, Y., (2009a); Ultra Light Weight Fuel Cell Electrical Vehicle (UL-FCV), Proc. of IEEE Int. Symp. on Industrial Electronics, pp.189-194.

Takahashi, Y., (2009b); Environmental System Education using Small Fuel Cell Electrical Vehicle, Journal of Fuel Cell Technology, vol.9, no.1, pp.128-131.

Yamamuro, S., (2003); Development of Fuel Cell Powered Wheelchair, Kuromoto Kihou, no.52, pp.40-44.

9

Surface-Barrier Solar Cells based on Monocrystalline Cadmium Telluride with the Modified Boundary

P.M. Gorley[1], V.P. Makhniy[1], P.P. Horley[1,2],
Yu.V. Vorobiev[3] and J. González-Hernández[2]
[1]*Science and Education Center "Semiconductor Material Science and Energy-Efficient Technology" at Yuri Fedkovych Chernivtsi National University,*
58012 Chernivtsi,
[2]*Centro de Investigación en Materiales Avanzados S.C., Chihuahua / Monterrey,*
31109 Chihuahua,
[3]*Centro de Investigación y de Estudios Avanzados del IPN, Unidad Querétaro,*
76230 Querétaro,
[1]*Ukraine*
[2,3]*México*

1. Introduction

Cadmium telluride is one of the most promising materials for solar cell (SC) applications due to its unique physical and chemical parameters. In the first place, it has the band gap E_g ≈ 1.5 eV (300 K) close to the optimal value for photovoltaic conversion (Fahrenbruch & Bube, 1983; Donnet, 2001). The highest temperature and radiation stability of CdTe in comparison with Si and GaAs (Ryzhikov, 1989; Korbutyak et al., 2000) permits to use SCs based on cadmium telluride under elevated temperatures and a considerable flux of ionizing radiation. The possible alternative to CdTe with the similar band gap – gallium arsenide and its solid solutions – are far more difficult to obtain and expensive due to the rarity of Ga (Mizetskaya et al., 1986; Kesamanly & Nasledov, 1973; Andreev et al., 1975).

Solar cells may use diode structures of different kind: p-n-junction, heterojunction (HJ) or surface barrier contact. Despite CdTe has a bipolar conductivity, creation of p-n junction cell based on it is impractical due to high resistivity of p-CdTe and technological difficulty to make ohmic contacts to this material. Therefore, heterojunctions offer more versatile solution by allowing larger parameter variation of junction components than those acceptable for p-n junctions (Alferov, 1998). Additionally, direct band gap of cadmium telluride allows to use this material in thin film form, which was confirmed experimentally for thin film junction nCdS/pCdTe (Sites & Pan, 2007). Alas, the crystalline parameters and coefficients of thermal expansion for CdS (as well as the other semiconductors with wider band gap) significantly differ from those of CdTe (Milnes & Feucht, 1972; Sharma & Purohit, 1979; Simashkevich, 1980), so that the resulting HJ would inherit a significant concentration of the defects at the junction boundary, which will decrease the performance of the solar cell. Moreover, despite the low cost of thin film cells comparing to those based on the bulk material, the technology for the thin-film CdTe has a significant perspectives to be improved (Chopra & Das, 1983; Britt & Ferekides, 1993). Therefore, it seems more appropriate to use monocrystalline cadmium telluride, which has well established and reliable technology (Ryzhikov, 1989; Korbutyak et al., 2000; Mizetskaya et al., 1986). It is preferable to design a technological method for manufacturing photovoltaic devices that could be easily adapted for creation of similar structures based on thin film form of CdTe after minute correction of the corresponding technological regimes.

For this type of applications, surface-barrier diode (SBD) is definitely a good candidate due to its advantages over other diode types (Strikha & Kil'chitskaya, 1992) – simple single-cycle technology necessary to create mono- and multi-element photodiodes of arbitrary area and topology over mono- or poly-crystalline substrates, as well as anomalously low temperatures of barrier contact deposition so that the parameters of the base substrates do not undergo any significant changes in the process. Additionally, substrates of any conductivity type are suitable for formation of SBDs, including those with pre-deposited ohmic contacts. The presence of a strong sub-surface electric field favors efficient separation of non-equilibrium carriers generated by the high-energy phonons. Finally, SBDs can have much lower values of series resistance R_0 comparing with the p-n junctions and hetero-junctions, as they have one semiconductor region in place of two.

However, many of the recent papers (Amanullah, 2003; Mason et al., 2004; Kim et al., 2009, 2009; Gnatyuk et al., 2005; Higa et al., 2007) are rather dedicated to ionizing radiation detectors based on high-resistive CdTe. In contrast with SC, here the surface effects are of far lower importance because ionizing particles penetrate deeper into the material. Independently on the application area of the SBD devices, they should have the largest possible height of the potential barrier. Analysis of literature sources points the impossibility to obtain SBDs with high barrier, minimal series resistance and surface recombination rate using the traditional technological methods (Milnes & Feucht, 1972; Strikha & Kil'chitskaya, 1992; Rhoderick, 1978; Valiev et al., 1981; Sze & Kwok, 2007) – thus, the new methodology should be involved.

One of the perspective ways to solve this problem involves technologies that modify the sub-surface properties of the base substrates, at the same time keeping parameters of bulk material free from significant changes. Here we analyze the experimental results concerning electrical, optical and photoelectric properties of n-CdTe substrates with the modified substrate and surface-barrier solar cells based on them.

2. Objects and methodology of investigations

The base substrates with dimensions $4 \times 4 \times 1$ mm^3 were cut from the bulk CdTe monocrystal, grown by Bridgeman method. The base material featured intrinsic defect electron conductivity $0.1 – 0.05$ $\Omega^{-1} \cdot$cm^{-1} at 300 K as was not doped during the growth process. The substrates were polished mechanically and chemically in the solution of $K_2Cr_2O_7$:H_2O:HNO_3 in proportion 4:20:10 with further rinsing in de-ionized water. As a result, the surface of the substrates gained a mirror-reflective look, and the samples featured a weak photo-luminescence (PL) at 300 K. The similar luminescence is also observed for the cleft surfaces, but it is completely absent in the mechanically-polished plates.

One of the largest sides of the plates with the mirror-reflective surface was deposited with indium ohmic contacts by soldering. Before the creation of a rectifying contact, which was formed with a semitransparent layer of gold deposited by vacuum sputtering, the contact-bearing side of the substrate was subjected to a different additional treatment. The first group of the samples was annealed in the air; further on, they will be referred to as $CdTe:O_2$ (Makhniy et al., 2009). The second group was processed in boiling aquatic suspension of base metal salts (Li_2CO_3, K_2CO_3 and Na_2CO_3), further addressed CdTe:AS (Makhniy & Skrypnyk, 2008). The third group of the samples was formed by chemically-etched substrates not subjected to any additional treatment; these will be further referred to as CdTe. The SBDs based on them served as reference material for comparative studies of modified surface diodes made of the samples belonging to $CdTe:O_2$ and CdTe:AS groups.

The dark current-voltage and capacitance-voltage curves (CVC and CpVC) were measured using the common methodology (Vorobiev et al., 1988; Batavin et al., 1985). The luminescence of the samples was excited with He-Ne and N_2-lasers (wave lengths 0.63 and 0.337 μm, respectively). The radiation, reflection and transmission spectra (N_ω, R_ω and T_ω) were obtained with a universal setup, allowing measurements in standard and differential modes (Makhniy et al., 2004). The spectra were registered automatically with a recording equipment KS-2, also allowing to obtain relaxation curves for the photoluminescence intensity. The light source for measurements of reflection and transmission spectra was a xenon lamp with a smooth spectrum in the investigated energy ranges. All the obtained spectra were corrected for non-linearity of the measuring system. PL spectra were plotted as a number of photons per unitary energy interval N_ω versus photon energy $\hbar\omega$.

The light source used for measuring of photoelectric characteristics was an incandescent lamp with a tungsten filament and a deuterium lamp. Fine tuning of the illumination level in the ranges of 4-5 orders of magnitude was done using a set of calibrated filters. The integral light and loading characteristics of the solar cell were measured with a common methodology (Koltun, 1985; Koltun, 1987). To study the spectral distribution of photosensitivity S_ω we used monochromator DMR-4 with energy dispersion 0.5 – 6.0 eV and precision 0.025 eV/mm. The Si and ZnSe photodiodes with known absolute current sensitivity were used as reference detectors.

The temperature measurements were performed in the ranges 300 – 450 K. The sample was deposited into a specially designed thermal chamber allowing appropriate illumination, quick variation and steady temperature maintenance with the precision of ±1 K. The photovoltaic efficiency of the solar cells was studied by comparing their photoelectric parameters with the reference ITO-Si cell, which under 300 K had the efficiency of 10% for AM2 illumination.

3. Optical properties

As it was shown by the previous studies (Makhniy et al., 2004), the surface modification of the n-CdTe substrates changes their optical properties in a different way depending on the annealing conditions. This approach also works for CdTe:AS and $CdTe:O_2$ samples featuring a sharp efficiency increase for the edge A-band of luminescence η (Fig. 1), which at 300 K can reach several percents. At the same time, for the substrates with mirror-reflecting surface this parameter does not exceed 0.01%. As the effective length of laser radiation in cadmium telluride is $\ell_{PL} \leq 10^{-5}$ cm, the photoluminescence takes place in a narrow sub-surface layer. Therefore, the luminescence intensity I_{PL} in the first approximation can be considered inversely proportional to the concentration of surface defects N_S. As surface modification decreases this concentration for more than two orders of magnitude, it makes a good motivation to use the substrates with modified surface for photodiode applications sensitive in short-wave region.

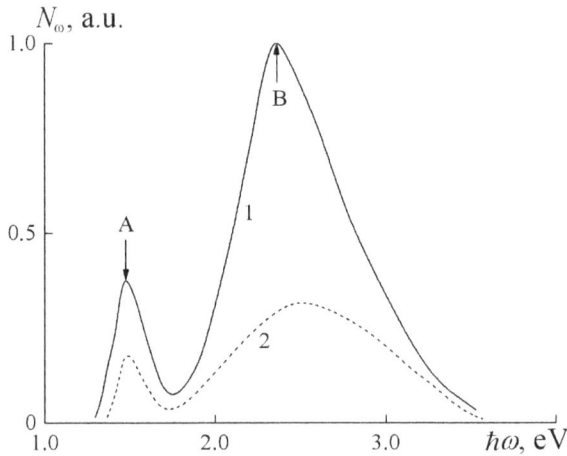

Fig. 1. PL spectra for the substrates $CdTe:O_2$ with 1) free surface and 2) covered with a golden film

To the contrast to CdTe:AS, the PL spectra of $CdTe:O_2$ samples feature a wide B-band situated in the intrinsic absorption area of CdTe at $\hbar\omega > E_g$, Fig. 1. The half-width $\Delta\hbar\omega_{1/2}$ of this band at 300 K is about 0.7 – 0.8 eV. It was found that the intensity of this high-energy band changes with time but that of A-band remains stable, also the peak of A-band $\hbar\omega_m$ shifts towards the low-energy region. It is important to mention that the stationary values of intensity I_{PL} and energy $\hbar\omega_m$ can be achieved several minutes after enabling the laser excitation (Fig. 2). Photoluminescence spectrum shown in Fig. 1, curve 1, was measured namely under such conditions. The experimental dependence of time constant for B-band $t_{1/2}(T)$ can be described by the Arrhenius law (inset to Fig. 2) with activation energy ~ 0.2 eV.

Our investigations revealed that the deposition of a semitransparent golden film onto the surface of $CdTe:O_2$ stabilizes I_{PL} (Fig. 2, curve 2). Despite the figure shows only the initial part of the $I_{PL}(t)$ curve, it remains unchanged not only after several hours, but also after switching laser illumination on and off for several times. This result is very important from the applied point of view, because semitransparent golden film works as an efficient barrier contact required for the optimal performance of SCs based on $CdTe:O_2$ substrates (Ciach et al., 1999).

The existence of high-energy B-band in the PL spectra of $CdTe:O_2$ samples can be explained by quantization of carrier energy caused by the presence of nano-scale structure formations, which is confirmed with the images obtained by the AFM Nanoscope-III in the periodic contact mode (Fig. 3). As one can see, the surface of annealed $CdTe:O_2$ samples is composed with granules some 10 – 50 nm in size (Fig. 3a), which may eventually join into a larger (100 – 500 nm) formations (Fig. 3b).

Annealing the samples under the optimal conditions (e.g., temperature and time) will result in optimization of A-band intensity corresponding to the edge luminescence (Makhniy et al., 2009). Using the relation $I_{PL} \sim N_S^{-1}$, one may come to the conclusion that such high-luminescent samples should have the minimal surface recombination rate. The AFM image of such substrates $CdTe:O_2$ reveals formation of nano-grains of various sizes (Fig. 3b). It is

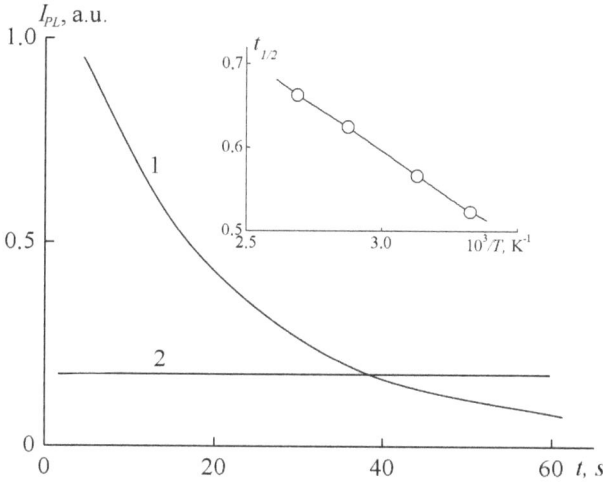

Fig. 2. Time dependence of B-band intensity at 300 K for CdTe:O_2 samples 1) with free surface and 2) covered with golden film. The inset shows the temperature dependence of time variable $t_{1/2}$

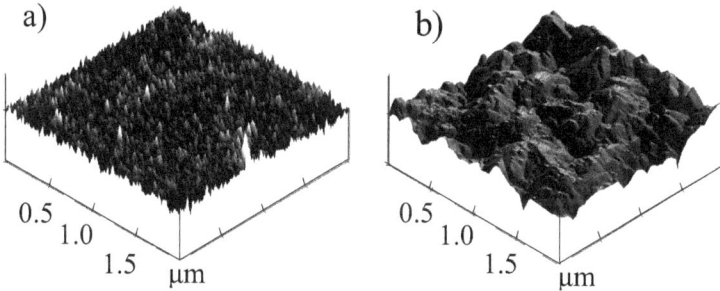

Fig. 3. AFM images of a) the original and b) air-annealed n-CdTe substrates

worth mentioning that large and small grains play a different role for the properties of surface-modified CdTe:O_2. The transition energy is defined by the quantization rule (Zayachuk, 2006; Pool & Owens, 2006)

$$\Delta E = \frac{n^2 \hbar^2}{2d^2} \left(\frac{1}{m_n^*} + \frac{1}{m_p^*} \right),$$ (1)

where m_n^* and m_p^* are the effective masses of electron and a hole; d is lateral dimension of nano-object responsible for the energy peak $\hbar\omega_m$ in the PL spectrum. The depth of the quantum well for such objects is

$$\Delta E = \hbar\omega_m - E_g .$$ (2)

Using the values $m_n^* = 0.11 \, m_0$; $m_p^* = 0.35 m_0$; $E_g = 1.5$ eV and $\hbar\omega_m \approx 2.5$ eV defined from the experiment, we applied expressions (1) and (2) to estimate nano-particle dimension d to be about 5 nm, which is two times smaller than the minimum observed size of the small grains ($d \approx 10$ nm) appearing in the AFM images. This controversy can be removed taking into account that the majority of the grains have a pyramidal shape (Fig. 3b), so that the B-band may be formed by contribution from their top parts, which are narrower than their base. The secondary proof of such possibility consists in presence of the photons with energy $\hbar\omega > \hbar\omega_m$, most probably caused by nano-objects with dimensions smaller than 5 nm.

It is important to emphasize that B-band can not be caused by the luminescence of CdO film that may be eventually formed during the annealing process. In the first place, we did not observe any visible radiation for the samples with CdO film created by photo-thermal oxidation over the substrates with mirror-smooth surface. Moreover, the differential reflection spectrum R_ω' of such samples has a peak corresponding to E_g of cadmium oxide, which is absent in the spectra of the modified substrates, Fig. 4.

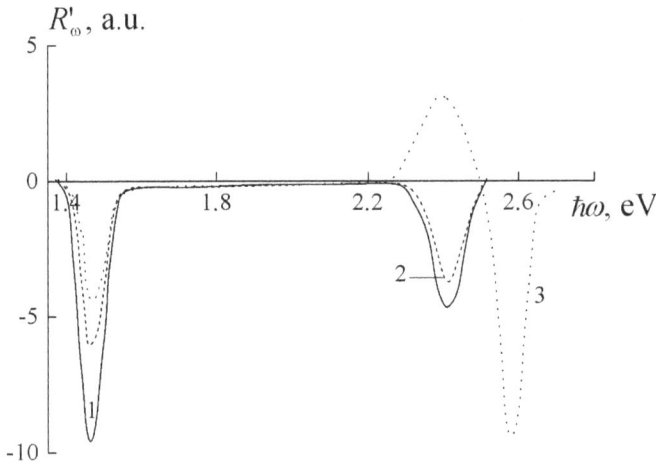

Fig. 4. Differential reflection spectra for various substrates: 1) CdTe, 2) CdTe:O$_2$ and 3) heterostructures CdO/CdTe at T = 300 K

The peaks located at 1.5 and 2.4 eV will correspond to the energy distance between the conduction band E_C and the edges of the valence band: the main sub-band E_{VA} and spin-orbital split band E_{VB}. The peak at $\hbar\omega_m \approx 2.6$ eV correlates with the band gap of CdO at 300 K (Madelung, 2004). The high-energy "tail" of B-band continues much further than it should if being solely defined by E_g(CdTe). It is worth noting that optical transmission / absorption of the samples is defined with the sample group. The "smooth" transmission curve T_ω for CdTe and CdTe:AS substrates has a sharp high-energy edge at $\hbar\omega_m \approx 1.5$ eV corresponding to the band gap of CdTe, Fig. 5.

In contrast, the samples annealed in the air feature a significant decrease of T_ω with red-shifted high-energy transmission edge intercepting the abscissa axis at the energy 1.3 eV, which is significantly lower than E_g of cadmium telluride. The detected peculiarities of transmission spectrum can be explained by the presence of super-grains (100 – 500 nm in size) at the surface of the samples. Such grains may cause light scattering and multiple reflections decreasing the absolute value of T_ω. As these processes intensify for the larger $\hbar\omega$, it may be the cause of the observed red-shift of T_ω curve. It is worth mentioning that in the

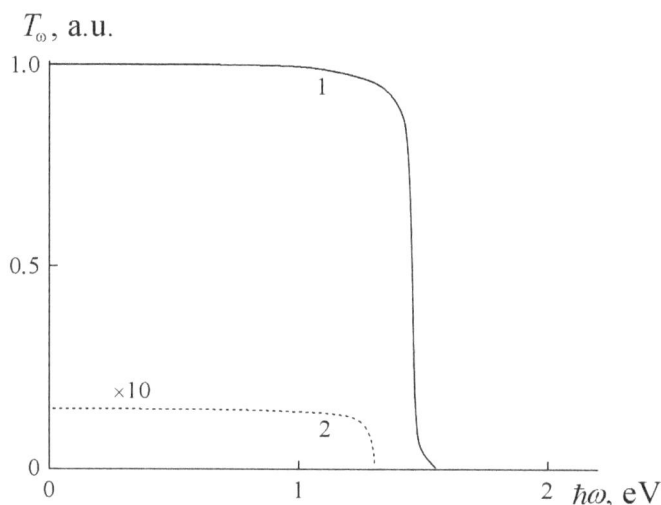

Fig. 5. Optical transmission spectra for different substrates: 1) CdTe and CdTe:AS; 2) CdTe:O$_2$ at 300 K.

photosensitivity spectra S_ω of Au- CdTe:O$_2$ diodes the sensitivity increases drastically for the energies exceeding the band gap of CdTe. It seems feasible to suggest that the observed quantum-scale surface texture is most probably formed by some kind of self-organization phenomena. This hypothesis is confirmed by a considerably narrow temperature and time intervals of the annealing process (Makhniy et al., 2003) that enable formation of the modified structure. Moreover, reproducing the same parameters under the different conditions (noble gas atmosphere or vacuum) does not lead to the desired effect, suggesting that one of the atmosphere components (in particular oxygen) plays an important role in the formation of a nano-crystalline surface structure. The definition of such dependence requires additional experimental and theoretical studies.

4. Electrical properties and carrier transport mechanisms

4.1 Potential barrier height

The height of the potential barrier φ_0 is one of the most important parameters of SBD: it limits open-circuit voltage U_{OC} of the photovoltaic device and enters the exponent in the expression describing over-barrier and tunneling currents, which with increase of φ_0 will yield a lower dark current. Also, higher barrier will favor a larger working temperature of a solar cell.

It was found out that the SBDs based on CdTe:AS and CdTe:O$_2$ feature significantly higher barrier than that achievable for the devices based on the non-modified CdTe. This point can be clearly seen from Fig. 6, where we plot forward-biased branches of current-voltage curves (CVCs) of the studied structures in their linearity region. It is important to highlight that the difference between the slopes of the segments for all three groups of diodes studied is insignificant, proving the close similarity of series resistance in the system diode base – deposited contacts. On the other hand, it also suggests that the contribution of the modified layer into the value of R_0 is negligibly small.

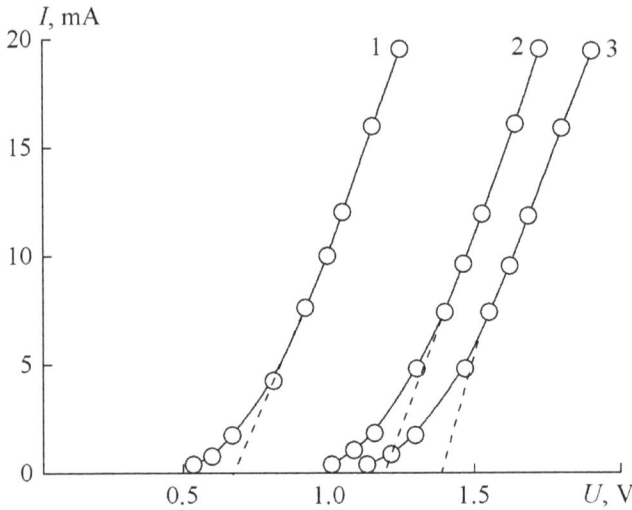

Fig. 6. Direct branches of CVC for the SBDs based on different substrates: 1) CdTe, 2) CdTe:AS, 3) CdTe:O_2 at 300 K.

The latter can be confirmed by the equal $I(U)$ dependence for the different diode types (Fig. 6). The same φ_0 and R_0 observable for the SBDs that differ by processing of their substrates (treatment in the solutions of Li, Na and K salts) suggest that various impurities have the same behavior in the resulting device. It is worth noting that the values of φ_0 determined from the direct CVC branches agree with the results obtained from the capacity measurements. The further analysis of the electrical properties revealed that the surface state apart from influencing φ_0 also change the character of physical processes in the SBDs, which is reflected in the dependences of forward / reverse currents as a function of voltage.

4.2 Direct current formation mechanisms in the surface barrier diodes

Figure 7 displays the initial segments of direct CVC branches measured for the modified surface diodes. As one can see from the figure, for $eU \geq 3kT$ the curves can be successfully approximated with the following expression (Rhoderick, 1978; Fahrenbruch & Bube, 1983)

$$I = I_0 \exp (eU / nkT) \tag{3}$$

where I_0 is a cut-off current at $U = 0$, and n is the non-ideality coefficient, which can be easily found as a slope of the straight CVC segments plotted in semi-logarithmic coordinates.

For Au-CdTe diodes the value of n is equal to the unity, suggesting the dominating over-barrier carrier transport for the SBDs made on the base of moderately-doped substrates (Rhoderick, 1978; Makhniy, 1992). It also means that the dielectric layer between metal and semiconductor is tunneling-transparent with the barrier height lower than $E_g/2$. The dependence $I_0(T)$ is mainly defined by the factor (Sze & Kwok, 2007)

$$I_0 \approx (- \varphi_0 / kT), \tag{4}$$

which gets an excellent confirmation in the experiment. Plotting $I_0(T)$ in $\ln I_0 - 1/T$ coordinates, one can achieve a good linear fit with the slope of 0.9 eV corresponding to φ_0 at 0 K.

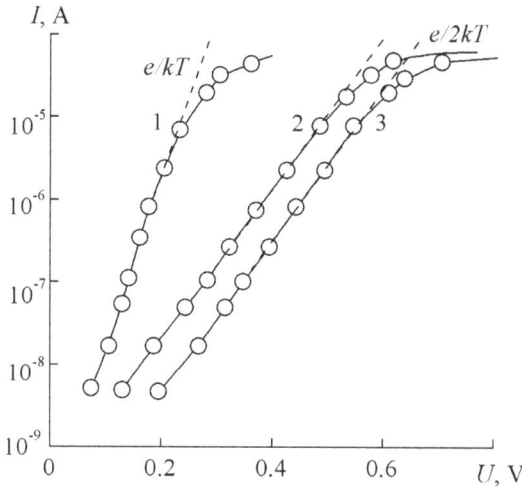

Fig. 7. The direct CVC branches of the diodes based on: 1) CdTe, 2) CdTe:AS and 3) CdTe:O_2 at 300 K.

In contrast to Au-CdTe contacts, the potential barrier height in the SBD with a modified surface will significantly exceed the value of $E_g(CdTe)/2 = 0.75$ eV at 300 K (Fig. 6). For such diodes the dominating carrier transport mechanism (Makhniy, 1992) is a recombination current in the space charge region (SCR) involving deep impurity levels. It can be also described by formula (3) with non-ideality coefficient $n = 2$. As one can see from Fig. 7, the initial direct CVC branches for Au-CdTe:AS and Au-CdTe:O_2 diodes fit excellently with the expression (3) for the temperature T = 300 K. However, it is also possible to approximate the experimental data with another exponential factor (Makhniy 1992)

$$I_0 \approx (-\varphi_0 / kT), \tag{5}$$

For this case, the measured $I_0(T)$ plotted in $\ln I_0 - 1/T$ coordinates will represent a straight line with a slope 1.6 eV corresponding to the band gap of cadmium telluride at 0 K.

Deviation of the experimental points from a straight line under the higher bias is caused by the voltage drop over series resistance of the diode, reducing the applied voltage U to the voltage at the barrier U_0 as

$$U_0 = U - IR. \tag{6}$$

Taking into account (6), one can re-write the expression (3) in the form (Makhniy 1992)

$$\ln I - \frac{eU}{kT} = \ln I_0 - \frac{eR_0}{nkT} I . \tag{7}$$

When the over-barrier current is dominating (which is correct for the voltages about φ_0) the non-ideality coefficient is equal to the unity and the direct branches of CVC can be efficiently plotted in the coordinates $\ln I_0 - eU/kT$. These experimental data can be nicely fit with expression (7) in the case of the diodes with modified surface (Fig. 8) with ordinate interception yielding the $\ln I_0$ for the given temperature.

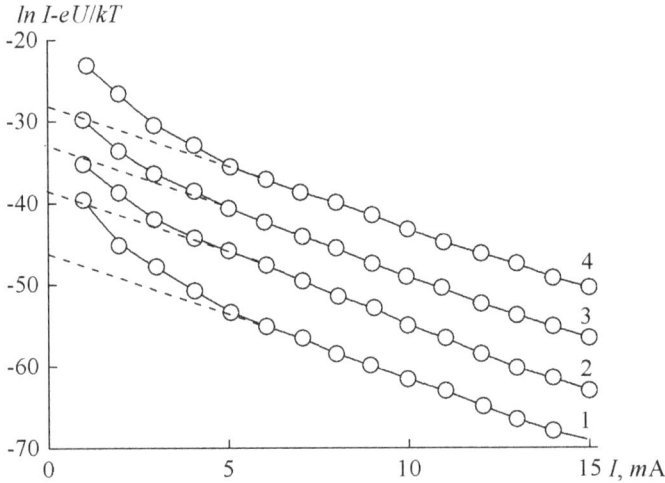

Fig. 8. Comparison of experimental data for direct CVC branches (Au-CdTe:AS diode) with theoretical formula (7) for the temperatures 1) 300 K, 2) 320 K, 3) 340 K and 4) 360 K

The dependence $I_0(T)$ obeys the expression (4) with $\varphi_0 = 1.6$ eV, which corresponds to the barrier height at 0 K. Taking into account the linear dependence $\varphi_0(T) = \varphi_0(0) - \gamma_\varphi T$ and using $\gamma_\varphi = 1.4 \cdot 10^{-3}$ eV/K one can estimate $\varphi_0 = 1.18$ eV at the room temperature, which correlates well with the experimental result 1.2 eV, Fig. 6.

While Fig. 8 presents the data for Au-CdTe:AS diode only, the similar results were obtained for all the structures studied. Therefore, the direct current transport in surface-barrier diodes based on CdTe has a predominant over-barrier character. In the diodes based on the substrates with a modified surface, the current is defined by carrier recombination in the space charge region (low direct bias) or over-barrier emission (high bias).

4.3 Reverse current in surface barrier diodes

It is important to emphasize that over-barrier and recombination cut-off currents I_0 at 300 K are always below 10^{-10} A (Fig. 7). Moreover, these currents in a theoretical model feature a weak dependence on the voltage, such as $U^{1/2}$ (Rhoderick, 1978; Sze & Kwok, 2007). However, in the experiments we obtained steeper curve of inverse current $I_{inv}(U)$, suggesting other current transport mechanisms.

It is logical to assume that in the SBDs based on wide-band semiconductors (which holds for CdTe) subjected to inverse bias tunneling of the carriers will become a dominating mechanism, including both inter-band tunneling and tunneling via the local levels (Makhniy, 1992). The validity of this hypothesis is supported by Fig. 9, where the inverse CVC exhibit a good fitting with the formula for tunneling current in an abrupt junction:

$$I_{inv} = a \exp\left(-\frac{b}{\sqrt{\varphi_0 - eU}}\right). \qquad (8)$$

Here a and b are the coefficients that can be found from the parameter substrate and the diode structure. The negative sign under the square root reflects the negative bias, and the voltage in the expression (8) should be substituted with the negative sign. The high slope of

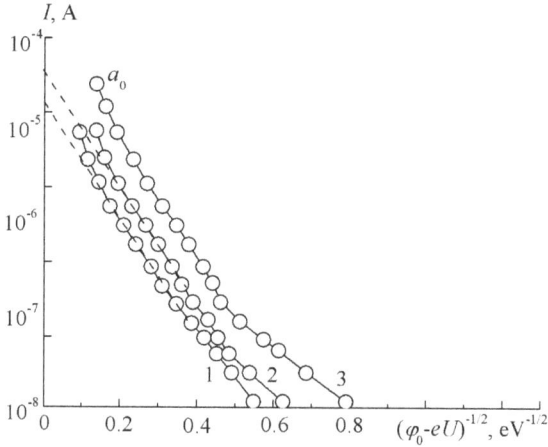

Fig. 9. Comparison of theoretical calculations according to (8) with the inverse CVC branches for the diodes based on: 1) CdTe, 2) CdTe:AS, and 3) CdTe:O_2 at 300 K.

the initial branches of inverse CVCs signals the parameter difference for the diode structures but not the substrates themselves. It is worth noting that the coefficient b depends on the height and width of the barrier, the depth of the local centers in the SCR and some other parameters (Makhniy, 1992), which requires more detailed study. The same can be commented concerning the dependence $I_{inv}(U)$ under high bias (Fig. 9), which in addition to tunneling may have also include the avalanche processes (Mizetskaya et al., 1986).

While the solution of the aforementioned tasks is important for the photodiodes, it is not that much crucial for solar cell applications that require a special attention to the parameters φ_0, n, I_0, and E_g appearing in the expression for the direct current.

5. Photoelectric properties of surface barrier diodes

5.1 Integral illuminated current voltage curves

The CVC of illuminated diode can be described as (Fahrenbruch & Bube, 1983; Koltun, 1987)

$$I = I_0 \left[\exp\left(\frac{eU}{nkT}\right) - 1 \right] - I_p \qquad (9)$$

with photocurrent I_p, dark current $I_0(U = 0)$ and non-ideality coefficient n, which is determined by the current transport mechanism. To determine the exact mechanisms involved in formation of CVCs for illuminated solar cells, we need introduce open circuit voltage (at $I = 0$) and short circuit current I_{SC} (for $U = 0$) that upon being substituted into (9) for a particular case $eU \geq 3kT$ would yield the expression

$$I_p = I_{SC} = I_{SC}^0 \exp\left(eU_L / nkT\right). \qquad (10)$$

Here I_{SC}^0 denotes the cut-off current for illuminated cell voltage $U_L = 0$, which may coincide with I_0 only in the case if the formation mechanisms for the light and dark currents are the same. Therefore, investigation of the integral light CVCs allow to determine the corresponding current formation mechanism, as well as to reveal both common and different traits between the electric and photoelectric properties of the materials studied. Analysis of $I_{SC}(U_{OC})$ curves shows that they are qualitatively similar, differing only in non-ideality value for the various SBD groups. Thus, for Au-CdTe diodes the value of $n = 1$ suggests over-barrier transport of photo-generated carriers, which can be also illustrated by fitting data presented in Fig. 10 (curve 1). It is worth mentioning that I_{SC}^0 is close to the cut-off current I_0, obtained for the dark CVC for the same diode at 300 K.

The energy slope of the segment $\ln I_{SC}^0 - 1/T$ yields the value of 0.9 eV that coincide with the potential barrier height φ_0 at 0 K, determined from the temperature dependence of the dark cut-off current I_0. In this way, one can conclude that dark and light currents of Au-CdTe contacts are formed by over-barrier emission of the carriers. To the contrary, the measured CVC of the illuminated SBDs with a modified surface will require non-ideality coefficient to be $n=2$, equation (10), to achieve the appropriate fitting illustrated in Fig. 10 (curve 2) for Au-CdTe:AS diode. As it was mentioned before, such dependence is characteristic for carrier recombination in the space charge region involving local centers, while the photosensitivity data for Au-CdTe suggests the dominating inter-band generation of the carriers.

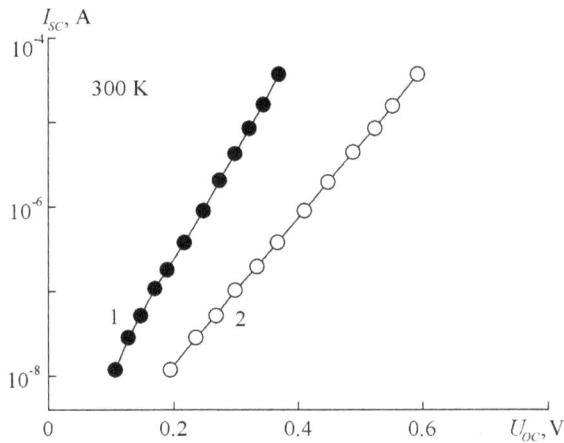

Fig. 10. Dependence of short circuit current on the open circuit voltage for Au-CdTe (1) and Au-CdTe:AS (2) diodes.

This controversy can be eliminated (Ryzhikov, 1989): under the stationary conditions the carrier generation rate G should be equal to the recombination rate R. However, if these phenomena are caused by distinct mechanisms, they will be described by the different analytical expressions. It is worth mentioning that the measurement of the dependences $I_{SC}(U_{OC})$ is performed in so-called compensation mode, when the photocurrent is equalized with dark current under the presence of direct bias. Due to this, one actually monitors recombination of photo-carriers, which yields the expression similar to that of the dark current. It is important that for the SBDs based on the substrates with a modified surface the dark current under the low direct bias is controlled by the recombination processes in space charge region (see Section 4). As carrier generation in the surface barrier diodes takes place in the space charge region due to the fundamental absorption of high-energy photons with $\hbar\omega > E_g$, it is more probable that they will recombine at the same device region taking advantage of the local impurity centers rather than via the inter-band recombination.

5.2 Dependence of I_{SC} and U_{OC} on illumination level and the temperature

For all the SBDs studied, the short circuit current depends linearly on the power of the incident light flux L, which is a consequence of linearity of photo-carrier generation as

$$I_{SC} = \beta L, \tag{11}$$

where β is a proportionality coefficient independent on L. As one can see from Fig. 11, the equation (11) holds well for illumination power spanning over several orders of magnitude. Using the expressions (10) and (11) it is easy to show that

$$U_{OC} = nkT\left(\ln\left(I_{SC} / I_{SC}^{0}\right)\right) = nkT\ln\left(\beta L / I_{SC}^{0}\right). \tag{12}$$

As one can see from Fig. 12, expression (12) describes well the experimental dependences of $U_{OC}(L)$ for the low illumination conditions. The tendency of $U_{OC}(L)$ curves to saturate under high L is caused by the potential barrier compensation. The higher value of U_{OC} for the SBDs with a modified surface in comparison with that of Au-CdTe diodes is explained by the significant difference of potential barrier heights for these rectifying structures (see Sec. 2).

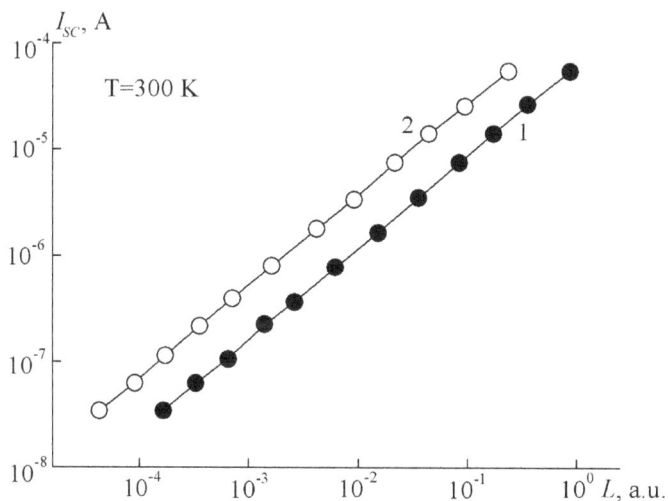

Fig. 11. Dependence of short circuit current on radiation power for Au-CdTe (1) and Au-CdTe:AS (2) diodes

Our investigations revealed that the temperature increase leads to a slight growth of short circuit current, which is caused by an insignificant decrease of series resistance R_0 characterizing the device. Under high illumination open circuit voltage tends to saturate, so that in the first approximation $U_{OC} \approx \varphi_0$ or E_g depending on formation mechanism of the light CVCs (see above). Therefore, under otherwise equal conditions, the open circuit voltage should diminish with increase of temperature, which was confirmed experimentally for all SBDs studied. It is important to emphasize that $U_{OC}(T)$ curve for SBDs with a modified surface features much weaker slope than that observable for the reference diodes. This can be explained by the different temperature dependence of variation coefficient for the band gap E_g and barrier height γ_φ (see Sec. 4).

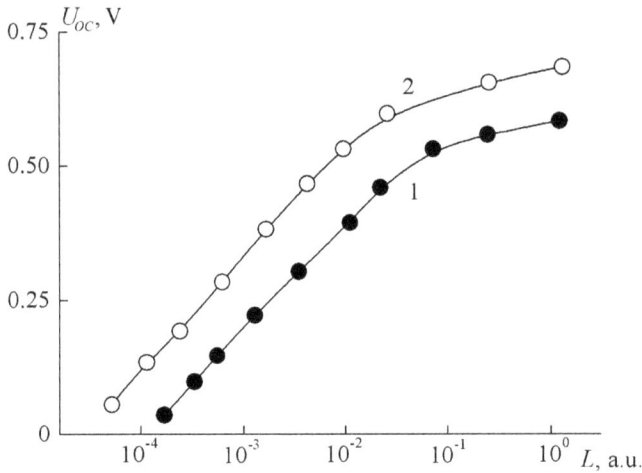

Fig. 12. Dependence of open circuit voltage on the illumination level for Au-CdTe (1) and Au-CdTe:AS (2) diodes

5.3 Spectral characteristics

The previous investigations of optical, electrical and photoelectrical properties of the SBDs predict several peculiar and promising characteristics of these structures. In the first place, the low energy limit $\hbar\omega_{min}$ of the photosensitivity spectrum S_ω is determined with high-energy edge of the transmission spectrum. According to the data presented in Fig. 5, the low-energy edge of S_ω curves for Au-CdTe and Au-CdTe:AS diodes will be about 1.5 eV, while for Au-CdTe:O_2 contacts it is somewhat lower, reaching 1.3 eV (Fig. 13).

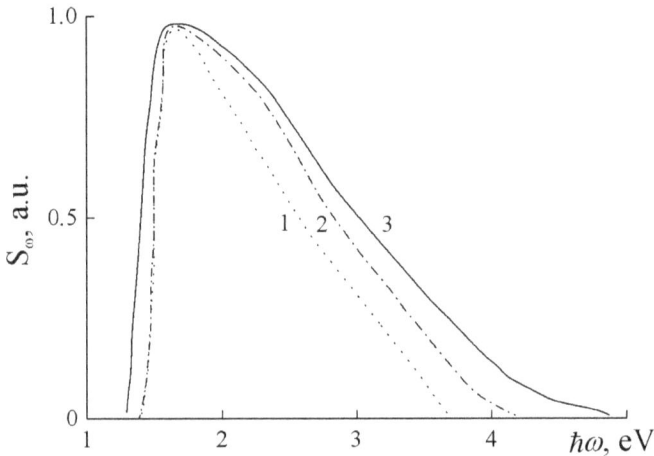

Fig. 13. Photosensitivity spectra for the SBD based on the substrates of CdTe (1), CdTe:AS (2) and CdTe:O_2 (3) at 300 K

High-energy photosensitivity, as it was discussed above, is defined by the surface recombination rate, which in turn depends on concentration of the defects N_S at the junction boundary. Thus, high short-wave sensitivity should be expected in SBDs based on the substrates with a maximum efficiency of edge luminescence band. This conclusion is confirmed by the data presented in Fig. 13, featuring wider photo-sensitivity peaks for Au-CdTe:O_2 diodes, which are based on the substrates with the most intensive A-band (see Section 3) reaching 10% for 300 K. At the same time, for case of CdTe:AS and CdTe substrates, the corresponding A-band values would be 1-2% and 0.01%, respectively, strongly influencing the behavior of S_ω spectra for photon energies $\hbar\omega > E_g$ (Fig. 13).

Investigation of the temperature dependence of the SBD spectra reveals that higher values of T mainly affect their low-energy "tail". The edge of S_ω under the temperature varying in the ranges 300-450 K almost replicates the dependence $E_g(T)$. In the contrast to this, the temperature dependence of the surface defects (*i.e.*, their concentration, energy distribution and scattering cross-sections) is negligibly weak, as one may conclude from the photosensitivity spectra for $\hbar\omega > E_g$.

Type of the surface barrier diode		Au-CdTe	Au-CdTe:AS	Au-CdTe:O_2
Parameters	φ_0, eV	0.7	1.2	1.35
	R_0, Ω	30	30	32
	U_{OC}, V	0.4	0.6	0.7
	I_{SC}, mA/cm^2	10	15	22
	Efficiency, %	5	9	13

Table 1. Main parameters of the solar cells

5.4 Main parameters of surface barrier solar cells

The diodes with the effective photosensitive area of 2·10^{-2} cm^2 were illuminated from the side of a semi-transparent golden contact by light equivalent to AM2 solar illumination conditions. All the measurements were performed at 300 K and the best results obtained were summarized in Table 1.

The values characterize photovoltaic devices in "as-obtained" form without anti-reflection coatings or any special optimization of solar cell design. Due to this, the values of fill factor *ff* are under 0.7 because of considerable series resistance (see Table 1). The reduction of the latter would allow increase of both I_{SC} and *ff*, improving the efficiency of the SCs.

6. Conclusions

The authors developed a special technological approach consisting in a special surface treatment of monocrystalline plates of cadmium telluride before deposition of the barrier contact, which significantly improves the electrical and photoelectric properties of the surface-barrier devices of metal-semiconductor junction type created on their base. The annealing of n-CdTe substrates ($\rho \sim$ 10-20 Ω cm) in the air or aquatic suspense of alkaline metals increases the potential barrier height φ_0 up to 1.2-1.4 eV in comparison with $\varphi_0 \approx 0.7$ eV for non-annealed substrates. The SBDs with modified surface features much better open circuit voltage and short circuit current. The efficiency of solar cells based on the diodes studied under AM2 illumination was within the ranges 8-13% at the room temperature. The developed technology for surface-barrier solar cells is simple, cheap and ecologically clean. In addition, the proposed technological principles can be used for producing photovoltaic devices based on thin films of cadmium telluride.

7. Acknowledgements

This research was supported in part by the Ministry for Education and Science (Ukraine) through the financing provided for the scientific projects carried out at the Department of Optoelectronics, Department of Energy Engineering and Electronics, as well as at the Science and Education Center "Semiconductor Material Science and Energy-Efficient Technology" at the Chernivtsi National University, Research Project #SU/447-2009 for the years 2009-2010. We also acknowledge the financial support of CONACYT (México) for the research performed in CINVESTAV-Querétaro, Research Project CB-2005-01-48792 for the years 2007-2010.

8. References

Alferov, Zh.I. (1998). The history and the future of semiconductor structures. *Fiz. Tekhn. Poluprov.*, Vol. 32, No. 1, 3-19

Amanullah, F.M. (2003). Effect of isochronal annealing on CdTe and the study of electrical properties of Au–CdTe Schottky devices. *Canadian Journal of Physics*, Vol. 81, No. 3, 617-624

Andreev, V.M.; Dolginov, L.M. & Tret'yakov. D.N. (1975). *Liquid epitaxy in technology of semiconductor devices*, Soviet radio, Moscow

Batavin, V.V.; Kontsevoj, Yu.A. & Fedorovich. Yu.V. (1985). *Measurement of the parameters of semiconductor materials and structures*, Radio i svyaz', Moscow

Britt, J. & Ferekides, C. (1993). Thin film CdS/CdTe solar cell with 15.8% efficiency. *Applied Physics Letters*, Vol. 62, 2851-2852

Chopra, K.L. & Das, S.R. (1983). *Thin film solar cells*, Springer, New York

Ciach, R.; Demich, M.V.; Gorley, P.M.; Kuznicki, Z.; Makhniy, V.P.; Malimon, I.V. & Swiatek, Z. (1999). Photo and X-ray sensitive heterostructures based on cadmium telluride, *J. Cryst. Growth*, Vol. 197, No. 3, 675-679

Donnet, D. (2001). Cadmium telluride solar sells. In: *Crean Electricity from Photovoltaic*. Archer, M.D. & Hill, R. (Eds.), 245-276, Imperial College Press

Fahrenbruch, A.L. & Bube, R.H. (1983). *Fundamentals of solar cells: photovoltaic solar energy conversion*. Academic Press, New York

Gnatyuk, V.A.; Aoki, T.; Hatanaka, Y. & Vlasenko, O.I. (2005). Metal–semiconductor interfaces in CdTe crystals and modification of their properties by laser pulses, *Applied Surface Science*, Vol. 244, 528-532

Higa, A.; Owan, I.; Toyama, H.; Yamazato, M.; Ohno, R. & Toguchi, M. (2007). Properties of Al Schottky Contacts on CdTe(111)Cd Surface Treated by He and H_2 Plasmas *Japanese Journal of Applied Physics*, Vol. 46, 2869-2872

Kesamanly, F.P. & Nasledov, D.N. (Eds.) (1973). *Gallium arsenide: production, properties and applications*, Nauka, Moscow

Kim, K.; Cho, S.; Suh, J. Won, J. Hong, J. & Kim. S. (2009). Schottky-type polycrystalline CdZnTe X-ray detectors. *Current Applied Physics*, Vol. 9, No. 2, 306-310

Koltun, M.M. (1985). *Optics and metrology of the solar cells*, Nauka, Moscow

Koltun, M.M. (1987). *Solar cells*, Nauka, Moscow

Korbutyak, D.V.; Mel'nychuk, S.V.; Korbut, Ye.V. & Borysyuk, M.M. (2000). *Cadmium telluride: impurity defect states and detector properties*, Ivan Fedoriv, Kyiv

Madelung, O. (2004). *Semiconductors: Data Handbook*, Springer

Makhniy, V.P. & Skrypnyk, M.V. (2008). Patent for the useful model UA №31891 published 25.04.2008.

Makhniy, V.P. (1992). *Physical processes in the diode structures based on wide-band A^2B^6 semiconductors*. Dissertation of the Doctor in Physics and Mathematics, Chernivtsi

Makhniy, V.P.; Demych, M.V. & Slyotov, M.M. (2003). Declaration patent UA №5010A published 22.04.2003.

Makhniy, V.P.; Skrypnyk, M.V. & Demych, M.V. (2009a). Patent for a useful model UA №40056 published 23.05.2009.

Makhniy, V.P.; Slyotov, M.M. & Skrypnyk, N.V. (2009b). Peculiar optical properties of modified surface of monocrystalline cadmium telluride, *Ukr. J. Phys. Opt.*, Vol. 10, No. 1, 54-60

Makhniy, V.P.; Slyotov, M.M.; Stets, E.V.; Tkachenko, I.V.; Gorley, V.V. & Horley, P.P. (2004). Application of modulation spectroscopy for determination of recombination center parameters. *Thin Solid Films*, Vol. 450, 222-225

Mason, W.; Almeida, L.A.; Kaleczyc, A.W. & Dinan, J.H. (2004). Electrical characterization of Cd/CdTe Schottky barrier diodes. *Applied Physics Letters*, Vol. 85, No. 10, 1730-1732

Milnes, A. & Feucht, D. (1972). *Heterojucntions and metal semiconductor junctions*, Academic Press, New York

Mizetskaya, I.B.; Oleynik, G.S.; Budennaya, L.D.; Tomashek, V.N. & Olejnik. N.D. (1986). *Physico-chemical bases for the synthesis of monocrystals of semiconductor solid solutions $A^{II}B^{VI}$*, Naukova dumka, Kiev

Pool, Ch. Jr. & Owens, F. (2006). *Nanotechnologies*, Tekhnosfera, Moscow

Rhoderick, E.H. (1982). *Metal-semiconductor contacts*, Clarendon Press, Oxford

Ryzhikov, V.D. (1989). *Scintillating crystals of the semiconductor compounds A^2B^6: Obtaining, properties, applications*, NIITEKhIM, Moscow

Sharma, B.L. & Purohit, R.K. (1979). *Semiconductor heterojunctions*, Soviet radio, Moscow

Simashkevich, A.V. (1980). *Heterojunctions based on semiconductor $A^{II}B^{VI}$ compounds*, Sthtiintsa, Kishinev

Sites, J. & Pan, J. (2007). Strategies to increase CdTe solar-cell voltage. *Thin Solid Films*, Vol. 515, No. 15, 6099-6102.

Strikha, V.I. & Kil'chitskaya, S.S. (1992). *Solar cells based on the contact metal-semiconductor*, Energoatomizdat, St. Petersburg

Sze, S.M. & Kwok, K.Ng. (2007). *Physics of semiconductor devices*, J. Willey & Sons, New Jersey

Valiev, K.A.; Pashintsev, Yu.I. & Petrov, G.V. (1981). *Using contacts metal-semiconductor in electronics*, Soviet radio, Moscow

Vorobiev, Yu.V.; Dobrovol'skiy, V.N. & Strikha, V.I. (1988). *Methods in semiconductor studies*, Vyshcha shkola, Kiev

Zayachuk, D.M. (2006). *Low-scale structures and super-lattices*, Lviv Polytechnic, Lviv

10

Efficiency of Thin-Film CdS/CdTe Solar Cells

Leonid Kosyachenko
Chernivtsi National University
Ukraine

1. Introduction

Over the last two decades, polycrystalline thin-film CdS/CdTe solar cells fabricated on glass substrates have been considered as one of the most promising candidates for large-scale applications in the field of photovoltaic energy conversion (Surek, 2005; Goetzberger et al., 2003; Romeo et al., 2004). CdTe-based modules have already made the transition from pilot scale development to large manufacturing facilities. This success is attributable to the unique physical properties of CdTe which make it ideal for converting solar energy into useful electricity at an efficiency level comparable to traditional Si technologies, but with the use of only about 1% of the semiconductor material required by Si solar cells.

To date, the record efficiencies of laboratory samples of CdS/CdTe solar cells and large-area modules are ~ 16.5 % and less than 10 %, respectively (Britt & Ferekides, 1993; Hanafusa et al., 1997; Meyers & Albright, 2000; Wu et al., 2001; Hanafusa et al., 2001; Bonnet, 2003). Thus, even the record efficiency of such type solar cells is considerable lower than the theoretical limit of 28-30% (Sze, 1981). Next challenge is to improve the performance of the modules through new advances in fundamental material science and engineering, and device processing. Further studies are required to reveal the physical processes determining the photoelectric characteristics and the factors limiting the efficiency of the devices.

In this chapter, we present the results of studying the losses accompanying the photoelectric conversion in the thin-film CdS/CdTe heterostructures and hence reducing the efficiency of modules on glass substrate coated with a semitransparent ITO or SnO_2 conducting layer. We discuss the main parameters of the material used and the barrier structure determining the photoelectric conversion efficiency in CdS/CdTe solar cell: (i) the width of the space-charge region, (ii) the lifetime of minority carriers, (iii) their diffusion length and drift length, (iv) the surface recombination velocity, and (v) the thickness of the CdTe absorber layer.

Among other factors, one of the important characteristics determining the efficiency of a solar cell is the spectral distribution of the quantum efficiency which accounts for the formation of the drift and diffusion components of the photocurrent and ultimately the short-circuit current density. In the paper particular attention is given to this aspect of solar cell. We demonstrate the possibility to describe quantitatively the quantum efficiency spectra of the thin-film CdS/CdTe solar cells taking into account the recombination losses at the CdS-CdTe interface and the back surface of the CdTe absorber layer.

Charge collection efficiency in thin-film CdS/CdTe solar cells are also discussed taking into consideration losses caused by a finite thickness of the p-CdTe layer, recombination losses at the front and back surfaces as well as in the space-charge region. The dependences of the drift and diffusion components of short-circuit current on the uncompensated acceptor concentration, charge carrier lifetime, recombination velocities at the interfaces are evaluated and discussed.

The mechanism of the charge transport in the CdS/CdTe heterostructure determining the other photoelectric parameters of the solar cell, namely, the open-circuit voltage and fill factor is also considered. It is shown that the above-barrier (diffusion) current of minority carriers is important only at high bias voltage, and the dominant charge transport mechanism is the generation-recombination occurring in the depletion layer. The observed *I–V* characteristics in the dark and the light are described mathematically in the context of the Sah-Noyce-Shockley theory.

2. Spectral distribution of quantum efficiency of CdS/CdTe heterostructure

In this section we will describe mathematically the spectral distribution of quantum efficiency of the thin-film CdS/CdTe solar cells taking into account the main parameters of the material used and the barrier structure, recombination in the space-charge region, at the CdS-CdTe interface and the back surface of the CdTe absorber layer.

Quantum efficiency η_{ext} is the ratio of the number of charge carriers collected by the solar cell to the number of photons of a given energy (wavelength λ) shining on the solar cell. Quantum efficiency relates to the response (A/W) of a solar cell to the various wavelengths in the spectrum. In the case of monochromatic radiation (narrow spectral range) $\eta_{ext}(\lambda)$ relates to the radiation power P_{opt} and the photocurrent I_{ph} by formula

$$\eta_{ext}(\lambda) = \frac{I_{ph}/q}{P_{opt}/h\nu},$$

(1)

where q is the electronic charge, $h\nu$ is the photon energy.

2.1 Experimental

Fig. 1(a) shows the quantum efficiency spectra of the CdS/CdTe solar cell taken at different temperatures. The substrates used for the development of thin film layers were glass plates coated with a semitransparent ITO (SnO_2 + In_2O_3) layer. The window layer CdS ($\sim$ 0.1 μm) was developed by chemical bath deposition (CBD); the absorber layer CdTe (1-3 μm) was deposited on top of CdS by close-space sublimation (CSS) (Mathew et al., 2007). Non-rectifying ohmic contact to the CdTe layer was fabricated by sputtering Ni in vacuum after bombarding the CdTe surface by Ar ions with energy $\sim$ 500 eV. The electrical characteristics of two neighboring Ni contacts on the CdTe surface were linear over the entire range of measured currents.

The spectral characteristics of the samples in the 300-900 nm range were recorded with a photoresponse spectral system equipped with a quartz halogen lamp. The spectral distribution of the photon flux at the outlet slit of the system was determined using a calibrated Si photodiode.

As can be seen from Fig. 1(a), compared with the literature data, the measured curves seem to reflect the most common features of the corresponding spectral curves for these devices (Sites et al., 2001; McCandless et al., 2003; Ferekides et al., 2004).

In the long-wavelength region, the spectra are restricted to the value λ_g corresponding to the band gap of CdTe which is equal to 1.46 eV at 300 K ($\lambda_g = hc/E_g = 845$ nm). In the short-

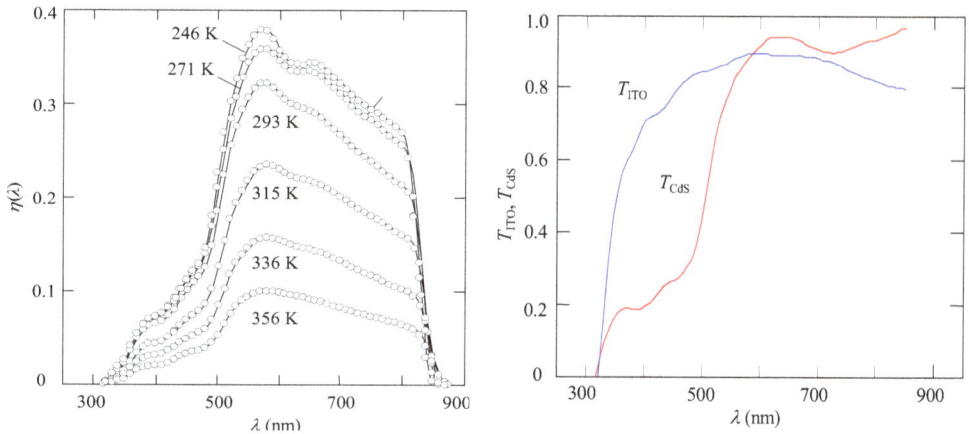

Fig. 1. (a) Spectral distribution of the quantum efficiency of CdS/CdTe device measured at different temperatures. (b) The transmission curves of the ITO coated glass (T_{ITO}) and the CdS layer (T_{CdS}) as a functions of the wavelength λ.

wavelength side, the quantum efficiency decays due to the lower transmission through the thin film layers: CdS in the range $\lambda < 500\text{-}520$ nm and ITO at $\lambda < 350$ nm (Fig. 1(b)).

The external quantum efficiency η_{ext} is related with the quantum efficiency of photoelectric conversion in the CdTe absorber layer, the transmission of the glass plate coated by ITO, T_{ITO}, and the transmission of the CdS layer, T_{CdS}, by the expression:

$$\eta_{ext} = T_{ITO}T_{CdS}\eta_{int} \qquad (2)$$

where η_{int} is the ratio of photogenerated carriers collected to the photon flux that arrives at the CdTe absorber layer.

In order to describe the external quantum efficiency spectrum η_{ext} we used the measured spectral dependences $T_{ITO}(\lambda)$ and $T_{CdS}(\lambda)$ shown in Fig. 1(b). The quantum efficiency η_{int} will be determined in the following by considering the photoelectric processes in the diode structure.

2.2 Width of the space-charge region and energy diagram of thin-film CdS/CdTe heterostructure

One of the parameters of a solar cell that determines the electrical and photoelectric characteristics is the width of the space-charge region W. It is known that in CdS/CdTe solar cells only the CdTe is contributing to the light-to-electric energy conversion and the window layer CdS absorbs light in the range $\lambda < 500\text{-}520$ nm thereby reducing the photocurrent. Therefore in numerous papers where the energy band diagram of a CdS/CdTe junction is discussed a band bending in the CdS layer (and hence a depletion layer) is not depicted (see, for example, Goetzberger et al, 2003; Birkmire & Eser, 1997; Fritsche et al., 2001). Analyzing the efficiency of CdS/CdTe solar cells, however, one is forced to assume the concentration of uncompensated acceptors in the CdTe layer to be $10^{16}\text{-}10^{17}$ cm^{-3} and even higher (a narrow depletion layer is assumed). It may appear that the latter comes into conflict with the commonly accepted model of CdS/CdTe as a sharply asymmetrical p-n heterojunction. In fact, this is not the case because the width of space-charge region of a diode structure is determined by the concentration of uncompensated impurity ($N_d - N_a$ for CdS) rather than by the free carrier concentration (n for n-CdS). These values coincide only in an uncompensated semiconductor with a shallow donor level whose ionization energy is less than the average thermal energy kT. However, the CdS layer contains a large number of background impurities (defects) of donor and acceptor types, which introduce in the band gap both shallow and deep levels. The effect of self-compensation is inherent in this material: a donor impurity introduces p-type compensating defects in just as high concentration as needed to annihilate virtually the electrical activity of donor (Desnica et al, 1999). As a result, CdS is a compensated semiconductor to a greater or lesser extent. In this case, the Fermi level is known to be pinned by the level (so-called pinning effect), which is partially compensated with the degree of compensation N_a / N_d close to 0.5 (if N_a / N_d = 0.5 the Fermi level exactly coincides with the deep impurity level) (Mathew et al., 2007). Assuming $N_d - N_a$ equal to n, one can make serious mistake concerning the determination of the space-charge region width in CdS. Thus, due to a large number of background impurities (defects), the uncompensated donor concentration in the CdS layer can be much higher than the electron concentration in the conduction band of CdS. When such is the case the depletion layer of the CdS/CdTe diode structure is virtually located in the p-CdTe layer even in the case of a CdS layer with comparable high-resistivity (Fig. 2). This is identical to the case of an asymmetric abrupt p-n junction or a Schottky diode; the width of the space-charge region in the CdS/CdTe heterojunction can be expressed as (Sze, 1981):

$$\varphi(x,V) = (\varphi_o - qV)\left(1 - \frac{x}{W}\right)^2 ,$$
$$(3)$$

$$W = \sqrt{\frac{2\varepsilon\varepsilon_o(\varphi_o - qV)}{q^2(N_a - N_d)}} ,$$
$$(4)$$

where ε_o is the electric constant, ε is the dielectric constant of the semiconductor, $\varphi_o = qV_{bi}$ is the barrier height at the semiconductor side (V_{bi} is the built-in potential), V is the applied voltage, and $N_a - N_d$ is the uncompensated acceptor concentration in the CdTe layer.

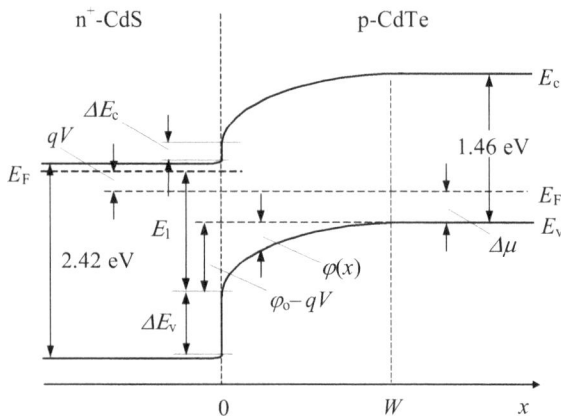

Fig. 2. Energy diagram of n-CdS/p-CdTe heterojunction for forward-bias condition. ΔE_c and ΔE_v show the discontinuities (offsets) of the conduction and valence bands, respectively.

2.3 Theoretical description of quantum efficiency

The quantum efficiency η_{int} can be found from the continuity equation which is solved using the boundary conditions. The exact solution of this equation with account made for the drift and diffusion components as well as surface recombination at the CdS/CdTe interface leads to rather cumbersome and non-visual expression for η_{int} (Lavagna et al, 1977):

$$\eta_{int} = \frac{1 + \dfrac{S}{D_n} \exp\left(-\dfrac{W^2}{W_o^2}\right) [A(\alpha) - D_1(\alpha)]}{1 + \dfrac{S}{D_n} \exp\left(-\dfrac{W^2}{W_o^2}\right) B} - \frac{\exp(-\alpha W)}{1 + \alpha L_n} - D_2(\alpha), \tag{5}$$

where S is the velocity of recombination at the front surface, D_n is the diffusion coefficient of electrons, and $L_n = (\tau_n D_n)^{1/2}$ is the electron diffusion length. The values A, B, D_1 and D_2 in Eq. (5) are the integral functions of the absorption coefficient α, the width of the depletion layer W and the effective Debye length $W_o = (\varepsilon\varepsilon_o kT/q^2(N_a - N_d))^{1/2}$.

Eq. (5) may be essentially simplified. At the boundary between the depletion and neutral regions ($x = W$), the photogenerated electrons are entrained by strong electric field and, hence, one may put $\Delta n(W) = 0$. This means that the terms D_1 and D_2 in Eq. (5) can be neglected. When calculating the values $A(\alpha)$ and B, one can replace the integration by multiplication of the maximum value of the integrands by their "half-widths". The half-widths are determined by the value of x at the point where the value of the integrand is smaller than the peak value by a factor of $e = 2.71$.

After such simplification, instead of Eq. (5) one can write (Kosyachenko et al., 1999):

$$\eta_{int} = \frac{1 + \dfrac{S}{D_n}\left(\alpha + \dfrac{2}{W}\dfrac{\kappa_o - qV}{kT}\right)^{-1}}{1 + \dfrac{S}{D_n}\left(\dfrac{2}{W}\dfrac{\varphi_o - qV}{kT}\right)^{-1}} - \frac{\exp(-\alpha W)}{1 + \alpha L_n}. \tag{6}$$

Comparison of the dependences $\eta_{int}(\lambda)$ calculated in a wide range of the parameters and the absorption coefficient α shows that equation (6) approximates the exact equation (5) very well (Kosyachenko, 2006).

It should be emphasized that Eqs. (5) and (6) do not take into consideration the recombination at the back surface of the CdTe layer (when deriving Eq. (5) the condition $\Delta n = 0$ at $x \to \infty$ was used) which can result in significant losses in the case of a thin CdTe layer with large diffusion length of the minority carriers. However, we can use the Eq. (5) to find the expression for the drift component of the photoelectric quantum yield. This can be done as follows.

In the absence of recombination at the front surface, equation (6) transforms into the known Gartner formula (Gartner, 1959)

$$\eta_{int} = 1 - \frac{\exp(-\alpha W)}{1 + \alpha L_n} \tag{7}$$

which ignores recombination at the back surface of the CdTe layer. In this case, the photoelectric quantum yield caused by processes in the space-charge region is equal to the absorptivity of this layer, that is, $1 - \exp(-\alpha W)$. Thus, subtracting the term $1 - \exp(-\alpha W)$ from the right side of Eq. (7), we obtain the expression for the diffusion component of the photoelectric quantum yield

$$\eta_{\text{diff}}^{\text{o}} = \exp(-\alpha W)\frac{\alpha L_n}{1 + \alpha L_n},\tag{8}$$

which, of course, ignores recombination at the back surface of the CdTe layer. Subtracting $\exp(-\alpha W)\alpha L_n/(1 + \alpha L_n)$ from the right side of Eq. (6) we come to the expression for the *drift* component of the photoelectric quantum yield taking into account surface recombination at the CdS-CdTe interface:

$$\eta_{\text{drift}} = \frac{1 + \dfrac{S}{D_n}\left(\alpha + \dfrac{2}{W}\dfrac{\varphi_o - qV}{kT}\right)^{-1}}{1 + \dfrac{S}{D_n}\left(\dfrac{2}{W}\dfrac{\varphi_o - qV}{kT}\right)^{-1}} - \exp(-\alpha W).\tag{9}$$

For the *diffusion* component of the photoelectric quantum yield that takes into account surface recombination at the back surface of the CdTe layer, we can use the exact expression obtained for the p-layer in a p-n junction solar cell (Sze, 1981)

$$\eta_{\text{dif}} = \frac{\alpha L_n}{\alpha^2 L_n^2 - 1}\exp(-\alpha W)\times$$

$$\times\left\{\alpha L_n - \frac{\dfrac{S_b L_n}{D_n}\left[\cosh\left(\dfrac{d - W}{L_n}\right) - \exp(-\alpha(d - W))\right] + \sinh\left(\dfrac{d - W}{L_n}\right) + \alpha L_n \exp(-\alpha(d - W))}{\dfrac{S_b L_n}{D_n}\sinh\left(\dfrac{d - W}{L_n}\right) + \cosh\left(\dfrac{d - W}{L_n}\right)}\right\}\tag{10}$$

where d is the thickness of the CdTe absorber layer, S_b is the recombination velocity at the back surface of the CdTe layer.

The *total* quantum yield of photoelectric conversion in the CdTe absorber layer is the sum of the two components:

$$\eta_{\text{int}} = \eta_{\text{drift}} + \eta_{\text{dif}}.\tag{11}$$

2.4 Comparison of calculation results with experiment

Fig. 3(a) shows the computed spectra of the external quantum efficiency $\eta_{\text{ext}}(\lambda)$ illustrating the effect of the uncompensated donor impurities in a CdS/CdTe heterojunction. In this calculation the absorption curve $\alpha(\lambda)$ was used from publication Toshifumi et al., 1993, the values of S and τ_n were taken as 10^7 cm/s and 10^{-10} s, respectively.

It can be seen from Fig. 3(a) that, as the uncompensated donor concentration varies, the shape of the $\eta_{\text{ext}}(\lambda)$ curves undergo significant changes. If the $N_a - N_d$ decreases from 10^{17} cm^{-3} to 10^{13} cm^{-3}, the external quantum efficiency increases first and then decreases. The increase in $\eta_{\text{ext}}(\lambda)$ is due to the expansion of the depletion layer and, hence, to more efficient collection of photogenerated carriers from the bulk of the film. However, if the depletion

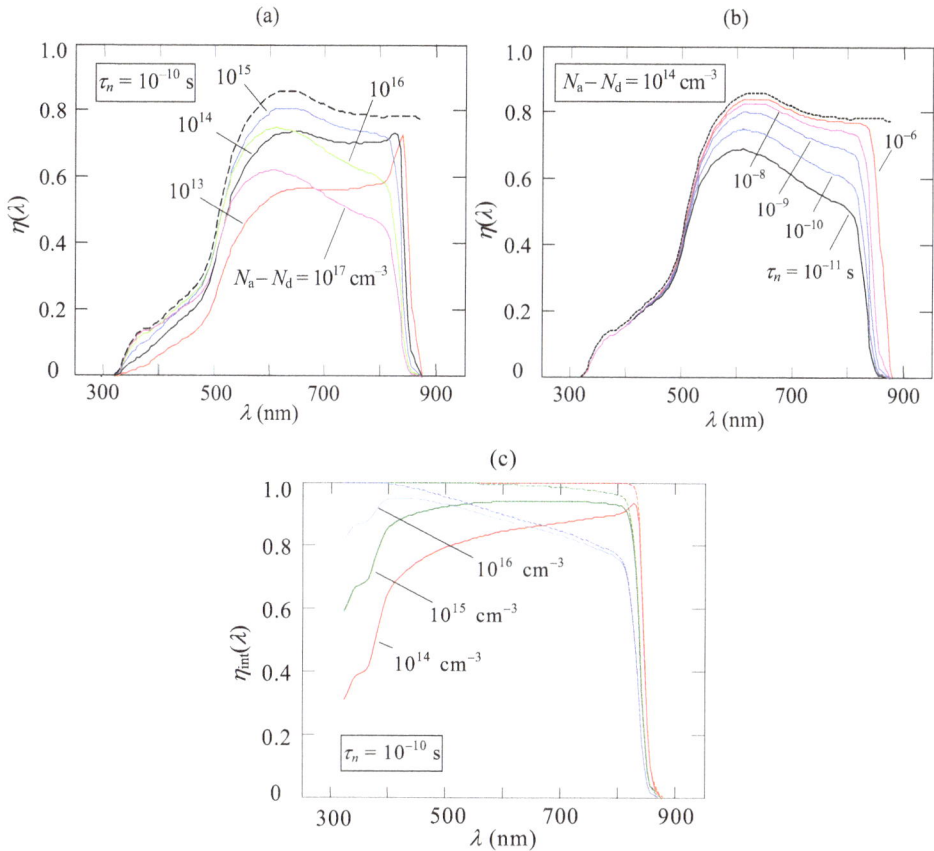

Fig. 3. (a) The external quantum efficiency spectra η_{ext} calculated using equations. (4),(9)-(11) at $\tau_n = 10^{-10}$ s and different uncompensated acceptor concentrations $N_a - N_d$, (b) at $N_a - N_d = 10^{16}$ cm^{-3} and different electron lifetimes τ_n and (c) the internal quantum efficiency spectra η_{int} calculated at different uncompensated acceptor concentrations and surface recombination velocity $S = 7 \times 10^7$ cm/s (solid lines) and $S = 0$ (dashed lines).

layer widens, the electric field becomes weaker which is favorable for the surface recombination. This effect is clearly demonstrated by the graphs for which $N_a - N_d$ is 10^{14} cm^{-3} and 10^{13} cm^{-3}: the surface recombination causes a significant decay with decreasing the wavelength. Evidently, the decay in photosensitivity in the $\lambda < 400$ nm region is also caused by absorption by the CdS layer and the conducting glass substrate. Absorption in the CdS layer masks the influence of surface recombination, however, that can be revealed in the spectra of "internal" quantum efficiency η_{int} shown in Fig. 3(c). As seen, the recombination losses are significant for rather low concentration of uncompensated acceptors 10^{14}-10^{15} cm^{-3}. As can be seen from Fig. 3(b), the variation in the carrier lifetime τ_n has practically no influence on the spectral curves of the device ($N_a - N_d = 10^{16}$ cm^{-3}) in the wavelength range $\lambda < 500$ nm. This is because in this spectral range, the depth of penetration of photons α^{-1} ($\alpha > 10^5$ cm^{-1}) is equal to or even smaller than the width of the space-charge region W.

Thus, only a small portion of the incident radiation is absorbed outside the space charge region and therefore the dependence of diffusion component of photocurrent on the electron lifetime is negligible.

On the other hand, in the wavelength range $\lambda > 500$ nm a considerable portion of the radiation is absorbed outside the space charge region and consequently as the electron lifetime increases the photoresponse also increases. When $N_a - N_d$ increases the effect of the electron lifetime increases and when $N_a - N_d$ decreases the effect of the electron lifetime becomes weaker so that at $N_a - N_d \leq 10^{14}$ cm^{-3} the photosensitivity is practically independent of the electron lifetime except the long-wavelength edge of the spectrum. It follows from Fig. 3(a) and (b) that, by varying the values of $N_a - N_d$ and τ, one can obtain the photosensitivity spectra of various shapes including those similar to the experimental curves shown in Fig. 1(a).

Fig. 4 illustrates the comparison of the calculated curves $\eta_{ext}(\lambda)$ using Eqs. (4),(9)-(11) with the measured spectrum taken at 300 K (we can not do it for other temperatures due to the lack of α values). The figure shows a quite good fit of the experimental data with the calculated values. Note that only the two adjustable parameters, the uncompensated acceptor concentration $N_a - N_d$ and electron lifetime τ_n, have been used to fit the calculation results with the experimental data and which were found to be 7×10^{16} cm^{-3} and 8×10^{-11} s, respectively. As can be seen from Fig. 3(c), the surface recombination velocity is not relevant at such high uncompensated acceptor concentration owing to the effect of a high electric field.

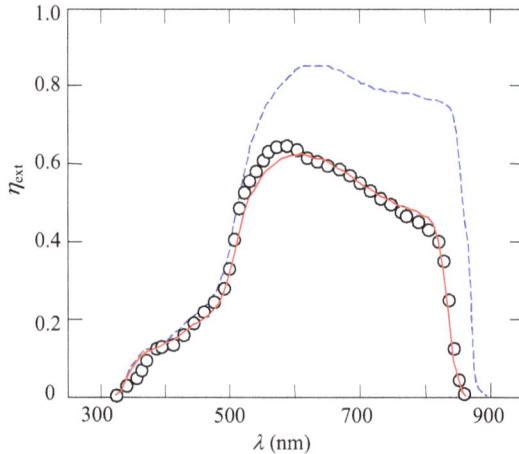

Fig. 4. Comparison of the measured (circles) and calculated (solid line) quantum efficiency spectrum η_{ext} using Eqs. (4), (9)-(11) at $N_a - N_d = 7 \times 10^{16}$ cm^{-3}, $\tau_n = 8 \times 10^{-11}$ s. The dashed line shows the spectrum of 100 % internal efficiency ($N_a - N_d = 10^{16}$ cm^{-3}, $\tau_n = 10^{-6}$ s, $d = 100$ μm).

3. Short-circuit current in CdS/CdTe heterostructure

The obtained expressions for quantum efficiency spectra can be used to calculate the short-circuit current density J_{sc} which is a quantitative solar cell characteristic reflecting the charge collection efficiency under radiation. The calculations will be done for AM1.5 solar radiation using Tables ISO 9845-1:1992 (Standard ISO, 1992). If Φ_i is the spectral radiation power density (in mW cm^{-2} nm^{-1}) and $h\nu$ is the photon energy (in eV), the spectral density of the incident photon flux is $\Phi_i / h\nu_i$ (in s^{-1}cm^{-2}), and then

$$J_{sc} = q \sum_i \eta_{int}(\lambda) \frac{\Phi_i(\lambda)}{h\nu_i} \Delta\lambda , \qquad (12)$$

where $\Delta\lambda_i$ is the wavelength range between the neighboring values of λ_i (the photon energy $h\nu_i$) in the table and the summation is over the spectral range $\lambda < \lambda_g = hc/E_g$.

3.1 The drift component of the short-circuit current

Let us first consider the drift component of the short-circuit current density J_{drift} using Eq. (12). Fig. 5 shows the calculation results for J_{drift} depending on the space-charge region width W. In the calculations, it was accepted $\varphi_o - qV = 1$ eV, $S = 10^7$ cm/s (the maximum possible velocity of surface recombination) and $S = 0$. The Eq. (9) was used for $\eta_{int}(\lambda)$.

Important practical conclusions can be made from the results presented in the figure.

If $S = 0$, the short-circuit current gradually increases with widening of W and approaches a maximum value of $J_{drift} = 28.7$ mA/cm^2 at $W > 10$ µm (the value $J_{drift} = 28.7$ mA/cm^2 is obtained from equation (12) at $\eta_{drift} = 1$).

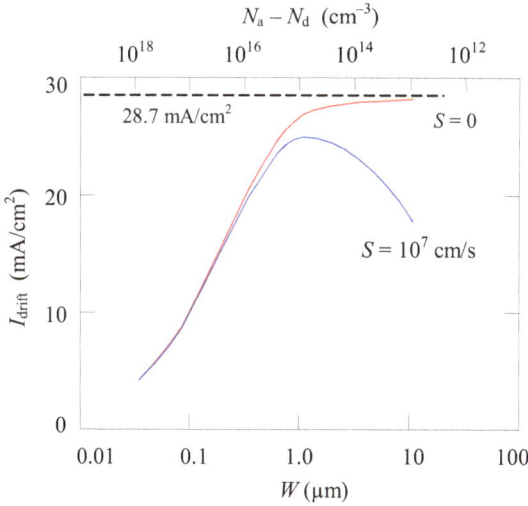

Fig. 5. Drift component of the short-circuit current density J_{drift} of a CdTe-based solar cell as a function of the space-charge region width W (the uncompensated acceptor concentration $N_a - N_d$) calculated for the surface recombination velocities $S = 10^7$ cm/s and $S = 0$.

Such result should be expected because the absorption coefficient α in CdTe steeply increases in a narrow range $h\nu \approx E_g$ and becomes higher than 10^4 cm^{-1} at $h\nu > E_g$. As a result, the penetration depth of photons α^{-1} is less than ~ 1 µm throughout the entire spectral range and in the absence of surface recombination, all photogenerated electron-hole pairs are separated by the electric field acting in the space-charge region.

Surface recombination decreases the short-circuit current only in the case if the electric field in the space-charge region is not strong enough. The electric field decreases as the space-charge region widens, i.e. when the uncompensated acceptor concentration $N_a - N_d$ decreases. One can see from Fig. 5 that the influence of surface recombination at $N_a - N_d = 10^{14}$-10^{15} cm^{-3} is quite significant. However, as $N_a - N_d$ increases and consequently the electric field strength becomes stronger, the influence of surface recombination becomes weaker, and at $N_a - N_d \geq 10^{16}$ cm^{-3} the effect is virtually eliminated. However in this case, the short-circuit current density decreases with increasing $N_a - N_d$ because a significant portion of radiation is absorbed outside the space-charge region.

It should be noted that the fabrication of the CdTe/CdS heterostructure is typically completed by a post-deposition heat treatment. The annealing enables grain growth, reduces defect density in the films, and promotes the interdiffusion between the CdTe and CdS layers. As a result, the CdS-CdTe interface becomes alloyed into the $CdTe_xS_{1-x}$-CdS_yTe_{1-y} interface, and the surface recombination velocity is probably reduced to some extent (Compaan et al, 1999).

3.2 The diffusion component of the short-circuit current

In order to provide the losses caused by recombination at the CdS-CdTe interface and in the space-charge region at a minimum we will accept in this section $N_a - N_d \geq 10^{17}$ cm^{-3}. On the other hand, to make the diffusion component of the short-circuit current J_{dif} as large as possible, we will set $\tau_n = 3\times10^{-6}$ s, i.e. the maximum possible value of the electron lifetime in CdTe. Fig. 6(a) shows the calculation results of J_{dif} (using Eqs. (10) and (12)) versus the CdTe layer thickness d for the recombination velocity at the back surface $S = 10^7$ cm/s and $S = 0$ (the thickness of the neutral part of the film is $d - W$).

One can see from Fig. 6(a) that for a thin CdTe layer (few microns) the diffusion component of the short-circuit current is rather small. In the case $S_b = 0$, the total charge collection in the neutral part (it corresponds to $J_{dif} = 17.8$ mA/cm^2 at $\eta_{dif} = 1$) is observed at $d = 15\text{-}20$ μm. To reach the total charge collection in the case $S_b = 10^7$ cm/s, the CdTe thickness should be 50 μm or larger. Bearing in mind that the thickness of a CdTe layer is typically between 2 and 10 μm, for $d = 10$, 5 and 2 μm the losses of the diffusion component of the short-circuit current are 5, 9 and 19%, respectively. The CdTe layer thickness can be reduced by shortening the electron lifetime τ_n and hence the electron diffusion length $L_n = (\tau_n D_n)^{1/2}$. However one does not forget that it leads to a significant decrease in the value of the diffusion current itself. This is illustrated in Fig. 6(b), where the curve $J_{dif}(\tau_n)$ is plotted for a thick CdTe layer (50 μm) taking into account the surface recombination velocity $S_b = 10^7$ cm/s. As it can be seen, shortening of the electron lifetime below 10^{-7}-10^{-6} s results in a significant lowering of the diffusion component of the short-circuit current density. Thus, when the space-charge region width is narrow, so that recombination losses at the CdS-CdTe interface can be neglected (as seen from Fig. 5, at $N_a - N_d > 10^{16}$-10^{17} cm^{-3}), the conditions for generation of the high diffusion component of the short-circuit current are $d > 25\text{-}30$ μm and $\tau_n > 10^{-7}$-10^{-6} s.

In connection with the foregoing the question arises why for total charge collection the thickness of the CdTe absorber layer d should amount to several tens of micrometers. The value d is commonly considered to be in excess of the effective penetration depth of the radiation into the CdTe absorber layer in the intrinsic absorption region of the semiconductor. As mentioned above, as soon as the photon energy exceeds the band gap of CdTe, the absorption coefficient α becomes higher than 10^4 cm^{-1}, i.e. the effective penetration depth of radiation α^{-1} becomes less than 10^{-4} cm = 1 μm. With this reasoning, the absorber layer thickness is usually chosen at a few microns. However, all that one does not take into the account, is that the carriers arisen outside the space-charge region, diffuse into the neutral part of the CdTe layer penetrating deeper into the material. Carriers reached the back surface of the layer, recombine and do not contribute to the photocurrent. Losses

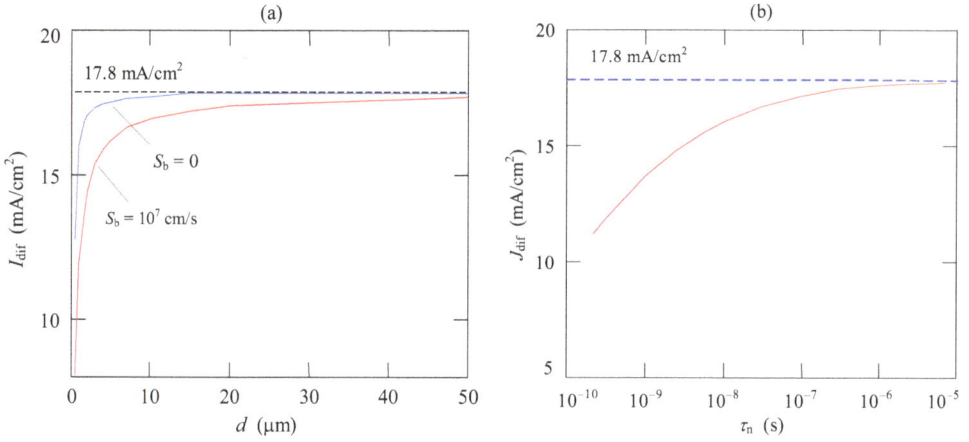

Fig. 6. Diffusion component of the short-circuit current density J_{dif} as a function of the CdTe layer thickness d calculated at the uncompensated acceptor concentration $N_a - N_d = 10^{17}$ cm^{-3}, the electron lifetime $\tau_n = 3\times10^{-6}$ s and surface recombination velocity $S_b = 10^7$ cm/s and $S_b = 0$ (a) and the dependence of the diffusion current density J_{dif} on the electron lifetime for the CdTe layer thickness $d = 50$ μm and recombination velocity at the back surface $S_b = 10^7$ cm/s (b).

caused by the insufficient thickness of the CdTe layer should be considered taking into account this process.

Consider first the spatial distribution of excess electrons in the neutral region governed by the continuity equation with two boundary conditions. At the depletion layer edge, the excess electron density Δn can be assumed equal zero (due to electric field in the depletion region), i.e.

$$\Delta n = 0 \text{ at } x = W. \tag{13}$$

At the back surface of the CdTe layer we have surface recombination with a velocity S_b:

$$S_b \Delta n = -D_n \frac{d\Delta n}{dx} \text{ at } x = d, \tag{14}$$

where d is the thickness of the CdTe layer.

Using these boundary conditions, the exact solution of the continuity equation is (Sze, 1981):

$$\Delta n = T(\lambda)N_o(\lambda)\frac{\alpha\tau_n}{\alpha^2 L_n^2 - 1}\exp[-\alpha W]\left\{\cosh\left(\frac{x-W}{L_n}\right) - \exp[-\alpha(x-W)] - \right.$$

$$\frac{\frac{S_b L_n}{D_n}\left[\cosh\left(\frac{d-W}{L_n}\right) - \exp[-\alpha(d-W)]\right] + \sinh\left(\frac{d-W}{L_n}\right) + \alpha L_n\exp[-\alpha(d-W)]}{\frac{S_b L_n}{D_n}\sinh\left(\frac{x-W}{L_n}\right) + \cosh\left(\frac{d-W}{L_n}\right)}\left.\times\sinh\left(\frac{x-W}{L_n}\right)\right\} \tag{15}$$

where $T(\lambda)$ is the optical transmittance of the glass/TCO/CdS, which takes into account reflection from the front surface and absorption in the TCO and CdS layers, N_o is the number of incident photons per unit time, area, and bandwidth (cm^{-2}s^{-1}nm^{-1}), $L_n = (\tau_n D_n)^{1/2}$ is the electron diffusion length, τ_n is the electron lifetime, and D_n is the electron diffusion coefficient related to the electron mobility μ_n through the Einstein relation: $qD_n/kT = \mu_n$.

Fig. 7 shows the electron distribution calculated by Eq. (15) for different CdTe layer thicknesses. The calculations have been carried out at $\alpha = 10^4$ cm^{-1}, $S_b = 7\times10^7$ cm/s, $\mu_n = 500$ cm^2/(V·s) and typical values $\tau_n = 10^{-9}$ s and $N_a - N_d = 10^{16}$ cm^{-3} (Sites & Xiaoxiang, 1996). As it is seen from Fig. 7, even for the CdTe layer thickness of 10 μm, recombination at back surface leads to a remarkable decrease in the electron concentration. If the layer thickness is reduced, the effect significantly enhances, so that at $d = 1$-2 μm, surface recombination "kills" most of the photo-generated electrons. Thus, the photo-generated electrons at 10^{-9} s are involved in recombination far away from the effective penetration depth of radiation (~ 1 μm). Evidently, the influence of this process enhances as the electron lifetime increases, because the non-equilibrium electrons penetrate deeper into the CdTe layer due to increase of the diffusion length. Calculation using Eq. (15) shows that if the layer thickness is large (~ 50 μm), the non-equilibrium electron concentration reduces 2 times from its maximum value at a distance about 8 μm at $\tau_n = 10^{-8}$ s, 20 μm at $\tau_n = 10^{-7}$ s, 32 μm at $\tau_n = 10^{-6}$ s.

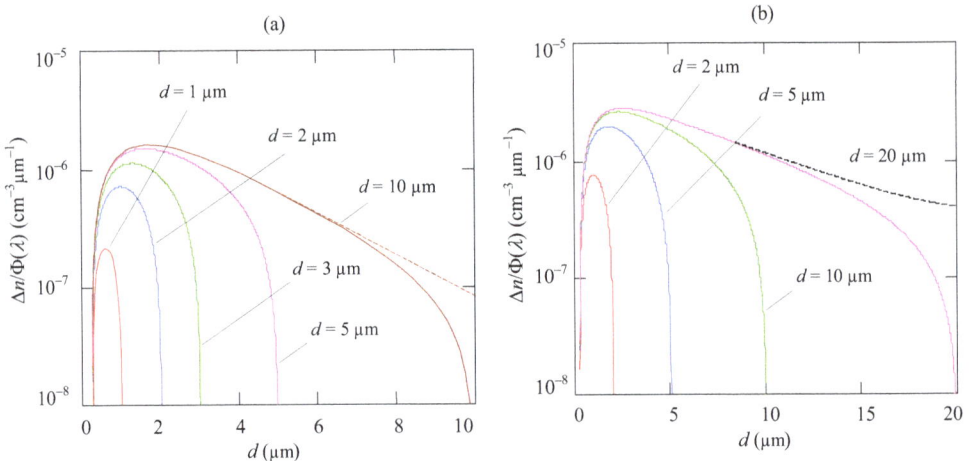

Fig. 7. Electron distribution in the CdTe layer at different its thickness d calculated at the electron lifetime $\tau_n = 10^{-9}$ s (a) and $\tau_n = 10^{-8}$ s (b). The dashed lines show the electron distribution for $d = 10$ and 20 μm if recombination at the back surface is not taken into account.

3.3 The density of total short-circuit current

It follows from the above that the processes of the photocurrent formation within the space-charge region and in the neutral part of the CdTe film are interrelated. Fig. 8 shows the *total* short-circuit current J_{sc} (the sum of the drift and diffusion components) calculated for different parameters of the CdTe layer, i.e. the uncompensated acceptor concentration, minority carrier lifetime and layer thickness. As the space-charge region is narrow (i.e., $N_a - N_d$ is high), a considerable portion of radiation is absorbed *outside* the space-charge region. One can see that when the film thickness and electron diffusion length are large enough (the top curve in Fig. 8(a) for $d = 100$ μm, $\tau_n > 10^{-6}$ s), practically the total charge collection takes place and the density of short-circuit current J_{sc} reaches its maximum value of 28.7 mA/cm² (note, the record experimental value of J_{sc} is 26.7 mA/cm² (Holliday et al, 1998)). However if the space-charge region is too wide ($N_a - N_d < 10^{16}$-10^{17} cm⁻³) the electric field becomes weak and the short-circuit current is reduced due to recombination at the front surface.

For $d = 10$ μm, the shape of the curve J_{sc} versus $N_a - N_d$ is similar to that for $d = 100$ μm but the saturation of the photocurrent density is observed at a smaller value of J_{sc}. A significant lowering of J_{sc} occurs after further thinning of the CdTe film and, moreover, for $d = 5$ and 3 μm, the short-circuit current even decreases with increasing $N_a - N_d$ due to incomplete charge collection in the neutral part of the CdTe film.

It is interesting to examine quantitatively how the total short-circuit current varies when the electron lifetime is shorter than 10^{-6} s. This is an actual condition because the carrier lifetimes in thin-film CdTe diodes can be as short as 10^{-9}-10^{-10} s and even smaller (Sites & Pan, 2007).

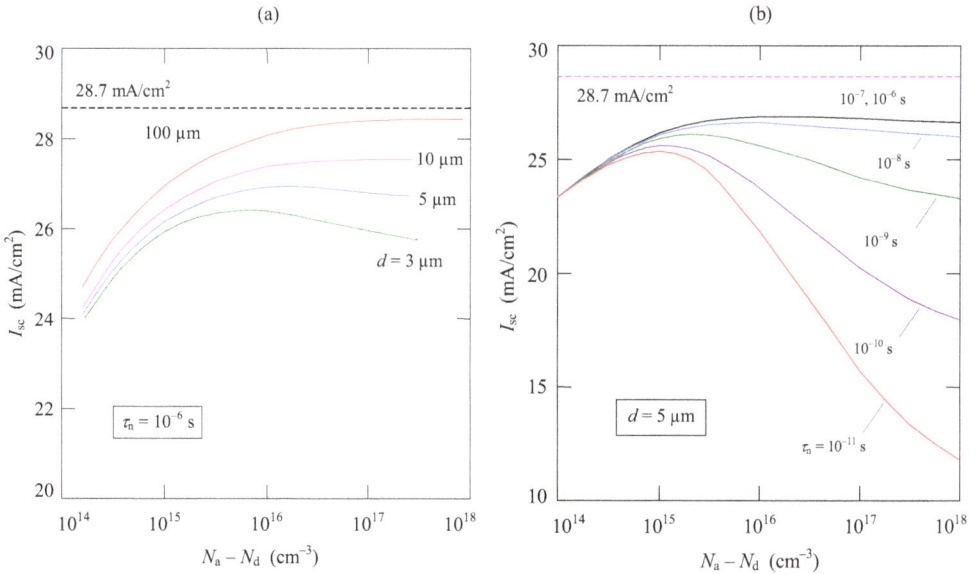

Fig. 8. Total short-circuit current density J_{sc} of a CdTe-based solar cell as a function of the uncompensated acceptor concentration $N_a - N_d$ calculated at the electron lifetime $\tau_n = 10^{-6}$ s for different CdTe layer thicknesses d (a) and at the thickness $d = 5$ μm for different τ_n (b).

Fig. 5(b) shows the calculation results of the total short-circuit current density J_{sc} versus the concentration of uncompensated acceptors $N_a - N_d$ for different electron lifetimes τ_n. Calculations have been carried out for the CdTe film thickness $d = 5$ μm which is often used in the fabrication of CdTe-based solar cells (Phillips et al., 1996; Bonnet, 2001; Demtsu & Sites, 2005; Sites & Pan, 2007). As it can be seen, at $\tau_n \geq 10^{-8}$ s the short-circuit current density is 26-27 mA/cm² when $N_a - N_d > 10^{16}$ cm⁻³. For shorter electron lifetime, J_{sc} peaks in the $N_a - N_d$ range $(1-3) \times 10^{15}$ cm⁻³. As $N_a - N_d$ is in excess of this concentration, the short-circuit current decreases since the drift component of the photocurrent reduces. In the range of the uncompensated acceptor concentration $N_a - N_d < (1-3) \times 10^{15}$ cm⁻³, the short-circuit current density also decreases, but because of recombination at the front surface of the CdTe layer. Anticipating things, it should be noted, that at $N_a - N_d < 10^{15}$ cm^{-3}, recombination in the space-charge region becomes also significant (see Fig. 9). Thus, in order to reach the short-circuit current density 25-26 mA/cm^2 when the electron lifetime τ_n is shorter than 10^{-8} s, the uncompensated acceptor concentration $N_a - N_d$ should be equal to (1-3)×10^{15} cm^{-3} (rather than $N_a - N_d > 10^{16}$ cm^{-3} as in the case of $\tau_n \geq 10^{-8}$ s).

4. Recombination losses in the space-charge region

In analyzing the photoelectric processes in the CdS/CdTe solar cell we ignored the recombination losses (capture of carriers) in the space-charge region. This assumption is based on the following considerations.

The mean distances that electron and hole travels during their lifetimes along the electric field without recombination or capture by the centers within the semiconductor band gap, i.e. the electron drift length λ_n and hole drift length λ_p, are determined by expressions

$$\lambda_n = \mu_n E \tau_{no}, \tag{16}$$

$$\lambda_p = \mu_p E \tau_{po}, \tag{17}$$

where E is the electric-field strength, μ_n and μ_p are the electron and hole mobilities, respectively.

In the case of uniform field ($E = $ const), the charge collection efficiency is expressed by the well-known Hecht equation (Eizen, 1992; Baldazzi et al., 1993):

$$\eta_c = \frac{\lambda_n}{W}\left[1 - \exp\left(-\frac{W-x}{\lambda_n}\right)\right] + \frac{\lambda_p}{W}\left[1 - \exp\left(-\frac{x}{\lambda_p}\right)\right]. \tag{18}$$

In a diode structure, the problem is complicated due to nonuniformity of the electric field in the space-charge region. However, due to the fact that the electric field strength decreases linearly from the surface to the bulk of the semiconductor, the field nonuniformity can be reduced to the substitution of E in Eqs. (16) and (17) by its average values $E_{(0,x)}$ and $E_{(x,W)}$ in the portion $(0, x)$ for electrons and in the portion (x, W) for holes, respectively:

$$E_{(x,W)} = \frac{(\varphi_o - eV)}{eW}\left(1 - \frac{x}{W}\right), \tag{19}$$

$$E_{(0,x)} = \frac{(\varphi_o - eV)}{eW}\left(2 - \frac{x}{W}\right). \tag{20}$$

Thus, with account made for this, the Hecht equation for the space-charge region of CdS/CdTe heterostructure takes the form

$$\eta_c = \frac{\mu_p E_{(x,W)}\tau_{po}}{W}\left[1 - \exp\left(-\frac{W-x}{\mu_p E_{(x,W)}\tau_{po}}\right)\right] + \frac{\mu_n E_{(0,x)}\tau_{no}}{W}\left[1 - \exp\left(-\frac{x}{\mu_n E_{(0,x)}\tau_{no}}\right)\right]. \tag{21}$$

Fig. 9(a) shows the curves of charge-collection efficiency $\eta_c(x)$ computed by Eq. (21) for the concentration of uncompensated acceptors 3×10^{16} cm^{-3} and different carrier lifetimes $\tau = \tau_{no} = \tau_{po}$. It is seen that for the lifetime 10^{-11} s the effect of losses in the space-charge region is remarkable but for $\tau \geq 10^{-10}$ s it is insignificant (μ_n and μ_n were taken equal to 500 and 60 cm^2/(V·s), respectively). For larger carrier lifetimes the recombination losses can be neglected at lower values $N_a - N_d$.

Thus, the recombination losses in the space charge-region depend on the concentration of uncompensated acceptors $N_a - N_d$ and carrier lifetime τ in a complicated manner. It is also seen from Fig. 9(a) that the charge collection efficiency η_c is lowest at the interface CdS-CdTe ($x = 0$). An explanation of this lies in the fact that the product $\tau_{no}\mu_n$ for electrons in CdTe is order of magnitude greater than that for holes. With account made for this, Fig. 9(b) shows the dependences of charge-collection efficiency on $N_a - N_d$ calculated at different carrier lifetimes for the "weakest" place of the space-charge region concerning charge collection of photogenerated carriers, i.e. at the cross section $x = 0$. From the results presented in Fig. 9(b), it follows that at the carrier lifetime $\tau \geq 10^{-8}$ s the recombination losses can be neglected at the uncompensated acceptor concentration $N_a - N_d \geq 10^{14}$ cm^{-3} while at $\tau = 10^{-10}$-10^{-11} s it is possible if $N_a - N_d$ is in excess of 10^{16} cm^{-3}.

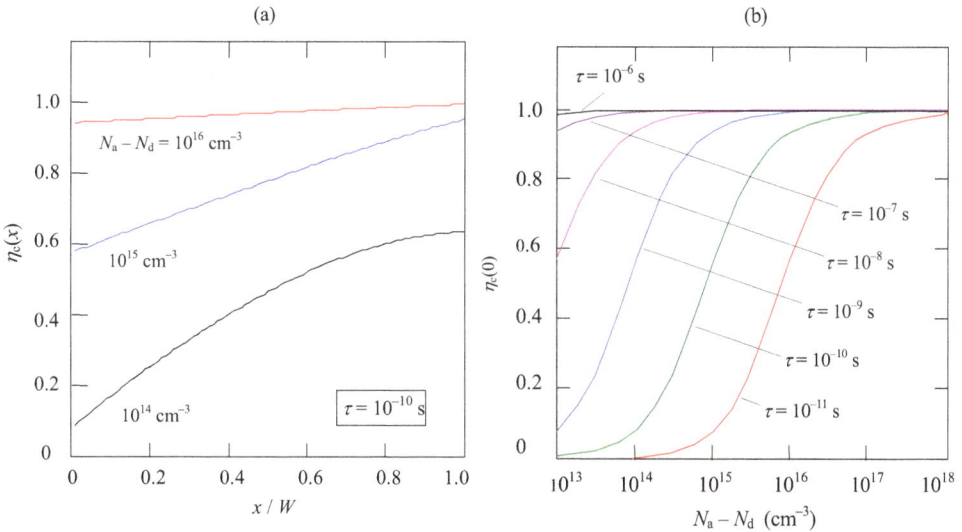

Fig. 9. (a) The coordinate dependences of the charge-collection efficiency $\eta_c(x)$ calculated for the uncompensated acceptor concentrations $N_a - N_d = 3\times10^{16}$ cm^{-3} and different carrier lifetimes τ. (b) The charge-collection efficiency η_c at the interface CdS-CdTe ($x = 0$) as a function of the uncompensated acceptor concentration $N_a - N_d$ calculated for different carrier lifetimes τ.

5. Open-circuit voltage, fill factor and efficiency of thin-film CdS/CdTe solar cell

In this section, we investigate the dependences of the open-circuit voltage, fill factor and efficiency of a CdS/CdTe solar cell on the resistivity of the CdTe absorber layer and carrier lifetime with the aim to optimize these parameters and hence to improve the solar cell efficiency. The open-circuit voltage and fill factor are controlled by the magnitude of the forward current. Therefore the *I-V* characteristic of the device is analyzed which is known to originate primarily by recombination in the space charge region of the CdTe absorber layer. The *I-V* characteristic of CdS/CdTe solar cells is most commonly described by the semi-empirical formulae which consists the so-called "ideality" factor and is valid for some cases. Contrary to usual practice, in our calculations of the current in a device, we use the recombi-nation-generation Sah-Noyce-Shockley theory developed for p-n junction (Sah et al., 1957) and adopted to CdS/CdTe heterostructure (Kosyachenko et al., 2005) and supplemented with over-barrier diffusion flow of electrons at higher voltages. This theory takes into account the evolution of the *I-V* characteristic of CdS/CdTe solar cell when the parameters of the CdTe absorber layer vary and, therefore, reflects adequately the real processes in the device.

5.1 *I-V* characteristic of CdS/CdTe heterostructure

The open-circuit voltage, fill factor and efficiency of a solar cell is determined from the *I-V* characteristic under illumination which can be presented as

$$J(V) = J_d(V) - J_{ph},$$ (22)

where $J_d(V)$ is the dark current density and J_{ph} is the photocurrent density.
The dark current density in the so-called "ideal" solar cell is described by the Shockley equation

$$J_d(V) = J_s \left[\exp\left(\frac{qV}{kT}\right) - 1 \right],$$ (23)

where J_s is the saturation current density which is the voltage independent reverse current as qV is higher than few kT.
An actual *I-V* characteristic of CdS/CdTe solar cells differs from Eq. (23). In many cases, a forward current can be described by formula similar to Eq. (23) by introducing an exponent index qV/AkT, where A is the "ideality" factor lied in the range 1 to 2. Sometimes, a close correlation between theory and experiment can be attained by adding the recombination component $I_o[\exp(qV/2kT) - 1]$ to the dark current in Eq. (23) (I_o is a new coefficient). Our measurements show, however, that such generalizations of Eq. (23) does not cover the observed variety of *I-V* characteristics of the CdS/CdTe solar cells. The measured voltage dependences of the forward current are not always exponential and the saturation of the reverse current is never observed. On the other hand, our measurements of *I-V* characteristics of CdS/CdTe heterostructures and their evolution with the temperature variation are governed by the generation-recombination Sah-Noyce-Shockley theory (Sah al., 1957). According to this theory, the dependence $I \sim \exp(qV/AkT)$ at $n \approx 2$ takes place only in the case where the generation-recombination level is placed near the middle of the band gap. If the level moves away from the midgap the coefficient A becomes close to 1 but only at low forward voltage. If the voltage elevates the *I-V* characteristic modified in the dependence where $n \approx 2$ and at higher voltages the dependence I on V becomes even weaker (Sah et al., 1957; Kosyachenko et al., 2003). At higher forward currents, it is also necessary to take into account the voltage drop on the series resistance R_s of the bulk part of the CdTe layer by replacing the voltage V in the discussed expressions with $V - I \cdot R_s$.

The Sah-Noyce-Shockley theory supposes that the generation-recombination rate in the section x of the space-charge region is determined by expression (Sah et al., 1957)

$$U(x,V) = \frac{n(x,V)p(x,V) - n_i^2}{\tau_{po}\left[n(x,V) + n_1\right] + \tau_{no}\left[p(x,V) + p_1\right]},$$ (24)

where $n(x,V)$ and $p(x,V)$ are the carrier concentrations in the conduction and valence bands, n_i is the intrinsic carrier concentration. The values n_1 and p_1 are determined by the energy spacing between the top of the valence band and the generation-recombination level E_t, i.e. $p_1 = N_v\exp(- E_t/kT)$ and $n_1 = N_c\exp[- (E_g - E_t)/kT]$, where $N_c = 2(m_nkT/2\pi\hbar^2)^{3/2}$ and $N_v = 2(m_pkT/2\pi\hbar^2)^{3/2}$ are the effective density of states in the conduction and valence bands, m_n and m_p are the effective masses of electrons and holes, τ_{no} and τ_{po} are the effective lifetime of electrons and holes in the depletion region, respectively.

The recombination current under forward bias and the generation current under reverse bias are found by integration of $U(x, V)$ throughout the entire depletion layer:

$$J_{gr} = q \int_0^W U(x,V)dx,$$ (25)

where the expressions for the electron and hole concentrations have the forms (Kosyachenko et al., 2003):

$$p(x,V) = N_c \exp\left[-\frac{\Delta\mu + \varphi(x,V)}{kT}\right],$$ (26)

$$n(x,V) = N_v\exp\left[-\frac{E_g - \Delta\mu - \varphi(x,V) - qV}{kT}\right].$$ (27)

Here $\Delta\mu$ is the energy spacing between the Fermi level and the top of the valence band in the bulk of the CdTe layer, $\varphi(x,V)$ is the potential energy of hole in the space-charge region. Over-barrier (diffusion) carrier flow in the CdS/CdTe heterostructure is restricted by high barriers for both majority carriers (holes) and minority carriers (electrons) (Fig. 2). For transferring holes from CdTe to CdS, the barrier height in equilibrium ($V = 0$) is somewhat lower than $E_{g\,CdS} - (\Delta\mu + \Delta\mu_{CdS})$, where $E_{g\,CdS} = 2.42$ eV is the band gap of CdS and $\Delta\mu_{CdS}$ is the energy spacing between the Fermi level and the bottom of the conduction band of CdS, $\Delta\mu$ is the Fermi level energy in the bulk of CdTe equal to $kT\ln(N_v/p)$, p is the hole concentration which depends on the resistivity of the material. An energy barrier impeding electron transfer from CdS to CdTe is also high but is equal to $E_{g\,CdTe} - (\Delta\mu + \Delta\mu_{CdS})$ at $V = 0$. Owing to high barriers for electrons and holes, under low and moderate forward voltages the dominant charge transport mechanism is recombination in the space-charge region. However, as qV nears φ_o, the over-barrier currents become comparable and even higher than the recombination current due to much stronger dependence on V. Since in CdS/CdTe junction the barrier for holes is considerably higher than that for electrons, the electron component dominates the over-barrier current. Obviously, the electron flow current is analogous to that occurring in a p-n junction and one can write for the over-barrier current density (Sze, 1981):

$$J_n = q \frac{n_p L_n}{\tau_n} \left[\exp\left(\frac{qV}{kT} \right) - 1 \right], \tag{28}$$

where $n_p = N_c \exp[- (E_g - \Delta\mu)/kT]$ is the concentration of electrons in the p-CdTe layer, τ_n and $L_n = (\tau_n D_n)^{1/2}$ are the electron lifetime and diffusion length, respectively (D_n is the diffusion coefficient of electrons).

Thus, according to the above discussion, the dark current density in CdS/CdTe heterostructure $J_d(V)$ is the sum of the generation-recombination and diffusion components:

$$J_d(V) = J_{gr}(V) + J_n(V) . \tag{29}$$

5.2 Comparison with the experimental data

The current-voltage characteristics of CdS/CdTe solar cells depend first of all on the resistivity of the CdTe absorber layer due to the voltage drop across the series resistance of the bulk part of the CdTe film R_s (Fig. 10(a)). The value of R_s can be found from the voltage dependence of the differential resistance R_{dif} of a diode structure under forward bias. Fig. 10 shows the results of measurements taken for two "extreme" cases: the samples No 1 and 2 are examples of the CdS/CdTe solar cells with low resistivity (20 $\Omega \cdot$cm) and high resistivity of the CdTe film (4×10^7 $\Omega \cdot$cm), respectively. One can see that, in the region of low voltage, the R_{dif} values decrease with V by a few orders of magnitude. However, at $V > 0.5$-0.6 V for sample No 1 and $V > 0.8$-0.9 V for sample No 2, R_{dif} reaches saturation values which are obviously the series resistances of the bulk region of the film R_s.

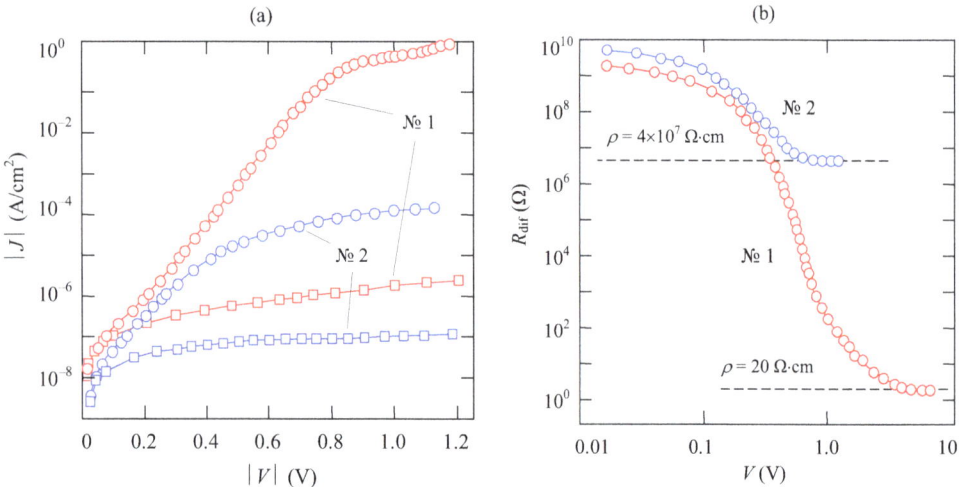

Fig. 10. *I-V* characteristics (a) and dependences of differential resistances R_{dif} on forward voltage (b) for two solar cells with different resistivities of CdTe layers: 20 and 4×10^7 $\Omega \cdot$cm (300 K).

Because the value of R_s for a sample No 1 is low, the presence of R_s does not affect the shape of the diode *I-V* characteristic. In contrast, the resistivity of the CdTe film for a sample No 2 is ~ 6 orders higher, therefore at moderate forward currents ($J > 10^{-6}$ A/cm²), the experimental points deviate from the exponential dependence which is strictly obeyed for sample No 1 over 6 orders of magnitude.

The experimental results presented in Fig. 11 reflect the common feature of the I-V characteristic of a thin-film CdS/CdTe heterostructure (sample No 1). The results obtained for this sample allow interpreting them without complications caused by the presence of the series resistance R_s. Nevertheless, in this case too, the forward I-V characteristic reveals some features which are especially pronounced. As one can see, under forward bias, there is an extended portion of the curve ($0.1 < V < 0.8$ V) where the dependence $I \sim \exp(qV/AkT)$ holds for $A = 1.92$. At higher voltages, the deviation from the exponential dependence toward lower currents is observed. It should be emphasized that this deviation is not caused by the voltage drop across the series resistance of the neutral part of the CdTe absorber layer R_s (which is too low in this case). If the voltage elevates still further (> 1 V), a much steeper increase of forward current is observed.

Analysis shows that all of varieties of the thin-film I-V characteristics are explained in the frame of mechanism involving the generation-recombination in the space-charge region in a wide range of moderate voltages completed by the over-barrier diffusion current at higher voltage.

The results of comparison between the measured I-V characteristic of the thin-film CdS/CdTe heterostructure (circles) and that calculated using Eqs. (25), (28) and (29) (lines) are shown in Fig. 11.

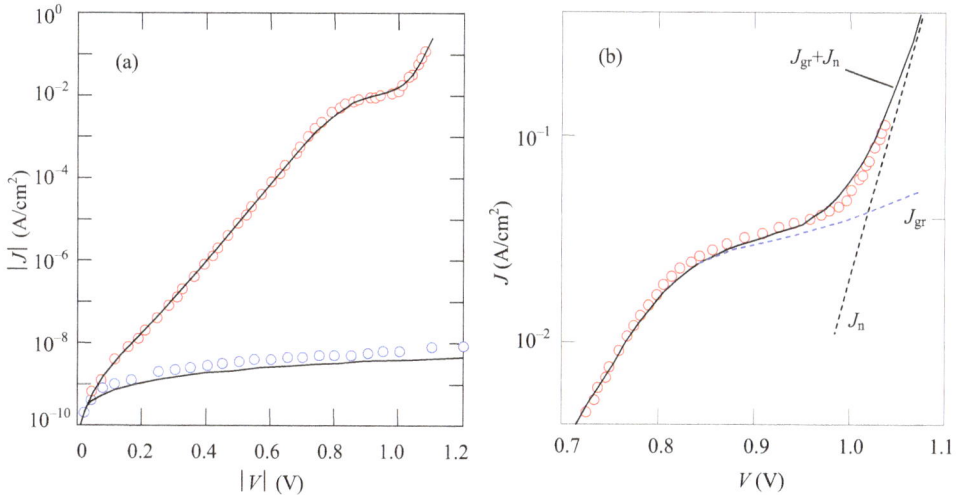

Fig. 11. (a) I-V characteristic of thin-film CdS/CdTe heterostructure. The circles and solid lines show the experimental and calculated results, respectively. (b) Comparison of the calculated and measured dependences in the range of high forward currents (J_{gr} and J_n are the recombination and diffusion components, respectively).

To agree the calculated results with experiment, the effective lifetimes of electrons and holes in the space-charge region were taken $\tau_{no} = \tau_{po} = \tau = 1.2 \times 10^{-10}$ s (τ determines the value of current but does not affect the shape of curve). The ionization energy E_t was accepted to be 0.73 eV as the most effective recombination center (the value E_t determines the rectifying coefficient of the diode structure), the barrier height φ_o and the uncompensated acceptor concentration $N_a - N_d$ were taken 1.13 eV and 10^{17} cm^{-3}, respectively. One can see that the I-V characteristic calculated in accordance with the above theory (lines) are in good agreement with experiment both for the forward and reverse connection (circles).

Attention is drawn to the fact that the effective carrier lifetime in the space charge region $\tau = (\tau_{n0}\tau_{p0})^{1/2}$ was taken equal to 1.5×10^{-8} s whereas the electron lifetime τ_n in the crystals is in the range of 10^{-7} s or longer (Acrorad Co, Ltd., 2009). Such a significant difference between τ and τ_n appears reasonable since τ_n is proportional to $1/N_t f$, where N_t is the concentration of recombination centers and f is the probability that a center is empty. Both of the values τ_{n0} and τ_{p0} in the Sah-Noyce-Shockley theory are proportional to $1/N_t$. At the same time, since the probability f in the bulk part of the diode structure can be much less than unity, the electron lifetime τ_n can be far in excess of the effective carrier lifetime τ in the space-charge region.

5.3 Dependences of open-circuit voltage, fill factor and efficiency on the parameters of thin-film CdS/CdTe solar cell

The open-circuit voltage V_{oc}, fill factor FF and efficiency η of a solar cell is determined from the I-V characteristic under illumination which can be presented as

$$J(V) = J_d(V) - J_{ph}, \tag{30}$$

where $J_d(V)$ and J_{ph} are the dark current and photocurrent densities, respectively.

Calculations carried out for the case of a film thickness $d = 5$ μm which is often used in the fabrication of CdTe-based solar cells and a typical carrier lifetime of 10^{-9}-10^{-10} s (Sites et al., 2007) in thin-film CdTe/CdS solar cells show that the maximum value of $J_{sc} \approx 25$-26 mA/cm^2 (Fig. 8(b)) is obtained when the concentration of noncompensated acceptors is $N_a - N_d = 10^{15}$-10^{16} cm^{-3}. Therefore, in the following calculations a photocurrent density $J_{sc} \approx 26$ mA/cm^2 will be used.

In Fig. 12(a) the calculated I-V characteristics of the CdS/CdTe heterojunction under illumination are shown. The curves have been calculated by Eq. (30) using Eqs. (25), (28), (29) for $\tau = \tau_{no} = \tau_{po} = 10^{-9}$ s, $N_a - N_d = 10^{16}$ cm^{-3} and various resistivities of the p-CdTe layer. As is seen, an increase in the resistivity ρ of the CdTe layer leads to decreasing the open-circuit voltage V_{oc}. As ρ varies, $\Delta\mu$ also varies affecting the value of the recombination current, and especially the over-barrier current. The shape of the curves also changes affecting the fill factor FF which can be found as the ratio of the maximum electrical power to the product $J_{sc}V_{oc}$ (Fig. 12(a)). Evidently, the carrier lifetime τ_n also influences the I-V characteristic of the heterojunction under illumination. In what follows the dependences of these characteristics on ρ and τ are analyzed.

The dependences of open-circuit voltage, fill factor and efficiency on the carrier lifetime calculated at different resistivities of the CdTe absorber layer are shown in Fig. 13. As is seen, V_{oc} considerably increases with lowering ρ and increasing τ. In the most commonly encountered case, as $\tau = 10^{-10}$-10^{-9} s, the values of $V_{oc} = 0.8$-0.85 V are far from the maximum possible values of 1.15-1.2 V, which are reached on the curve for $\rho = 0.1$ Ω·cm and $\tau > 10^{-8}$. A remarkable increase of V_{oc} is observed when ρ decreases from 10^3 to 0.1 Ω·cm.

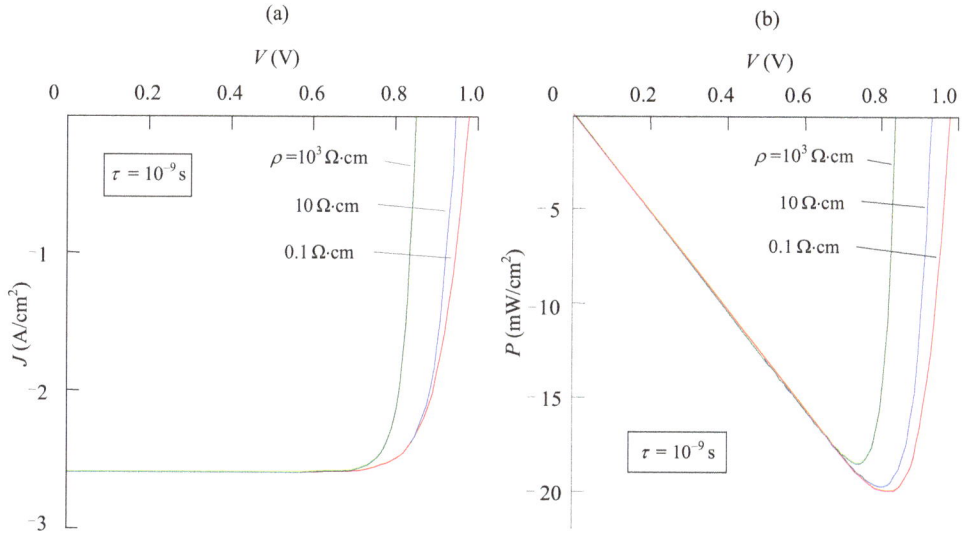

Fig. 12. *I-V* characteristics (a) and voltage dependence of the output power (b) of CdS/CdTe heterojunction under AM1.5 solar irradiation calculated for J_{sc} = 26 mA/cm², τ = 10⁻⁹ s and different resistivities ρ of the CdTe absorber layer.

Fig. 13(b) illustrates the dependence of the fill factor $FF = P_{max}/(J_{sc}\cdot V_{oc})$ on the parameters of the CdS/CdTe heterostructure within the same range of ρ and τ (P_{max} is the maximal output power found from the illuminated *I-V* characteristic). As it is seen, the fill factor increases from 0.81-0.82 to 0.88-0.90 with the increase of the carrier lifetime from 10⁻¹¹ to 10⁻⁷ s. The non-monotonic dependence of FF on τ for ρ = 0.1 Ω·cm is caused by the features of the *I-V* characteristics of the CdS/CdTe heterostructures, namely, the deviation of the *I-V* dependence from exponential law when the resistivity of CdTe layer is low (see Fig. 11, $V > 0.8$ V).

Finally, the dependences of the efficiency $\eta = P_{out}/P_{irr}$ on the carrier lifetime τ_n calculated for various resistivities of the CdTe absorber layer are shown in Fig. 13(c), where P_{irr} is the AM 1.5 solar radiation power over the entire spectral range which is equal to 100 mW/cm² (Standard IOS, 1992). As it is seen, the value of η remarkably increases from 15-16% to 21-27.5% when τ and ρ changes within the indicated limits. For τ = 10⁻¹⁰-10⁻⁹ s, the efficiency lies near 17-19% and the enhancement of η by lowering the resistivity of CdTe layer is 0.5-1.5% (the shaded area in Fig. 13(c)).

Thus, assuming τ = 10⁻¹⁰-10⁻⁹ s, the calculated results turn out to be quite close to the experimental efficiencies of the best samples of thin-film CdS/CdTe solar cells (16-17%).

The conclusion followed from the results presented in Fig. 13(c) is that in the case of a CdS/CdTe solar cell with CdTe thickness 5 μm, enhancement of the efficiency from 16-17% to 27-28% is possible if the carrier lifetime increases to $\tau \geq 10^{-6}$ s and the resistivity of CdTe reduces to $\rho \approx 0.1$ Ω·cm. Approaching the theoretical limit η = 27-28% requires also an increase in the short-circuit current density. As it is follows from section 3.3, the latter is possible for the thickness of the CdTe absorber layer of 20-30 μm and more.

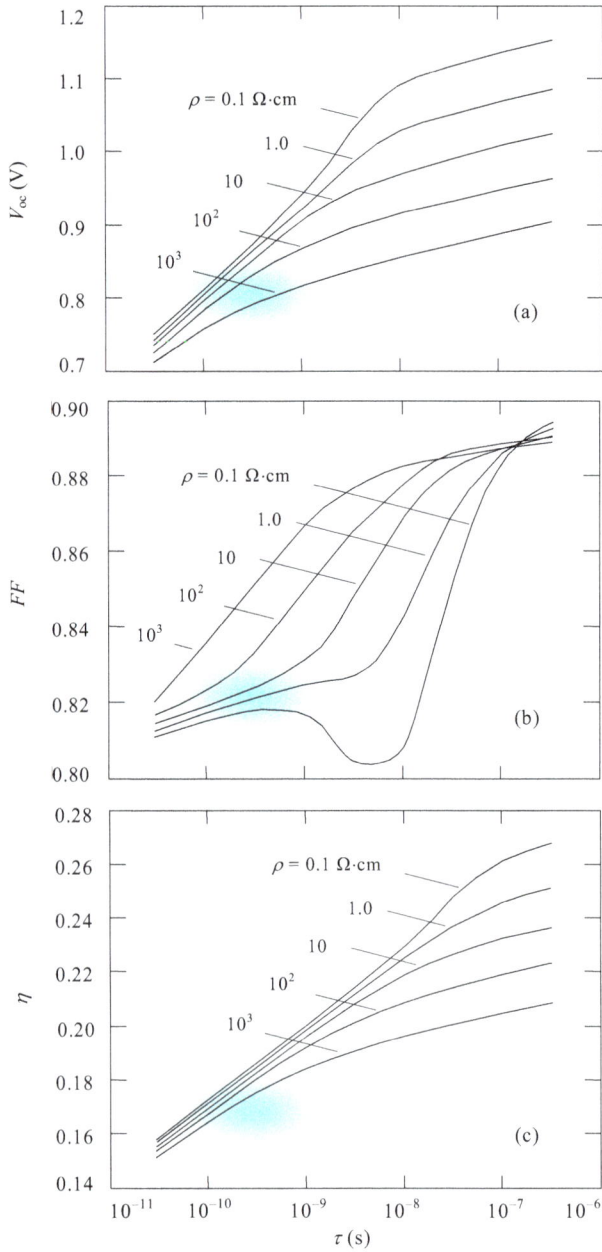

Fig. 13. Dependences of the open-circuit voltage V_{oc} (a), fill factor FF (b) and efficiency η (c) of CdS/CdTe heterojunction on the carrier lifetime τ calculated by Eq. (30) using Eqs. (24)-(29) for various resistivities ρ of the CdTe layer. The experimental results achieved for the best samples of thin-film CdS/CdTe solar cells are shown by shading.

6. Conclusion

The findings of this paper give further insight into the problems and ascertain some requirements imposed on the CdTe absorber layer parameters in a CdTe/CdS solar cell, which in our opinion could be taken into account in the technology of fabrication of solar cells.

The model taking into account the drift and diffusion photocurrent components with regard to recombination losses in the space-charge region, at the CdS-CdTe interface and the back surface of the CdTe layer allows us to obtain a good agreement with the measured quantum efficiency spectra by varying the uncompensated impurity concentration, carrier lifetime and surface recombination velocity. Calculations of short-circuit current using the obtained efficiency spectra show that the losses caused by recombination at the CdTe-CdS interface are insignificant if the uncompensated acceptor concentration $N_a - N_d$ in CdTe is in excess of 10^{16} cm^{-3}. At $N_a - N_d \approx 10^{16}$ cm^{-3} and the thickness of the absorbing CdTe layer equal to around 5 μm, the short-circuit current density of 25-26 mA/cm^2 can be attained. As soon as $N_a - N_d$ deviates downward or upward from this value, the short-circuit current density decreases significantly due to recombination losses or reduction of the photocurrent diffusion component, respectively. Under this condition, recombination losses in the space-charge region can be also neglected, but only when the carrier lifetime is equal or greater than 10^{-10} s.

At $N_a - N_d \geq 10^{16}$ cm^{-3}, when only a part of charge carriers is generated in the neutral part of the p-CdTe layer, *total* charge collection can be achieved if the electron lifetime is equal to several microseconds. In this case the CdTe layer thickness d should be greater than that usually used in the fabrication of CdTe/CdS solar cells (2-10 μm). However, in a common case where the minority-carrier (electron) lifetime in the absorbing CdTe layer amounts to 10^{-10}–10^{-9} s, the optimum layers thickness d is equal to 3–4 μm, i.e., the calculations support the choice of d made by the manufacturers mainly on an empirical basis. Attempts to reduce the thickness of the CdTe layer to 1–1.5 μm with the aim of material saving appear to be unwarranted, since this leads to a considerable reduction of the short-circuit's current density J_{sc} and, ultimately, to a decrease in the solar-cell efficiency. If it will be possible to improve the quality of the absorbing layer and, thus, to raise the electron lifetime at least to 10^{-8} s, the value of J_{sc} can be increased by 1–1.5 mA/cm^2.

The Sah-Noyce-Shockley theory of generation-recombination in the space-charge region supplemented with over-barrier diffusion flow of electrons provides a quantitative explanation for all variety of the observed *I-V* characteristics of thin-film CdS/CdTe heterostructure. The open circuit voltage V_{oc} significantly increases with decreasing the resistivity ρ of the CdTe layer and increasing the effective carrier lifetime τ in the space charge region. At $\tau = 10^{-10}$–10^{-9} s, the value of V_{oc} is considerably lower than its maximum possible value for $\rho \approx 0.1$ Ω·cm and $\tau > 10^{-8}$ s and the calculated efficiency of a CdS/CdTe solar cell with a CdTe layer thickness of 5 μm lies in the range 17-19%. An increase in the efficiency and an approaching its theoretical limit (28-30%) is possible in the case when the electron lifetime $\tau_n \geq 10^{-6}$ s and the thickness of CdTe absorber layer is 20-30 μm or more. The question of whether an increase in the CdTe layer's thickness is reasonable under the conditions of mass production of solar modules can be answered after an analysis of economic factors.

7. Acknowledgements

I thank X. Mathew, Centro de Investigacion en Energia-UNAM, Mexico, for the CdS/CdTe thin-film heterostructures for measurements, V.M. Sklyarchuk for sample preparation to study, V.V. Motushchuk and E.V. Grushko for measurements carried out, and all participants of the investigation for helpful discussion. The study was supported by the State Foundation for Fundamental Investigations (Ministry of Education and Science, Ukraine) within the Agreement Φ14/259-2007.

8. References

Acrorad Co, Ltd., 13-23 Suzaki, Gushikawa, Okinawa 904-2234, Japan. Available: http://www.acrorad.jp/us/cdte.html.

Aramoto T., Kumazawa S., Higuchi H., Arita T., Shibutani S., Nishio T., Nakajima J., Tsuji M., Hanafusa A., Hibino T., Omura K., Ohyama H. &, Murozono M., (1997). 16.0% Efficient Thin-Film CdS/CdTe Solar Cells. *J. Appl. Phys.* 36, 6304-6305.

Baldazzi G., Bollini D., Casali F., Chirco P., Donati A., Dusi W., Landini G., Rossi M. & Stephen J. B., (1993). Timing response of CdTe detectors, *Nucl. Instr. and Meth.* A326 319-324.

Birkmire, R.W. & Eser, E. (1997). Polycrystalline thin film solar cells: Present status and future potential, *Annu. Rev. Mater. Sc.* 27, 625.

Britt, J. & Ferekides, C., (1993), Thin-film CdS/CdTe solar cell with 15.8% efficiency. Appl. Phys. Lett. 62, 2851-2853.

Bonnet, D. (2001). Cadmium telluride solar cells. In: Clean Electricity from *Photovoltaic*. Ed. by M.D. Archer, R. Hill. Imperial College Press, pp. 245-276.

Bonnet, D. (2003). CdTe thin-film PV modules, In: *Practical Handbook of Photovoltaic: Fundamentals and Applications.* Ed. by T. Makkvart and L. Castaner. Elseivier, Oxford.

Compaan A.D., Sites J.R., Birkmire R.W., Ferekides C.S. and Fahrenbruch A.L. (1999). Critical Issues and Research Needs for CdTe-Based Solar Cells, *Proc. 195th Meeting of the Electrochemical Society),* PV99-11, Seattle, WA, pp. 241-249.

Demtsu S.H. & Sites J.R., (2005). Quantification of losses in thin-film CdS/CdTe solar cells, *Proc. 31rd IEEE Photovoltaic Specialists Conf.* pp. 3-7, Florida, Jan. 347-350.

Desnica, U.V., Desnica-Frankovic I.D., Magerle R., Burchard A. & Deicher M.. (1999). Experimental evidence of the self-compensation mechanism in CdS, *J. Crystal Growth*, 197, 612-615.

Eizen Y. (1992). Current state-of-the-art applications utilizing CdTe derectors, *Nucl. Instr. and Meth.* A322, 596-603.

Ferekides, C.S., Balasubramanian, U., Mamazza, R., Viswanathan, V., Zhao, H. & Morel, D.L. (2004) CdTe thin-film solar cells: device and technology issues, *Solar Energy* 77, 823-830.

Fritsche, J., Kraft, D., Thissen, A., Mayer, Th., Klein & A., Jaegermann W. (2001). Interface engineering of chalcogenide semiconductors in thin film solar cells: CdTe as an example, *Mat. Res. Soc. Symp.* Proc. , 668, 601-611.

Gartner W.W., (1959). Depletion-layer photoeffects in semiconductors, *Phys. Rev.* 116, 84-87.

Goetzberger, A. , Hebling, C. & Schock, H.-W. (2003). Photovoltaic materials, history, status and outlook, Materials Science and Engineering R40, 1-46.

Grasso, C., Ernst, K., R. Könenkamp, Lux-Steiner, M.C. & Burgelman, M. (2001). Photoelectrical Characterisation and Modelling of the Eta-Solar Cell. *Proc. 17th European Photovoltaic Solar Energy Conference*, vol. 1. pp. 211-214, Munich, Germany, 22-26 October.

Hanafusa, A., Aramoto, T., Tsuji, M., Yamamoto, T., Nishio, T., Veluchamy, P., Higuchi, H., Kumasawa, S., Shibutani, S., Nakajima, J., Arita T., Ohyama, H., Hibino T., Omura & K. (2001). Highly efficient large area (10.5%, 1376 cm2) thin-film CdS/CdTe solar cell, *Solar Energy Materials & Solar Cells*. 67, 21-29.

Holliday D. P., Eggleston J. M. and Durose K., (1998). A photoluminescence study of polycrystalline thin-film CdTe/CdS solar cells. *J. Cryst. Growth*. 186, 54-549.

Kosyachenko, L.A., Sklyarchuk, V.M. , Sklyarchuk, Ye.F. & Ulyanitsky, K.S. (1999). Surface-barrier p-CdTe-based photodiodes, *Semicond. Sci. Technol.*, 14, 373-377.

Kosyachenko L.A., Rarenko I.M., Zakharuk Z.I., Sklyarchuk V.M., Sklyarchuk Ye.F., Solonchuk I.V., Kabanova I.S. & Maslyanchuk E.L. (2003). Electrical properties of CdZnTe surface-barrier diodes. *Semiconductors*. 37, 238-242.

Kosyachenko L.A., Mathew X., Motushchuk V.V. & Sklyarchuk V.M. (2005). The generation-recombination mechanism of charge transport in a thin-film CdS/CdTe heterojunction, *Semiconductors*, 39, 539–542.

Kosyachenko, L.A. (2006). Problems of Efficiency of Photoelectric Conversion in Thin-Film CdS/CdTe Solar Cells, Semiconductors. 40, 710-727.

Lavagna, M., Pique, J.P. & Marfaing, Y. (1977). Theoretical analysis of the quantum photoelectric yield in Schottky diodes, *Solid State Electronics*, 20, 235-240.

Mathew, X., Kosyachenko, L.A., Motushchuk, V.V. & Sklyarchuk. O.F. (2007). Requirements imposed on the electrical properties of the absorbed layer in CdTe-based solar cells. *J. Materials Science: Materials in Electronics*. 18, 1021-1028.

McCandless, B.E., Hegedus, S.S., Birkmire, R.W. & Cunningham, D. (2003). Correlation of surface phases with electrical behavior in thin-film CdTe. devices. *Thin Solid Films* 431–432, 249-256.

Meyers, P.V. & Albright, S.P. (2000). Photovoltaic materials, history, status and outlook. Prog. Photovolt.: Res. Appl. 8, 161- 168.

Phillips J.I., Brikmire R.W., McCandless B.E., Mayers P.V. & Shaparman W.N., (1996). Polycrystalline heterojunction solar cells: a device perspective. *Phys. Stat. Sol.* (b) 31, 31-39.

Reference solar spectral irradiance at the ground at different receiving conditions. Standard of International Organization for Standardization ISO 9845-1:1992.

Romeo, N., Bosio, Canevari, A. V. & Podesta A., (2004). Recent progress on CdTe/CdS thin film solar cells, *Solar Energy* , 77, 795-801.

Sah C., Noyce R. & Shockley W. (1957). Carrier generalization recombination in p-n junctions and p-n junction characteristics, *Proc. IRE*. 45, 1228–1242.

Sites, J.R. & Xiaoxiang Liu, (1996). Recent efficiency gains for CdTe and CuIn$_{1-x}$Ga$_x$Se2 solar cells: What has changed? *Solar Energy Materials & Solar cells*. 41/42 373-379.

Sites J.R. & Pan J., (2007). Strategies to increase CdTe solar-cell voltage. *Thin Solid Films*, 515, 6099-6102.

Surek, T. (2005). Crystal growth and materials research in photovoltaics: progress and challenges, Journal of Crystal Growth. 275, 292-304.

Sze, S. (1981). *Physics of Semiconductor Devices*, 2nd ed. Wiley, New York.

Toshifumi, T., Adachi, S., Nakanishi, H. & Ohtsuka M. (993). K. Optical constants of Zn1-xCdxTe Ternary alloys: Experiment and Modeling. *Jpn. Appl. Phys.* 32, 3496-3501.

Wu, X., Keane, J.C., Dhere, R.G., Dehart, C., Albin, D.S., Duda, A., Gessert, T.A., Asher, S., Levi, D.H. & Sheldon, P. (2001). 16.5%-efficient CdS/CdTe polycrystalline thin-film solar cell, In: *Proceedings of the 17th European Photovoltaic Solar Energy Conference*, Munich, Germany, October 2001, p. 995-1000.

11

Control of a 3KW Polar-Axis Solar Power Platform with Nonlinear Measurements

John T. Agee and Adisa A. Jimoh
Tshwane University of Technology, Pretoria,
South Africa

1. Introduction

Environmental concerns and the finiteness of fossil fuels have engendered a global embrace for alternative energy systems. Botswana is a country blessed with abundant solar energy resources: having a mean solar day of 8.8 hours and 320 days of clear sunshine in a year (Anderson & Abkenari, 1999; Botswana Energy Report, 2003). It also experiences an excellent mean solar radiation intensity of 5.8/KW.m² (Anderson & Abkenari, 1999). Given that the country currently meets about 70% of her electricity needs through imports from the Southern African Power Pool (SAPP) (Botswana Power Corporation, 2004; SADC, 2004; Matenge & Masilo, 2004), the country is well motivated to integrate solar power into its energy generation base.

Earlier attempts at integrating solar power generation into the national energy mix in Botswana in the eighties, and, in fact, up to the nineties, advocated the use of solar power installations of a few Watts' capacity for lighting in small rural communities. Moreover, such solar power projects employed static installations. Static solar power installations generally have lower daily and seasonal efficiencies than sun-tracking installations. Compared to tracking systems, the lower efficiencies of static solar installations often mean that additional photovoltaic panels must be mounted to meet the required output capacity, thus raising the over-all cost of the facility. Consequently, the above-mentioned solar power philosophy imploded by reason of two shortcomings: customers' perception that initial installation costs were unduly high; and the sentiments of financial institutions that the business value of such small capacity installations was insignificant. Experiences around the SADC countries generally show that such integration philosophies have always not been sustainable (Geche & Irvine, 1996; Mogotsi, 2002; Lasschuit *et al*, 2009). For Botswana, as well as for several other countries in the SADC (Southern African Development Council) region, the high initial costs of solar power installations have been a major hindrance to the massive adoption of solar energy for rural communities (BPC, 2005; Solarie 2005).

The current development of solar power equipment for use in Botswana, and the possible subsequent extension to other SADC countries benefited from the findings reported above. First of all, the current efforts concentrate on the development of solar power equipment that could support rural entrepreneurial activities, in addition to basic lighting needs. This approach is rooted in the understanding that sustainability could be enhanced with the stimulation of economic activities. This approach is also thought to be supported by the fact that rural individuals could be motivated to acquire solar power systems that might enhance their economic wellbeing; and also that, such solar power installations with added economic value would attract the support of local financial institutions. Tables 1 to 3 show examples of typical rural enterprises around Botswana, and the example power requirements (SADC Report, 2004; de Lazzer, 2005). Additional scenarios that could be considered include cold storage facilities for anti-retroviral drugs, rural guest houses for tourism, battery charging and welding businesses. In all studied cases, the power requirements could be considered to be in the range 1-3KW; this forming the basis for the 3KW rating of the solar power platform studies presented in the chapter. Secondly, the current approach for equipment development uses tracking solar power systems, as opposed to the earlier approach that utilized static solar power systems. An extensive discussion comparing tracking solar power systems is presented elsewhere (de Lazzer, 2005; Agee et al 2006a). Suffice it to state that, polar-axis tracking systems present an option capable of producing 97.5% the output power of two-axis solar power systems; and this at acquisition and maintenance costs similar to those of the cheaper single-axis installations. This comparative economic advantage informed our choice of the polar-axis tracking solar power systems for the study reported in this chapter.

For the rest of the chapter, the physical structure and the data of the solar platform system is presented in section two. Dynamic modelling of the platform and model studies is presented in section three. Sensor characteristics modelling and validation form the contents of section four. Two controllers are investigated for the enhancement of the dynamic performance of the polar-axis solar power platform. The design and comparative analysis and discussions of these controllers is presented in section five of the chapter. Section six contains the conclusions and recommendations for further research. A list of references is included in section seven, to conclude the chapter.

Equipment	Power Consumption (W)
Refrigerator with freezer	550
Lighting bulb 60W (x2)	120
Television 51cm color	80
Radio portable	6
Fan	250
TOTAL	**1006**

Table 1. Energy requirements of a rural bar

Equipment	Power Consumption (W)
4 computers	1200
light	60
Radio-cassette	6
TV color	80
TOTAL	**1346**

Table 2. Energy requirements of a rural internet café.

2. Description of the 3KW polar-axis solar tracker hardware

Ten, 300KW solar panels were required to realize the 3KW design power level. Thus, the platform carries ten Shott 300W solar panels. In addition, two smaller Shell SQ 80W solar panels are provided to compensate for the energy looses and the power required in the electrical drive system (Alternative Energy Store, 2005; Shell.com, 2005). The detailed design of the 3KW solar power platform is presented in (de Lazzer, 2005). The weight of each of the 300W solar panels is 46.6Kg. The weight of the 80W solar panel is 7.5Kg /panel. The total weight is ≈500 Kg. The platform is 7 meters long by 3.8 meters wide. Therefore, the area of the platform is 26.4m². For the 300W solar panels, the weight is distributed on all the frame perimeter. The concentrated force value is 0.1 N/mm. For the 80W solar panels, the weight is distributed on the frame length only and its value is 0.03 N/mm. The solar panels are fixed symmetrically with respect to the beams of the platform. It is assumed that the structure experiences no dynamic effects. The load can therefore be classified as static. The arrangement of solar panels is shown in Figure 1. The standing 3KW platform is shown in Figure 2. The drive system consist of a d.c motor linked to the platform through a gear train having a gear ratio of 800. Additional provision was made for manual control for the purposes of field experimentation in the Botswana environment. This manual provision for the seasonal adjustment of the longitudinal inclination of the platform is visible from Figure 2, where it appears as a knob on the stem supporting the platform.

Equipment	Power (W)
4 refrigerators with freeze	2100
3 lights	120
Hi Fi	180
2 TV 51cm color	160
TOTAL	**2560**

Table 3. Energy requirements for a rural clinic

Parameters and their values		
$R_a=5\Omega$	$L_a=0.003H$	$B=3.95.10^{-6}$ Kg.ms^{-1}
$K_b=0.0636V/$ rad/s	$K_m=0.00711$ Kgm/A	$K=0.01Kg$ m²/ s²
$J_M=7.72.10^{-6}$ Kg m²	$J_L=970Kgm^2$	$N=1/n=1$ /800

Table 4. System parameters

The platform is a sensor-based tracking solar power platform. As suggested by the name, this type of tracking solar power system employs two photosensitive detectors to determine the position of the sun. Usually, two sensors are positioned on an imaginary line parallel to the east-west axis passing through the centre of the array of solar panels. They are arranged so that they produce a differential output whenever the active surface of the solar panels is not aligned perpendicular to the direction of sunrays. When the incident solar radiation is perpendicular to the plane of the array of PV cells, both sensors generate equal amount of current. If however, the incident solar radiation is not perpendicular to the array, then one of the sensors produces an output current greater than that of the second sensor. The differential output current from the sensor arrangement is a current whose magnitude depends of the angle of misalignment of the panels. The sign of the resultant current indicates the direction of the sun. The control system utilises the output of the sensor arrangement to control the motor that rotates the platform of solar panels until the electrical signal in each of the light detectors becomes equal.

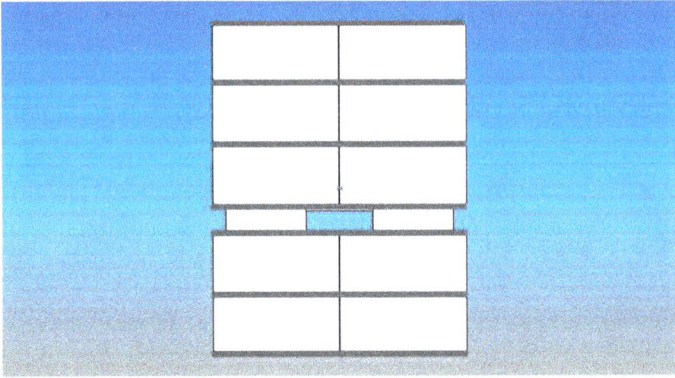

Fig. 1. Arrangement of solar panels on the platform.

For the purpose of investigating the dynamic performance of the platform, as well as for controller design , a model of the platform is developed, based on the application of relevant physical laws. The platform modeling is presented in the next section, section three of the chapter.

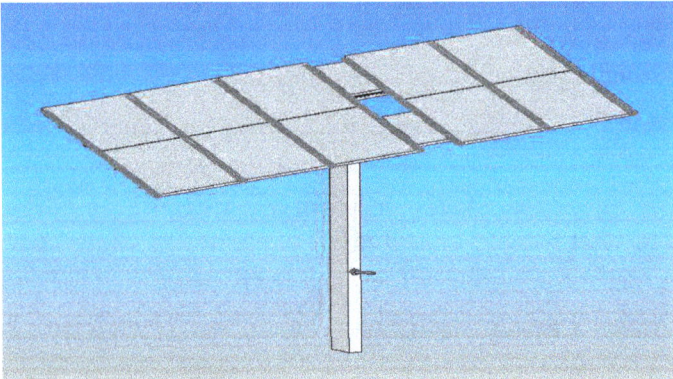

Fig. 2. The 3KW platform viewed from above

3. Mathematical modelling and analysis of the dynamics of the basic platform

The subsequent modeling presented in this section concerns the east-west motion of the platform. The block diagram representation of the platform in the east-west direction is shown in Figure 3.

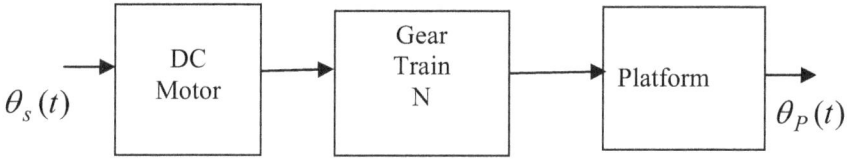

Fig. 3. Block diagram of 3KW solar power platform

Where: $\theta_s(t)$ is the instantaneous direction of sunlight and $\theta_p(t)$ the instantaneous position of the platform.

Hence, the model of the solar tracker is the system of dynamic equations linking the separately excited dc motor to the platform through the gear train as derived in the sequel.

3.1 Modelling of the separately excited DC motor

The typical equivalent circuit arrangement for a separately excited DC motor is shown in Figure 4. An applied armature voltage e_a creates an armature current i_a given by (Kuo & Golnaraghi, 2003):

$$e_a = R_a i_a + L_a \frac{di_a}{dt} + K_b \frac{d\theta_m}{dt} \qquad (1)$$

where $e_a(t)$: armature voltage (V); $i_a(t)$: armature

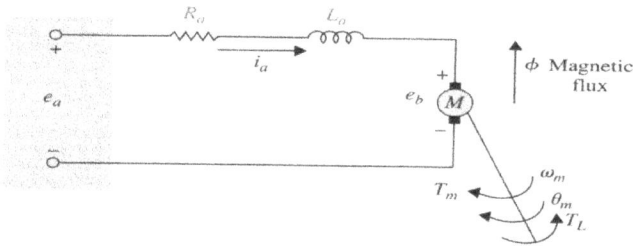

Fig. 4. Schematic diagram of a separately excited DC motor.

current (A); R_a: armature resistance (Ω); L_a: armature inductance (H); K_b: back-emf constant (V/rad/s) and $\theta_m(t)$: rotor displacement (rad.). This current causes a torque

$$T_m = K_m i_a \qquad (2)$$

where $T_m(t)$ is torque(N.m.) and K_m the torque constant (N.m/A).

The torque of the DC motor is coupled to drive the platform through the motor shaft and a gear train. The torque causes an angular displacement of the rotor θ_m, given by (Kuo & Golnaraghi, 2003):

$$T_m = J_t \frac{d^2\theta_m}{dt^2} + B\frac{d\theta_m}{dt} + K\theta_m \qquad (3)$$

where $J_t = J_m + N^2 J_l$ and J_m: moment of inertia of the motor ($kg.m^2$); J_l: moment of inertia of the solar power platform ($kg.m^2$); N: gear-train ratio between motor and platform; B: viscous-friction coefficient of the motor ($kg.m.s^{-1}$); K: spring constant ($kg.m^2.s^{-2}$). After taking the Laplace transform of equations (1)-(3), it is straightforward to obtain the open-loop transfer function for the system G(s)=θ_m/V$_a$ as shown in equation (4).

$$G(s) = \frac{K_m / L_a J}{s^3 + \left\{\dfrac{R_a J_t + L_a B}{L_a J_t}\right\} s^2 + \left\{\dfrac{R_a B + K L_a + K_b K_m}{L_a J_t}\right\} s + \dfrac{R_a K}{L_a J_t}} \tag{4}$$

Note further that, the angular position of the platform θ_p is related to the motor angular position θ_m through the gear ratio:

$$\theta_p / \theta_m = N = 1 / 800 \tag{5}$$

The gear ratio was decided by comparing similar applications (Kuo & Golnaraghi, 2003). The parameters of the open-loop platform system are given in the Table 4. A substitution of these parameters in equation (4) results in the open-loop transfer function:

$$G(s) = \frac{1559.2}{s^3 + 1666.7s^2 + 109.87s + 10965} = \frac{b}{s^3 + a_3 s^2 + a_2 s + a_1} \tag{6}$$

In subsection 3.2, the dynamic behaviour of this open loop system was investigated with the view to determining what controller would be most suitable for improving the platform dynamic performance.

3.2 Simulation of the open-loop platform system
The analysis of the dynamic performance of the platform was simulated using MATLAB. The following simulations were carried out.

3.2.1 Time-domain characterisation of the open-loop platform system
The system was simulated for a unit step increase in the input voltage. The results are shown in Figure 5. The performance of the system, from Figure 5, could be summarized as in Table 5. *Settling Time*: It is evident from Figure 5 and the summary in Table 5 that the settling time of 105 seconds for the platform well exceeds one minute. This is too long and would not be suitable for the successful tracking of sunlight, since the direction of the sun rays is likely to change significantly before the platform settles down to the last command. Improvements in the settling time would be required. The peak overshoot is 96% of the final value. This is much higher than the maximum 17% overshoot acceptable in literature (Kuo & Golnaragh, 2003). The maximum overshoot must be reduced. For a third order system, the damping ratio is not strictly defined (Kuo & Golnaraghi, 2003). However, the contribution of the root s_3=-1670 in the transient response is negligible. The complex pole-pair s_1, s_2=-0.031±j2.56 are the significant poles of the system. Their damping factor ξ is equal to 0.0121. This is much less than the damping factor of ≈ 0.707 required for optimum plant performance (Norman, 2004). An appropriate control strategy should enhance the damping of the system.

Settling time	105 Sec.
Peak overshoot	96%
Steady state error	0.875%
Eigenvalues	-1670, -0.031±j2.56
Damping factor	0.0121

Table 5. Summary of the dynamic performance of the platform

Steady State Error: The steady state error of 0.875% is well less than the typical tolerance band values of either 2% or 5% and does not need any further improvement.

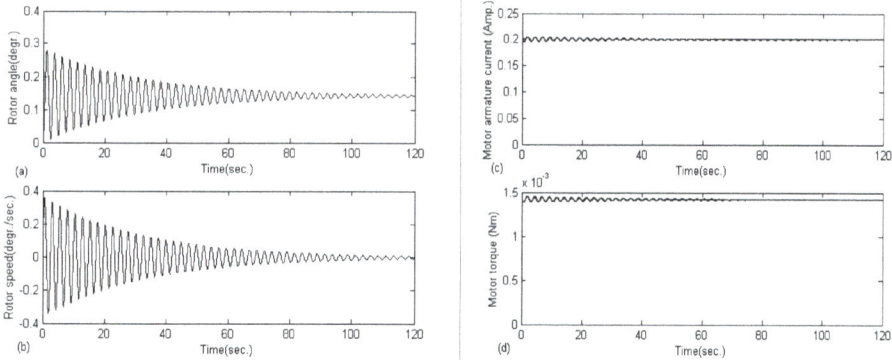

Fig. 5. Dynamic response of solar platform without feedback control

3.2.2 Frequency analysis of the open loop system
The frequency response of the open system is shown in Figure 6. From the Bode plot, the Gain and Phase margins are read. The Gain margin is 40.86dB and the phase margin is 10.4 deg. While the gain margin seems adequate, the phase margin is too low; and is hence due

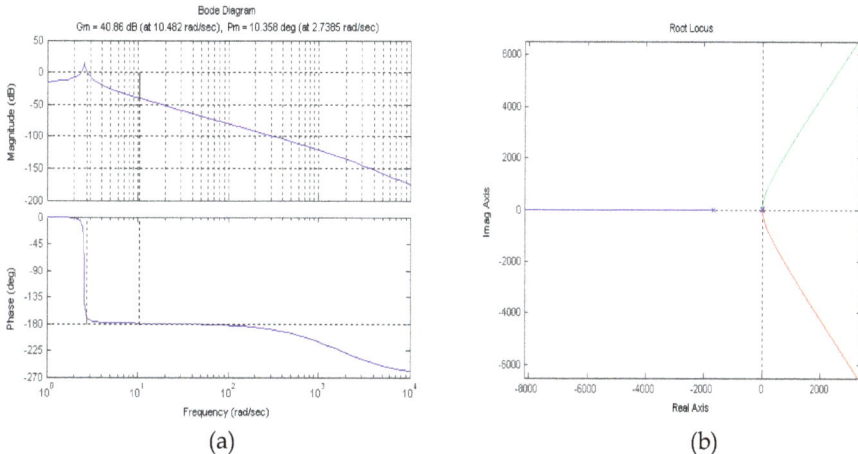

Fig. 6. (a) Bode plot of open- loop platform (b) The root locus of open-loop platform

for further improvements. More over, the roots-locus plots (Figure 6.b) show that even for very small increases in the forward gain of the system, the dominant pole pair crosses the imaginary axis into the right half of the s-plane. It could be concluded that a control strategy that only increases the forward gain of the system, such as proportional control, would aid instability. In the next section of the paper, we present the design of a controller that avoids increases in the forward gain of the system; and which replaces the existing dominant pole-pair of the plant with ones that have the optimum damping factor.

3.3 Summary results from dynamic simulations of the basic platform system

The open-loop simulations of the platform model show that the steady state error of the open-loop dynamics is satisfactory, hence needs no further improvements. The settling time of 105 seconds is well in excess of practical requirements for such systems; and needs to be substantially reduced. The controller design shall specify a target settling time of 2 seconds. The settling time of the open-loop platform system is, in fact, due to the poor damping factor ξ= 0.0121. Alternatively, the poor settling time could be interpreted as a consequence of the dominant pole-pair s_1, s_2=-0.031±j2.56 existing close to the origin of the s-plane. It therefore seems plausible to explore the use of a controller or controller-types that will cancel the existing dominant pole-pair and replace them with those that optimise the damping of the platform. Linear systems are considered to be optimally damped if the damping factor $\xi \approx \frac{1}{\sqrt{2}}$ (Kuo & Golnaragh, 2003).

4. Sensor modelling and characterisation

For the purpose of effecting feedback control of the platform, a sensor required to measure the relative orientation of the platform axis $\theta_p(t)$ for feedback was required. As indicated in section two of the chapter, a photovoltaic position sensor was employed for the measurement of the misalignment α(t) between the orientation of the platform axis and the direction of sun rays. Hence the sensor measurements are a function of the variable $\alpha(t) = \theta_s - \theta_p(t)$. The sensor consists of two photocells. The derivation of the sensor output relationship begins with a recollection of the theory of the photovoltaic cell.

4.1 The photovoltaic cell

The photovoltaic cell is a two terminal device which consists of a photodiode. The photodiode may be a p-n junction or p-i-n structure. When the cell absorbs light, mobile electrons and positively charged holes are created. If the absorption occurs within the junction's depletion region, or one diffusion length away from it, these carriers are swept from the junction by the built-in field of the depletion region, producing a photocurrent (Nelson, 2003). A detailed representation of the current phenomenon in a photocell, accounting for internal diode current is given as:

$$i_0 = I_{PH} - I_D = I_{PH} - I_S \cdot \{\exp(V / mV_T) - 1\}. \tag{7}$$

Where: I_{PH}: Photo current; I_D: Diode current; I_S: Diode reverse saturation current; m: Diode "ideally factor" m= 1-5V_T; Thermal voltage: $V_T = kT / e$; k: constant of Boltzmann; T: absolute temperature; e: charge of an electron. Internal voltage drop in practical photocells is accounted for by the addition of a series resistor R_S. Also leakage currents could be observed in photocells, which may be described as due to a parallel resistor R_p ; from these we could write the photocurrent of the photocell as:

$$0 = I_{PH} - I_D - I_P - i_0 \tag{8}$$

And

$$I_P = \frac{V_D}{R_P} = \frac{V + i_0 R_S}{R_P} \tag{9}$$

Hence,

$$i_0 = I_{PH} - I_S \left\{ \exp\left(\frac{V + i_0 R_S}{m V_T} \right) - 1 \right\} - \frac{V + i_0 R_S}{R_P} \tag{10}$$

4.2 The photovoltaic position sensor

Two photovoltaic cells were required to measure the position of the solar power platform, relative to the axis of sunlight. A photocell each was installed on either side of a line running perpendicular to the direction of rotation of the platform. Let these cells be "A" and "B" respectively, as shown in Fig. 7. The opening of the enclosure has a width of W. Each cell has the breadth D and height C. The cells are located a distance L behind the opening. Each cell generates a current, I_A and I_B, respectively, proportional to the intensity of its incident radiation. The difference between I_A and I_B is conditioned to generate an error voltage which could then be used to control the motion of the platform, as explained in the next subsection (Kuo & Golnaragh, 2003). It shall be shown below, the relationship between the cell currents I_A, I_B, and angle of misalignment α, of the platform.

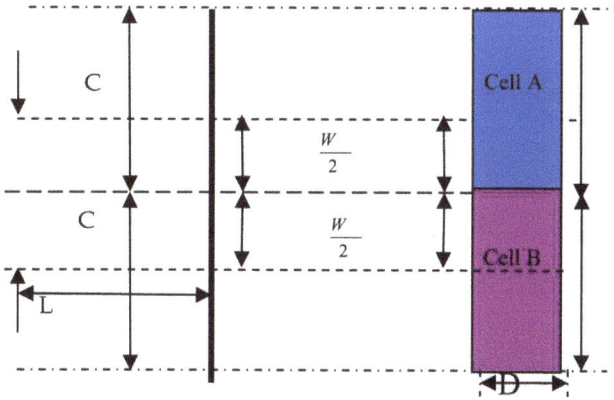

Fig. 7. Arrangement of photocells in the photovoltaic position sensor

4.3 Relating the light-intensity, exposed area of the cell and cell photocurrent

Consider the ray geometry of the photovoltaic position sensor with two identical photocells as shown in Fig. 8. With light incident perpendicular to both cells (angle of misalignment $\alpha=0$), the area of each cell, exposed, is given by

$$A_A = A_B = (W/2) * D \tag{11}$$

Both cells intercept the same amount of incident radiation, producing the same amount of current. In such a case, the difference between the two current values is zero and no error signal is produced. Now, consider the situation in which an angular misalignment of $\alpha \neq 0^0$, causes the incident light to enter the enclosure of the sensor at an angle of inclination α to the axis of the two cells. Then, from Fig. 8, the exposed areas of cells A and B are respectively given by:

$$A_A = D(\frac{W}{2} + h); A_B = D(\frac{W}{2} - h) \tag{12}$$

Also, from the geometry in Fig. 8,

$$h = L \tan \alpha \therefore A_A = D(\frac{W}{2} + L \tan \alpha); A_B = D(\frac{W}{2} - L \tan \alpha) \tag{13}$$

With the intercepted light energy being proportional to the light-sensing surface area of the intercepting cell,

$$P_{RA} = I_R D(\frac{W}{2} + L \tan \alpha); P_{RB} = I_R D(\frac{W}{2} - L \tan \alpha) \tag{14}$$

The photocurrent I_{PH} of a cell is known to be proportional to the light energy, therefore:

$$I_0 \propto P_R \tag{15}$$

Therefore, from Eq. (14), and Eq. (15), one can write:

$$i_A = K_1 P_{RA} = K_1 I_R D(\frac{W}{2} + L \tan \alpha);$$
$$i_B = K_1 P_{RB} = K_1 I_R D(\frac{W}{2} - L \tan \alpha) \tag{16}$$

Substituting Eq. (16) into Eq. (10) yields:

$$i_A = K_1 I_R D \frac{W}{2} + K_1 I_R DL \tan \alpha - I_s (e^{\frac{V_O \log(1+I_R)}{mV_T}} - 1)$$
$$- \frac{V_O \log(1+I_R) + I_A R_S}{R_P} \tag{17}$$

and,

$$i_B = K_1 I_R D \frac{W}{2} - K_1 I_R DL \tan \alpha - I_s (e^{\frac{V_O \log(1+I_R)}{mV_T}} - 1)$$
$$- \frac{V_O \log(1+I_R) + I_B R_S}{R_P} \tag{18}$$

Now, the differential output current for the cells would now be approximated as:

$$i_A - i_B = 2K_1 I_R DL \tan(\theta_s - \theta_p) \tag{19}$$

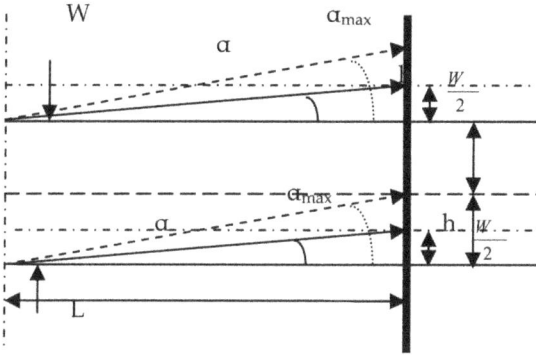

Fig. 8. Ray geometry in the sensor

4.4 The feedback signal

The circuit for feedback signal generation is shown in Figure 9. As derived in (Agee et al, 2009), the resultant output current, representing the difference between the direction of the sun rays θ_s and the axis of the rotor θ_m is given by (20):

$$i_A - i_B = 2K_1 I_R DLan(\theta_S - \theta_P) \tag{20}$$

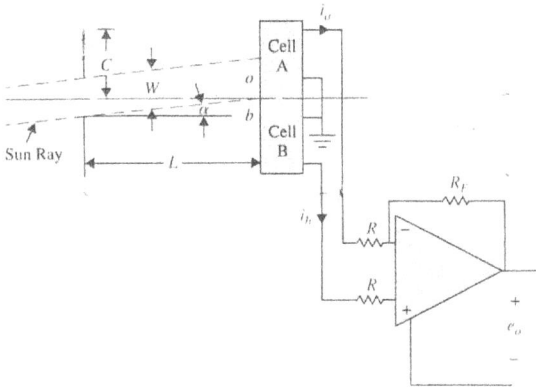

Fig. 9. Sensor signal conditioning for feedback

where i_A, i_B are cell currents from cells A and B respectively. The error voltage signal e_o is straight away derived from Figure 9, to be:

$$e_o = -R_F(i_A - i_B) = -2K_1 I_R DLR_F \tan(\theta_S - \theta_P) \tag{21}$$

However, with the introduction of a unity-gain inverting amplifier in the follow-up stage, we have

$$e_{o1} = -e_o = R_F(i_A - i_B) = 2K_1 I_R DLR_F \tan(\theta_S - \theta_P) \tag{22}$$

Thus with the measurements included for feedback control, the armature voltage V_a could now be expressed as

$$V_a = K_F(e_{o1}) = 2K_FK_1I_RDLR_F\tan\alpha; \qquad \alpha = \theta_S - \theta_P \qquad (23)$$

where, K_F is a constant of proportionality. For linear controller design, a linearised representation of sensor characteristics is required. By approximating $\tan(\theta_S - \theta_P) \cong \theta_S - \theta_P$, the linear sensor feedback model takes the form:

$$V_a = K_F(e_{o1}) = 2K_FK_1I_RDLR_F\alpha \qquad (24)$$

4.5 Validation of sensor model

The data in Table 6, together with equations (17) and (18) were used for the numerical validation of sensor model. Simulations were done in MATLAB. Fig. 10 and Fig. 11, respectively, confirm the nonlinear relationships between I_A, I_B and α. Figure 12 show the nonlinear relationship between α and the differential output cutrrent of the sensor.

PARAMETER	VALUE	PARAMETER	VALUE
Light intensity I_R	12 [mW/cm²]	R_S	5 [Ω]
Sensitivity of sensor K_1	0.55 [AW⁻¹]	D	5 [cm]
m, the ideal diodes factor	1	L	0.8 [cm]
Temperature voltage V_T	25.7 [mV] at T = 25°C	W	3.5 [cm]
Reverse saturation current	I_S = 2x10⁻⁶[A]	C	5 [cm]
Photo-voltage V_O	0.5 [V]	Area	12.5 [cm²]
Parallel resistance R_P	1 [KΩ]		

Table 6. Data used for the simulation of the photovoltaic position sensor

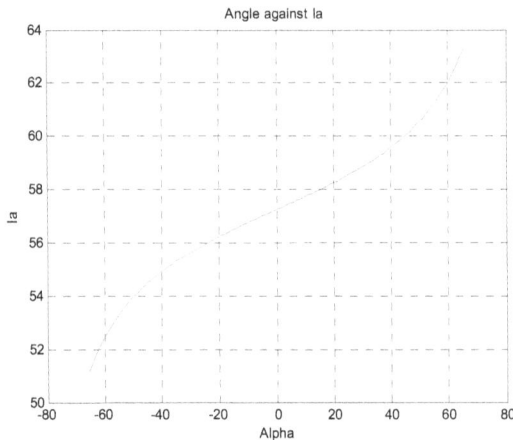

Fig. 10. Current I_A versus angle α

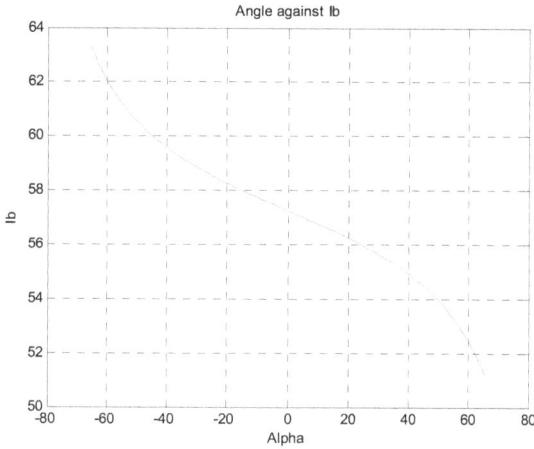

Fig. 11. Current I_B versus angle α

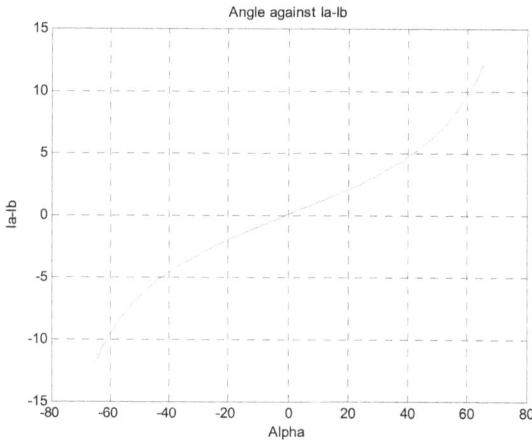

Fig. 12. I_A – I_B related with angle α.

5. Controller design

The structure for the platform system with controller is shown in Figure 13. This shows the embedding of the photosensor into the feedback network.

5.1 Performance specification

The controller was required to modify the response of the system in such a manner as to achieve the following performance specification:

- Settling time for 2% tolerance band = 2 seconds.
- Damping factor = 0.71

First, the pole-cancellation controller is designed and validated. A critical appraisal of the performance of the controller in the presence parameter variations and nonlinearities is presented. Finally, the design of the nonlinear controller, and a comparative analysis of its performance with respect to the pole-cancellation controller are presented.

5.2 Design of the pole cancellation controller for the platform

The structure for the system with controller is shown in Figure 13. The Figure shows the position of the sensor presented in section (4) of the chapter. It has been explained earlier that, both the settling time and the damping of the platform system need improvement. Our solution was to use a controller whose transfer function zeros cancelled the undesirable poles of the platform transfer function, G(s). Then, the poles of the controller were placed so as to achieve the desired closed-loop dynamic performance. Accordingly, the structure of the controller is specified to be the notch filter with the following transfer function:

$$G_C(s) = \frac{K_1 s^2 + \alpha_1 s + \beta_1}{K_2 s^2 + \alpha_2 s + \beta_2} = \frac{N_1(s)}{D_1(s)} \tag{25}$$

as shown in Figure 14 (Agee *et al*, 2006).

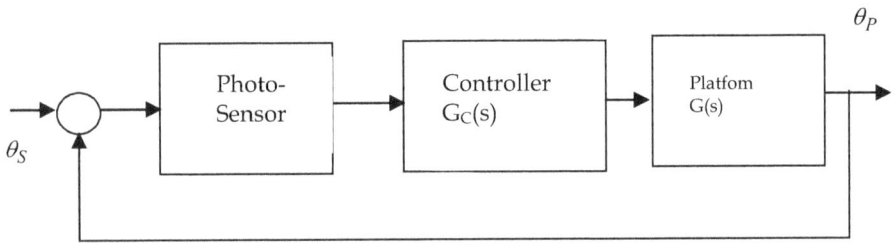

Fig. 13. Block diagram of controlled platform

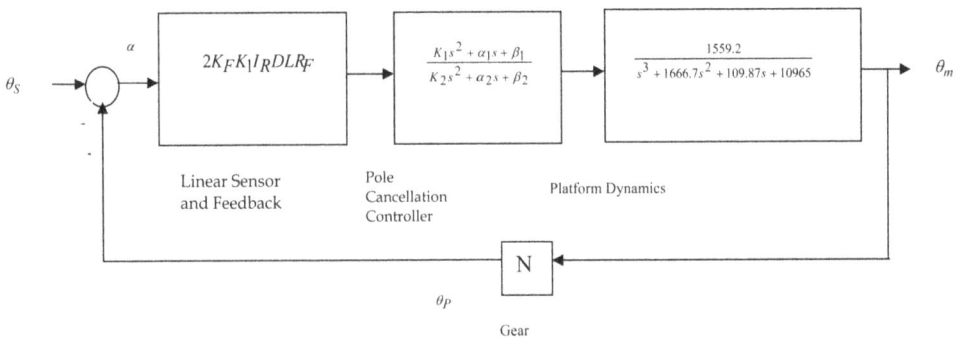

Fig. 14. Closed-loop block diagram of platform with a linear sensor

5.2.1 Tuning of pole-cancellation controller

The controller design entails the determination of the six coefficients $K_1, \alpha_1, \beta_1, K_2, \alpha_2$ and β_2. Note that the zeros of the controller are required to cancel the dominant pole-pair of the open-loop transfer function of the platform system. With the pole-cancellation controller included as in Figure 14, the closed-loop transfer function of the platform becomes equation (25). To determine K_1, α_1 and β_1 for the controller, consider the closed loop transfer function for the controlled system with linear sensor characteristics. Then, using the fact that the dominant pole-pair of the open loop transfer function is : -0.031±j2.56;

$$C(s) = \frac{(2K_F K_1 I_R DLNR_F G_C(s)G(s)}{1 + (2K_F K_1 I_R DLNR_F)G_C(s)G(s)} \qquad (26)$$

An equivalent second-order polynomial for the zeros of the notch filter is obtained. Hence, $K_1 = 1$; $\alpha_1 = 0.062$ and $\beta_1 = 6.5546$. Now, let $2K_F K_1 I_R DL = 1$, such that,

$$G_C(s)G(s) = \frac{(s + 0.031 - 2.56j)(s + 0.031 + 2.56j).1559.2}{D_1(s)(s + 1670)(s + 0.031 - 2.56j)(s + 0.031 + 2.56j)}$$

or,

$$G_C(s)G(s) = \frac{1559.2}{(K_2 s^2 + \alpha_2 s + \beta_2).(s + 1670)}$$

Then, for the closed-loop platform with pole-cancellation,

$$C(s) = \frac{1559.2 NR_F}{(K_2 s^2 + \alpha_2 s + \beta_2).(s + 1670) + 1559.2 NR_F} \qquad (27)$$

$$C(s) = \frac{1559.2 NR_F / K_2}{s^3 + (\frac{\alpha_2}{K_2} + 1670).s^2 + (\frac{\beta_2}{K_2} + 1670\frac{\alpha_2}{\beta_2})s + (1670\frac{\beta_2}{K_2} + 1559.2\frac{NR_F}{K_2})} \qquad (28)$$

Also, for a damping factor of 0.71, the settling time is given by: $t_{s_{1\%}} = 4.6 / \xi\omega_n$ (Kuo & Golnaraghi, 2003). Thus, for a settling time of 2 seconds, we deduce $\omega_n = 3.24$ rad/s. The new dominant pole-pair is given by $s_{1,2} = -\xi\omega_n \pm \omega_n\sqrt{1-\xi^2}$, yielding the dominant second-order factor $s^2 + 2\xi\omega_n + \omega_n^2$. With the third-pole of the closed-loop system situated at s= -d, we have the characteristic equation of the controlled system to be

$$s^3 + (d + 2\xi\omega_n)s^2 + (\omega_n^2 + 2\xi\omega_n d)s + d\omega_n^2 = 0 \qquad (29)$$

Compare equations (28) and (29) to obtain the following relations:

$$160y + z = d\omega_n^2$$
$$\omega_n^2 + 2\xi\omega_n d = y + 1670x \qquad (30)$$
$$d + 2\xi\omega_n = 1670 + x$$

Where,

$$x = \frac{\alpha_2}{K_2}, z = \frac{1.949 R_F}{K_2}; y = \frac{\beta_2}{K_2} \tag{31}$$

The simultaneous solution of equation (30) yields:

$$\omega_n = 1670\xi \pm \sqrt{697225\xi^2 - 2788900 + z / \Delta} \tag{32}$$
$$\Delta = d - 1670$$

It becomes a matter of substitutions to verify that for the given settling time of 2 seconds and damping factor $\xi = 0.71$, equation (32) is satisfied for

$$z = 2893.664, \Delta - 0.0001 \tag{33}$$

and further that the closed loop pole d=-1670.0001, $\alpha_2 / K_2 = 4.6008; \beta_2 / K_2 = 8.7648$. By setting K$_2$=1, obtain the complete transfer function of the controller as:

$$G_C(s) = \frac{s^2 + 0.0620s + 6.5546}{s^2 + 4.6008s + 8.7648} \tag{34}$$

and $R_F = z / 1.949 = 1.485 K\Omega$

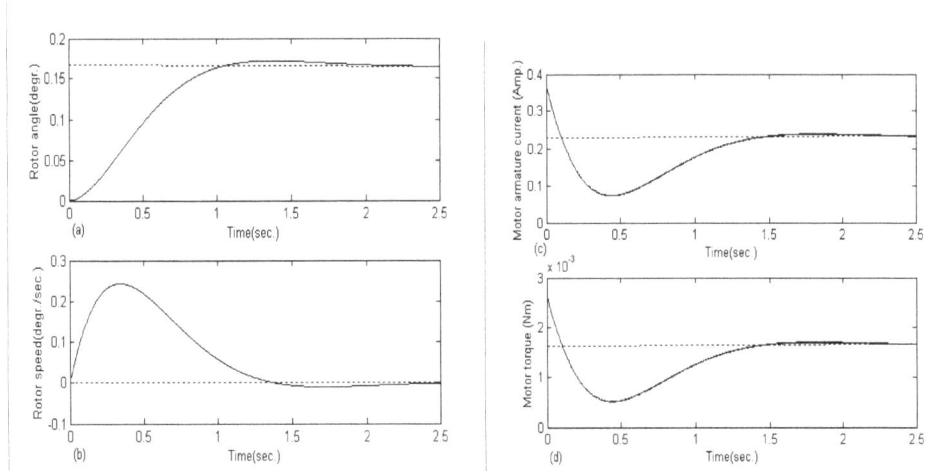

Fig. 15. Analysis of platform system controlled by pole-cancellation

5.3 Performance of the platform equipped with pole-cancellation controller
The transfer function of the closed-loop system with controller becomes:

$$C(s) = \frac{1665855}{s^3 + 1674s^2 + 7688s + 17505} \tag{35}$$

Both the transient response and the frequency response of the controlled system were simulated. The results are shown in Figure 15. It can be seen that, with the controller included, the system does not oscillate like it did before the introduction of the controller.

The overshoot is 4.3%. Overshoot peak has been reduced by 95%, as compared to the case of the uncontrolled system. The damping factor is improved to 0.71. The settling time is now reduced to 2 seconds. The steady state error is only 0.01 rads. This is much better than for the uncontrolled system. In fact, the steady state error has been reduced by 99%. The system is stable. The gain margin is improved by 75% to about 73dB while the phase margin is now infinity. The peak armature current has been reduced by 25% from 0.2A to 0 0.16 A. The peak torque required in the drive system increased from 1.5mN.m to 2.7mNm. It could therefore be concluded that the pole cancellation control strategy improved significantly the dynamic performance of the 3KW solar power platform; improving the damping factor from 0.0121 to 0.71; reducing the settling time from 105 sec. to 2 sec. Peak overshoot of the rotor angular position was reduced by 95%. Both the gain margin and the phase margin were also substantially improved. The peak torque requirement however was doubled. The controller structure is simple and may be easily implemented. However, the design neglected nonlinear sensor characteristics which may restrict the usefulness of the control strategy presented here. These effects on the performance of the pole-cancellation controller are discussed in details in subsection (5.4).

5.4 A Critical analysis of the shortcomings of the pole-cancellation control strategy for platform control

In practical systems, the efficacy of the pole-cancellation control strategy is limited by the effects of parameter variation/uncertainty and the nonlinear sensor behaviour. These two effects are discussed further in the following two subsections.

5.4.1 Effect of parameter uncertainty on performance of platform controlled by pole cancellation

Consider the effect of addititive parameter change on the open-loop platform model of equation (6). The resulting platform representation will admit the form in equation (36). Hence,

$$G(s) + \Delta G(s) = \frac{b + \Delta b}{s^3 + (a_3 + \Delta a_3)s^2 + (a_2 + \Delta a_2)s + (a_1 + \Delta a_1)} \tag{36}$$

Or, more specifically:

$$G(s) + \Delta G(s) = \frac{1559.2 + \Delta b}{s^3 + (1666.7 + \Delta a_3)s^2 + (109.87 + \Delta a_2)s + (10965 + \Delta a_1)} \tag{37}$$

Numerical investigation of the platform systems with variable parameters in MATLAB yielded the roots variations shown in Table 7. It is evident from Table 7 that the open-loop poles of the platform change significantly with variation in system parameters. Under such circumatances, a basic pole-cancellation control strategy will be ineffective, except where on-line adaption is introduced. Such additional complexities in the structure of the controller would make the above control strategy more expensive and hence, less attractive.

Moreover, the increased complexity in the systems would lead to the more complex platform block diagram shown in Figure 16, with uncancelled dynamics, where:

Parameter	Variation	s_1	s_2	s_3
Δa_3	+5%	-1750	-0.035+2.5i	-0.035–2.5i
Δa_3	+10%	-1833.3	-0.036+2.4i	-0.036–2.4i
Δa_2	+5%	-1670	-0.037+2.6i	-0.037–2.6i
Δa_2	+10%	-1670	-0.040+2.6i	-0.040–2.6i
Δa_1	+5%	-1670	-0.039+ 2.6i	-0.039–2.6i
Δa_1	+10%	-1670	-0.041 +2.7i	-0.041-2.7i

Table 7. Dependence of platform system poles on parameter variation

$$\eta(s) = \frac{\{s^3 + (1667.7 + \Delta a_3)s^2 + (109.87 + \Delta a_2)s + (10965 + \Delta a_1)\}\{s^3 + 1666.75s^2 + 109.87s + 10965\}}{\Delta b(s^3 + 1666.75s^2 + 109.87s + 10965) - 1559.2(s^2 + \Delta a_2 s + \Delta a_1)} \quad (38)$$

This, in turn, will modify the closed-loop response as shown in Figure 17-18. Note from Figure 17.b, that with a 10% increase in a₃, a non-positive defenite beahiour of the closed-loop platform is indicated. Figure 18 shows that, under some variations in a₂, the pole-cancellation controller loses function in steady state, and sustained oscillation of the plant is produced.

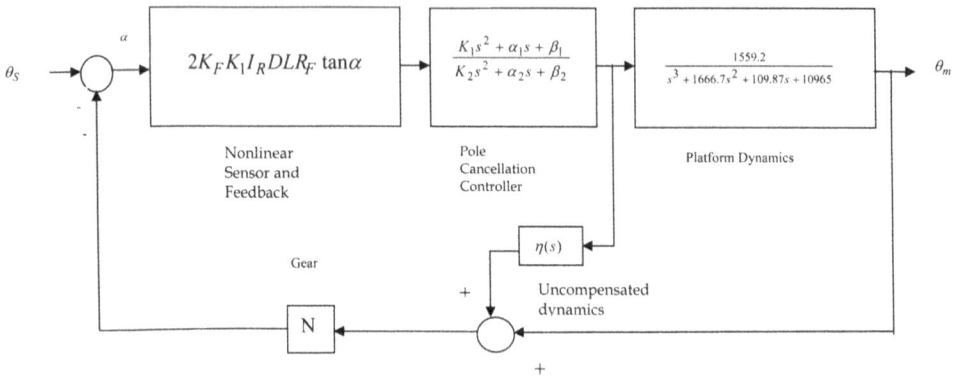

Fig. 16. Complex block diagram of platform due to failure of pole-cancellation

5.4.2 Effect of nonlinear measurement on linear controller performance

Recall that, while the approximate sensor characteristics was used for the design of the pole-cacellation controller, exact sensor characteristics are nonlinear, as in equation (23). In this subsection, a few comments are made as to how the true sensor characteristics affect the dynamics of the platform under pole-cancellation control.

Figure 19 compares the exact tangent characteristics with the linear approximation $\tan\alpha \approx \alpha$; together with the third-order polynomial approximation, $\tan\alpha \approx \alpha + \alpha/3!$. It is evident that none of the approximate representation is useful beyond $\alpha \geq 26°$. Hence, the linear model is only valid within a very narrow window of the complete domain of operation of the solar power opearation. This will lead to significant model mismatches, for which the performance of the pole-cancellation controlle would be significantly inadequate.

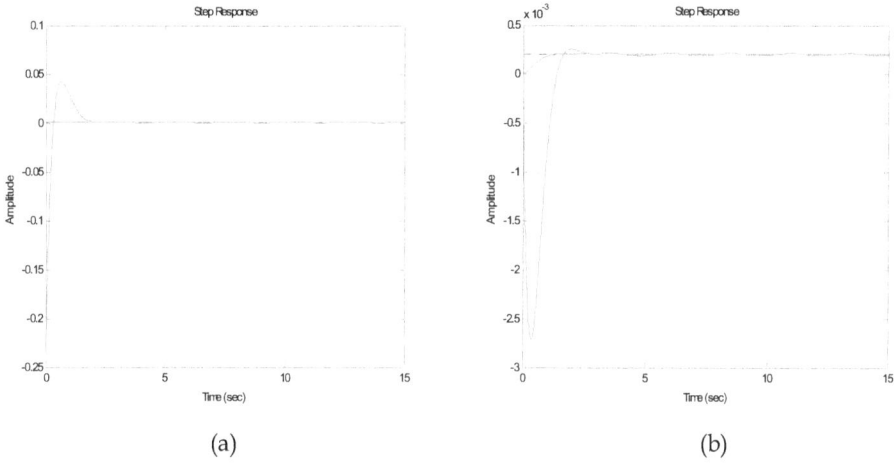

(a)

(b)

Fig. 17. (a) Step response with a 5% variation of a_3 (b) Step response with 10% variation of a_3.

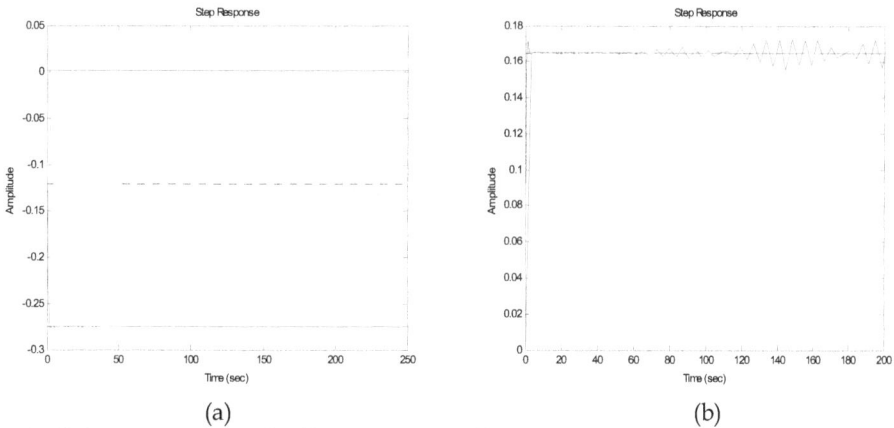

(a)   (b)

Fig. 18. (a) Step response with 5% increase in a_2 (b) Step response with 10% increase in a_2

5.4.3 A summary perspective on the linear control of solar power platform

It is evident from the critical analysis of the effect of parameter variations and nonlinearity, that the suitability of linear control strategies for the polar-axis solar power platform with nonlinearities is limited. In particular, whereas the structure of the pole-cancellation controller makes controller implementation simple, the viablility of this strategy in the practical environments of parameter variation and nonlinearities is not guaranteed. Because of the very high forward gains of the platform, linear control was also not robust. On the other hand, the linear controlled system offers a suitable reference model for the comparison of the performance of possible nonlinear control strategies that may be employed on the platform system. In the rest of the chapter, the design and simulation of a nonlinear controller, the feedback-linearised control of the platform, is presented. Results are compared with those from the linear systems as presented above.

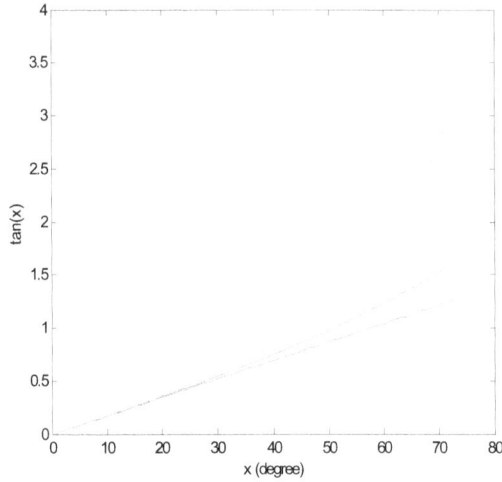

Fig. 19. Approximations of sensor tangent characteristics

5.5 Nonlinear control of platform

Feedback measurements used for the control of the tracking system lead to an overall nonlinear behaviour in the platform. This resulting nonlinear dynamics is much richer in complexity than the dynamics of the linear platform system. Consequently, nonlinear control strategies may be required to optimize the dynamic performance of the platform. The design of the nonlinear controller for the platform is presented in this section of the chapter. Controller discussions make comparisons of the nonlinear controller with the pole cancellation controller earlier designed for the tracking system.

5.5.1 The state space model of open-loop solar platform

Combining (1)-(3), it is straightforward to obtain the state-space representation for the open-loop systems, as shown in (39). Notice from the equation that, the basic open loop-loop platform, without feedback measurements is here again confirmed to be linear.

$$
\begin{aligned}
\dot{\theta}_1 &= \theta_2 \\
\dot{\theta}_2 &= \theta_3 \\
\dot{\theta}_3 &= -a_1\theta_1 - a_2\theta_2 - a_3\theta_3 + bV_a \\
y &= \theta_1 = \theta_m
\end{aligned}
\tag{39}
$$

where,

$$
a_1 = \frac{KR_a}{L_a J_t}, a_2 = \frac{KL_a + BR_a + K_b K_m}{L_a J_t}, a_3 = \frac{BL_a + J_t R_a}{L_a J_t}, b = \frac{K_m}{L_a J_t}
\tag{40}
$$

and $\theta = [\theta_1, \theta_2, \theta_3]^T \subset \Re^3 = [\theta_m, \dot{\theta}_m, \ddot{\theta}_m]^T$.

5.5.2 Nonlinear state-space model of platform with feedback measurements

The over all plant model, including the feedback measurement, will modify (39) to yield (41):

$$\dot{\theta}_1 = \theta_2$$
$$\dot{\theta}_2 = \theta_3 \tag{41}$$
$$\dot{\theta}_3 = -a_1\theta_1 - a_2\theta_2 - a_3\theta_3 + 2bK_FK_1I_RDLR_F\tan(\theta_s - \theta_m))$$

It is thus evident from (35) that where as the basic system has a linear model, the measurements used for control make the overall system nonlinear

5.5.3 The non-linear input-state feedback control of polar-axis solar power platform

For the linearisation of the platform systems using feedback, a function of state $x_1 = \psi(\theta_1, \theta_2, \theta_3)$ is required; where the $r \leq n$ derivatives of x_1 exist; and ψ is invertible such that all states of the platform $\theta_i; i = 1, 2, ..r$ and its input are functions of x_1 and its r derivatives. Recollect the fact that, for linear systems expressible in the controllable canonical form (as in equation (39), the input-state feedback linearising variable x_1 is the output of the canonical plant representation (Kuo & Golnaragh, 2003). It is possible to write that:

$$\theta_1 = x_1$$
$$\theta_2 = \dot{x}_1$$
$$\theta_3 = \ddot{x}_1 \tag{42}$$
$$V_a = (1/b)(\dddot{x}_1 + a_3\ddot{x}_1 + a_2\dot{x}_1 + a_1x_1); b \neq 0$$

Consequently, the model of the nonlinear platform could now be written in terms of the linearising variable as

$$\theta_1 = x_1$$
$$\theta_2 = \dot{x}_1$$
$$\theta_3 = \ddot{x}_1 \tag{43}$$
$$\theta_s - \theta_m = \tan^{-1}\left\{\left[\frac{1}{2bK_FDLI_RR_F}\right](\dddot{x}_1 + a_3\ddot{x}_1 + a_2\dot{x}_1 + a_1x_1)\right\}; b \neq 0$$

Let,

$$x_3 = v \tag{44}$$

Then

$$\dot{x}_1 = x_2$$
$$\dot{x}_2 = x_3 \tag{45}$$
$$\dot{x}_3 = -K_I(x_1 - x_1^*) - K_P(x_2 - x_2^*) - K_D(x_3 - x_3^*) = v(t)$$

where x_1^*, x_2^*, x_3^* are the respective steady states of x_1, x_2, x_3 and

$$v = -a_1 x_1 - a_2 x_2 - a_3 x_3 + 2K_F K_1 I_R DLR_F b K_F \tan(\theta_s - \theta_m) \qquad (46)$$

The nonlinear error measurements could now be written as:

$$\mu = \theta_s - \theta_m = \tan^{-1}\left\{\frac{(\rho + \sigma)}{2bK_F K_1 LI_R R_F}\right\}$$

$$\rho = -K_I(x_1 - x_1^*) - K_P(x_2 - x_2^*) - K_D(x_3 - x_3^*) \qquad (47)$$

$$\sigma = a_1 x_1 + a_2 x_2 + a_3 x_3$$

5.5.4 Tuning of the parameters of the nonlinear controller

For the evaluation of the v in equation (45), the controller parameters, K_I, K_P, K_D are chosen such that the stability of the linear system

$$\dot{x}_3 = -K_I(x_1 - x_1^*) - K_P(x_2 - x_2^*) - K_D(x_3 - x_3^*) \qquad (48)$$

is guaranteed.
Now, define the error between the states and their references in the following manner:

$$\dot{x}_3(t) = \dddot{e}(t)$$

$$x_3(t) - x_3^*(t) = \ddot{e}(t) \qquad (49)$$

$$x_2(t) - x_2^*(t) = \dot{e}(t)$$

$$x_1(t) - x_1^*(t) = e(t)$$

And recast equation (48) in the form of equation (50):

$$(s^3 + K_D s^2 + K_P s + K_I)E(s) = 0 \qquad (50)$$

Thus, select the tuning parameters K_I, K_P, K_D to ensure the asymptotic elimination of the error, such that.

$$\dot{x}_3(\infty) \to 0$$

$$x_3(\infty) \to x_3^*(\infty) \qquad (51)$$

$$x_2(\infty) \to x_2^*(\infty)$$

$$x_1(\infty) \to x_1^*(\infty)$$

Here we proceed to chose K_I, K_P, K_D applying the Routh-Hurtwits criterion to the equivalent s polynomial:

$$s^3 + K_D s^2 + K_P s + K_I = 0 \qquad (52)$$

Where

The nonlinear controller given by (43)-(47) shall be simulated and its impact on the dynamics of the platform compared with that of the linear controller.

5.6 Comparative simulation of the impact of nonlinear controller

Simulations comparing the performance of the platform under the impact of the feedback-linearised controller, with that under the impact of the pole-cancellation controller are shown in Figure 20 to Figure 22. The nonlinear system brings the system to rest within two seconds, as would the reference linear model. Overshoots are virtually eliminated in the dynamics of θ_m. Whereas the pole-cancellation strategy also brought the system to rest within two seconds, the accelerations observed in other system variables were very high. Overshoots remained significant. The nonlinear controller being reported in this paper achieves the same settling time and eliminates overshoots without causing excessive accelerations.

6. Conclusions and recommendations

6.1 Conclusions

The high overshoots associated with the dynamic response of the pole-cancellation controller could significantly add to the hardware costs of the tracking systems. Moreover, parameter uncertainty/variations could significantly compromise the performance of the platform under linear control. Feedback-linearised control of the platform yield exact linearization of platform dynamics, by feedback, without any approximations. It is demonstrated that this exact linearization leads to a better performance of the platform under the action of the feedback-linearised controller.

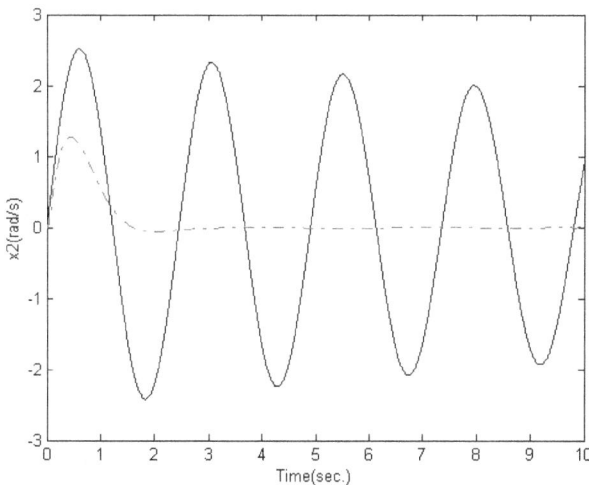

Fig. 20. Platform position under nonlinear control (_._) and linear control (---)

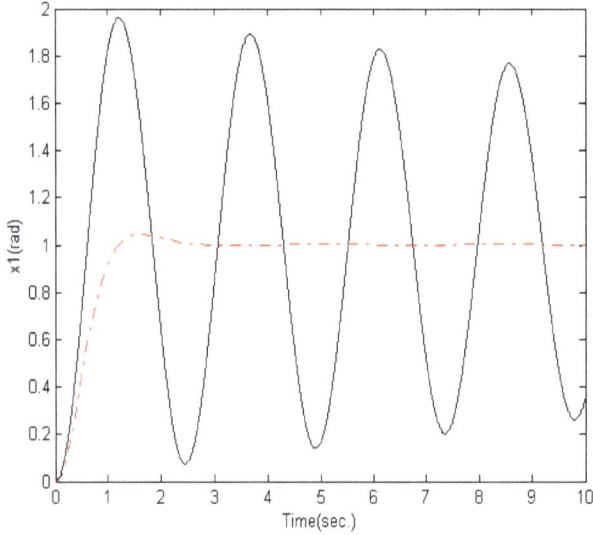

Fig. 21. Platform velocity under nonlinear control (_._) and linear control (----)

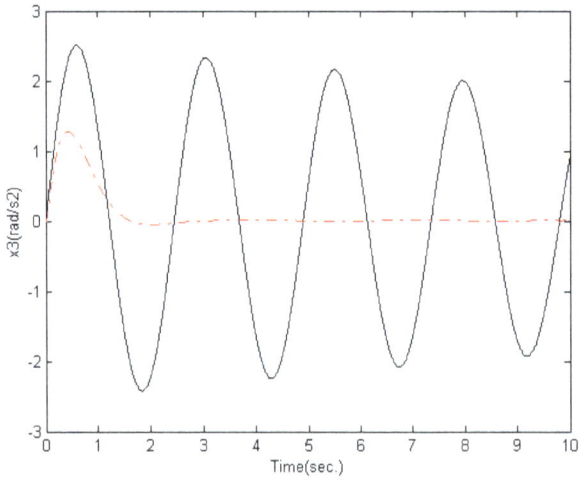

Fig. 22. Platform acceleration under nonlinear control (_._._) and linear control (----)

6.2 Recommendation for further studies

Further research directions here would consider the methods and costs of controller implementation. The feedback-linearised control strategy leads to a nonlinear controller whose structure is more complex than that of the linear pole-cancellation controller. It is recommended to explore neural networks for the implementation of the nonlinear controller. The basic structure of the solar tracker system is linear. The subsequent nonlinearity in the system is due to the measurement. It is recommended to further explore the use of a linear sensor or even sensor-less tracking strategies to simplify controller design.

7. References

Anderson, G. O. & Abkenari, M. H., (1990) Application of solar energy technology in Botswana. *Proceeding of IASTED Int. Conf. on Power and Energy Systems.* Las Vegas, Nevada, 1999, 141-150.

Agee, J. T., Masupe, S, Jeffrey, M. & Jimoh A. A.Enhancing the Output Characteristics of a Photovoltaic Position Snesor Using a feed-Forward Neural network, Advance Materials Research, Vol. 62-62, pp. 506-511. 2009

J. T. Agee, A. A. Jimoh (2007) Feedback linearised Control of a Solar Power Platform. IEEE Africon 2007. Namibia.

J. T. Agee, M. de Lazzer an M. K. Yanev, "A Pole cancellation strategy for stabilising a 3KW solar power platform. *Int. Conf. Power and Energy Systems(EuroPES 2006)*, Rhodes, Greece. June 26-28.

J. T. Agee, S. Obok-Opok and M. de lazzer, "Solar tracker technologies: market trends and field applications. *Int. Conf. on Eng. Resaerch and Devlopemnt: Impact on Industries.* 5-7th September, 2006.

Alternative Energy Store, 300W Shott Solar panel specifications. http://shop.altenergy store.com, 2005.

Archer, M. and Hill, R. (2001), Clean Electricity from Photovoltaics Botswana Power. http://www.wn.com /s/ Botswana power . 2004.

Southern Africa and the Southern African Development Community. http://www. eia.doe.gov /meu/cabs/sadc .hmtl. 2005.

Consultancy on Identifying and Overcoming Barriers to Widespread Adoption of Renewable Energy – Based Rural Electrification in Botswana: *Final report.* 2003.Customer Tips http://www.bpc.bw 10-2005

Daily energy consumption http://www.epsic.ch/pagesperso/ schneiderd/ Apelm /Sources/Solaire.htm 10-2005

De Lazzer, M. (2005). A positioning System for an Array of Solar Panels. Unpublished M.Sc Thesis University of Botswana.

Geche, J. & Irvine J. (1996). Photovoltaic Lighting in Rural Botswana: A Pilot Project. Renewable Energy for Development, Vol. 9, No. 2, pp. http://www.sei.se/red /red9609e.html 21st July, 2009

B. C. Kuo & F. Golnaraghi, *Automatic Control Systems* (eight edition, John Wiley and Sons, Inc., 2003).

Lasschuit, P, Westra, C. & van Roekel G., (xxxx). Financial Suatainability of PV Implementation in Swaziland. http://roo.undp.org/gef/solarpv/docs/bgmaterial /Misc%20PV%20Papers/ECN%20-%20Financial%20sustainability%20of%20PV %20implementation %20in%20Swaziland.pdf. 21st July, 2009

N. Matenge & V. Masilo, *Feasibility Study of Botswana Electricity Generation*(HND Project, *Power Point Presentation*, University of Botswana, 2004).

Mogotsi, B. (2002). Energy and Sustainable Development in Botswana. Sustainable Energy Watch Report. HELIO-Botswana.   Helio International.   http://www.helio-international.org /reports/2002/botswana.cfm. 21st July, 2009

Nelso, J (2003), The Physics of Solar Cells. Barnes and Noble.

N. S. Norman, *Control Systems Engineering* (John Willey and Sons, Inc. USA , 2004).

Shell SQ 80W solar panel specifications. http://www.shell.com, 2005.

TRACSTAR (Small Power systems) Solar Tracking for Architects Source: http://www. pacificsites.com/ sps/trackforarc.html, 2006.

Permissions

The contributors of this book come from diverse backgrounds, making this book a truly international effort. This book will bring forth new frontiers with its revolutionizing research information and detailed analysis of the nascent developments around the world.

We would like to thank all the contributing authors for lending their expertise to make the book truly unique. They have played a crucial role in the development of this book. Without their invaluable contributions this book wouldn't have been possible. They have made vital efforts to compile up to date information on the varied aspects of this subject to make this book a valuable addition to the collection of many professionals and students.

This book was conceptualized with the vision of imparting up-to-date information and advanced data in this field. To ensure the same, a matchless editorial board was set up. Every individual on the board went through rigorous rounds of assessment to prove their worth. After which they invested a large part of their time researching and compiling the most relevant data for our readers.

The editorial board has been involved in producing this book since its inception. They have spent rigorous hours researching and exploring the diverse topics which have resulted in the successful publishing of this book. They have passed on their knowledge of decades through this book. To expedite this challenging task, the publisher supported the team at every step. A small team of assistant editors was also appointed to further simplify the editing procedure and attain best results for the readers.

Apart from the editorial board, the designing team has also invested a significant amount of their time in understanding the subject and creating the most relevant covers. They scrutinized every image to scout for the most suitable representation of the subject and create an appropriate cover for the book.

The publishing team has been an ardent support to the editorial, designing and production team. Their endless efforts to recruit the best for this project, has resulted in the accomplishment of this book. They are a veteran in the field of academics and their pool of knowledge is as vast as their experience in printing. Their expertise and guidance has proved useful at every step. Their uncompromising quality standards have made this book an exceptional effort. Their encouragement from time to time has been an inspiration for everyone.

The publisher and the editorial board hope that this book will prove to be a valuable piece of knowledge for researchers, students, practitioners and scholars across the globe.

List of Contributors

Saïdou Madougou and Mohamadou Kaka
University Abdou Moumouni of Niamey, Niamey, Niger

Gregoire Sissoko
Université Cheikh Anta Diop de Dakar, Senegal

G. S. Aglietti, S. Redi, A. R. Tatnall, T. Markvart and S.J.I. Walker
University of Southampton, United Kingdom

Marco Aurélio dos Santos Bernardes
Centro Federal de Educação Tecnológica de Minas Gerais – CEFET-MG, Brazil

Silvia Colodrero, Mauricio E. Calvo and Hernán Míguez
Instituto de Ciencia de Materiales de Sevilla, Consejo Superior de Investigaciones Científicas-Universidad de Sevilla, Spain

Dr. Luis Jaime Caballero
Isofoton S.A., Spain

Dragos Ronald Rugescu
University of California at Davis, U.S.A.

Radu Dan Rugescu
University Politehnica of Bucharest, Romania E.U.

Rolf Stangl, Caspar Leendertz and Jan Haschke
Helmholtz-Zentrum Berlin für Materialien und Energie, Institut für Silizium Photovoltaik, Kekule-Str., Berlin, Germany

Yoshihiko Takahashi, Syogo Matsuo and Kei Kawakami
Department of Mechanical System Engineering, Department of Vehicle System Engineering, Kanagawa Institute of Technology, Japan

P.M. Gorley and V.P. Makhniy
Science and Education Center "Semiconductor Material Science and Energy-Efficient Technology" at Yuri Fedkovych Chernivtsi National University, Chernivtsi, Ukraine

J. González-Hernández
Centro de Investigación en Materiales Avanzados S.C., Chihuahua / Monterrey, Chihuahua, México

Yu.V. Vorobiev
Centro de Investigación y de Estudios Avanzados del IPN, Unidad Querétaro, Querétaro, México

P.P. Horley
Science and Education Center
"Semiconductor Material Science
and Energy-Efficient Technology" at
Yuri Fedkovych Chernivtsi National
University, Chernivtsi, Ukraine
Centro de Investigación en Materiales
Avanzados S.C., Chihuahua /
Monterrey, Chihuahua, México

Leonid Kosyachenko
Chernivtsi National University,
Ukraine

John T. Agee and Adisa A. Jimoh
Tshwane University of Technology,
Pretoria, South Africa

Index

www.ingramcontent.com/pod-product-compliance
Lightning Source LLC
Chambersburg PA
CBHW070738190326
41458CB00004B/1226